Semiconductor Device Technology

Other titles in Electrical and Electronic Engineering

J.C. Cluley: Electronic Equipment Reliability, second edition
A.R. Daniels: Introduction to Electrical Machines
W. Gosling: A First Course in Applied Electronics
B.A. Gregory: An Introduction to Electrical Instrumentation and Measurement Systems, second edition
L.F. Lind and J.C.C. Nelson: Analysis and Design of Sequential Digital Systems
Paul A. Lynn: An Introduction to the Analysis and Processing of Signals, second edition
A.G. Martin and F.W. Stephenson: Linear Microelectronic Systems
A. Potton: An Introduction to Digital Logic
Trevor J. Terrell: Introduction to Digital Filters
G. Williams: An Introduction to Electrical Circuit Theory

Semiconductor Device Technology

Malcolm E. Goodge

Senior Lecturer,
School of Electronic Engineering
and Computer Science,
Kingston Polytechnic

M

© Malcolm E. Goodge 1983

All rights reserved. No part of this publication may be reproduced or transmitted, in any form or by any means, without permission.

First published 1983 by
THE MACMILLAN PRESS LTD
London and Basingstoke
Companies and representatives
throughout the world

Typeset in 10/12 Times by
Photo-Graphics, Yarcombe, Nr. Honiton, Devon

Printed in Hong Kong

ISBN 0 333 29501 3
ISBN 0 333 29506 4 pbk

The paperback edition of the book is sold subject to the condition that it shall not, by way of trade or otherwise, be lent, resold, hired out, or otherwise circulated without the publisher's prior consent in any form of binding or cover other than that in which it is published and without a similar condition including this condition being imposed on the subsequent purchaser.

Contents

Preface vii

1 Semiconductor Physics 1

    1.1 Development of Electrical Models of Materials 2
    1.2 Semiconductors 14
    1.3 Free Carriers in Semiconductors 28
    1.4 Space Charge 47
    1.5 Surface Conditions 48
    1.6 Junctions 51
    1.7 Junction Breakdown 83
    1.8 Junction Modelling 86
    References and Further Reading 88
    Tutorial Questions 89

2 Discrete Semiconductor Devices 93

    2.1 Semiconductor Diodes 93
    2.2 Bipolar Junction Transistors (BJTs) 107
    2.3 Field-effect Transistors (FETs) 212
    2.4 Switching Devices 250
    2.5 Devices Used in Signal Generation 262
    2.6 Optical Devices 271
    References and Further Reading 300
    Tutorial Questions 302

3 Semiconductor Device Fabrication — 309

Introduction — 309
3.1 Planar Fabrication Process — 310
3.2 Current Developments in Fabrication Technology — 320
3.3 Diffusion Process — 323
References and Further Reading — 328
Tutorial Questions — 328

4 Integrated Circuit Technology — 329

Introduction — 329
4.1 Isolation — 330
4.2 Monolithic IC Elements — 335
4.3 Monolithic IC Fabrication Techniques — 355
4.4 Monolithic IC Design — 357
4.5 Scale of Integration — 365
4.6 Superintegration — 368
4.7 Monolithic IC Technologies in Practice — 369
4.8 LSI in Practice — 388
4.9 Film Circuits — 410
References and Further Reading — 416
Tutorial Questions — 417

Appendix A   *Terminal Parameter Modelling of Device Characteristics* — 419

Appendix B   *Nomenclature and Terminology* — 426

Appendix C   *Constants, Conversions, Unit Multiples* — 441

Appendix D   *Electromagnetic Spectrum* — 444

Appendix E   *Device Numbering Systems* — 445

Appendix F   *Component Coding* — 450

Appendix G   *Preferred (E-series) Component Values* — 453

Appendix H   *Manufacturers' Data Sheets of Selected Devices* — 455

Index — 482

Preface

This book is concerned with the *fabrication, operation* and *performance* of modern semiconductor devices, the term 'device' being used here to encompass both discrete devices such as transistors and integrated circuits. The material presented has developed from a series of lectures given to students of electronics in the School of Electronic Engineering and Computer Science at Kingston Polytechnic over the past few years.

To become a competent engineer in hardware electronics, that is, to be able to use electronic devices in practice to produce reliable equipment capable of meeting required performance specifications, it is essential to have an adequate appreciation of the properties and limitations of available devices. Also, to make the circuit design process efficient, it is necessary to be able to analyse circuit performance theoretically so that the effect of modifications can be reliably predicted rather than relying on a haphazard 'trial and error' approach. An understanding of the fundamental physical processes responsible for the terminal properties of devices together with knowledge of the construction and method of manufacture provides a foundation for such work.

The objective during the preparation of the book was to provide comprehensive information on the operation and performance of semiconductor devices supported by consideration of the physical processes involved, to provide ways of representing the performance quantitatively in the form of a model enabling analysis of circuits using the devices and to describe the construction and fabrication processes.

Chapter 1 is devoted to the fundamental physical processes involved in semiconducting materials and at junctions between materials. Widespread use is made of the particle atomic *model* of the structure of material and the point is emphasised that in reality the composition of material is not known, discussion of properties in terms of atomic structure, energy and charge being ways of *representing* or *modelling* the situation.

Chapter 2 deals with the wide range of discrete semiconductor devices. Diodes, bipolar transistors and field-effect transistors are considered in general while specialised devices are grouped according to application as switching devices, devices used in signal generation and optical devices. For each device, the physical construction and mechanism of operation is discussed, typical terminal characteristics are given and, where applicable, the consideration concludes with the formulation of a network model which represents the performance of the device in network form.

The planar fabrication process used to manufacture the vast majority of devices is described in detail in chapter 3.

Circuit integration, both in monolithic and film form, is considered in chapter 4. For monolithic ICs, the fabrication of individual circuit elements is described together with discussion of the design process, that is, the selection of the system to be implemented in integrated form, the choice of the technology to be used, the circuit design and the chip layout design. After this general consideration, the specific technologies are described and compared and their application at the various levels of integration discussed.

General information on network modelling techniques, terminology, preferred component values and device numbering and coding is given in the appendixes. A comprehensive list of symbols and abbreviations used in the text is also included together with manufacturers' device data sheets for some typical devices.

Experience with the courses at Kingston has shown which aspects of the subject present the greatest difficulties to students and during the preparation of the text I have tried to answer the questions that are invariably asked during the course and also to dispel the misconceptions that often occur.

An attempt is made throughout to bring the reader up to date with modern device structures, the corresponding improvements in performance, the use of new materials and trends in development. If the reader requires a deeper academic treatment of a topic, a comprehensive list of appropriate literature is given. Tutorial questions at the end of each chapter provide the opportunity to apply the work to numerical problems.

The material is intended to be covered in the first two years of an undergraduate or diploma course, the subject having a formal timetable allocation of two to three hours per week with students expected to devote an approximately equal period each week to private study. It is considered that treatment in the order presented in the text is the most logical; thus it is intended that the first year is devoted to semiconductor physics and discrete diodes and transistors which leads on from introductory courses at school or at TEC Diploma/Certificate level. The second year of the course can then deal with integrated circuits and specialised discrete devices used in switching, high-frequency signal generation and optical systems. Discussion of the planar fabrication process can be included in the first-year syllabus or left until the second year as time permits. The detail presented is far in excess of what can be included in the corresponding lecture course but the book is intended to act as a reference to help students to clarify points mentioned only briefly in lectures and to provide the answers to questions left unanswered at the end of lectures.

ACKNOWLEDGEMENTS

I should like to acknowledge the assistance of the staff and students of the School of Electronic Engineering and Computer Science at Kingston Polytechnic during the development of this work, in particular Professor John Coekin and Dr Don Pedder for their helpful comments during the preparation of the manuscript, and the degree and HND students over a number of years who have given their opinions of the devices content of their courses, with particular reference to its practical use in laboratory and industrial work, enabling modifications to be made.

I am grateful to Ferranti Electronics Limited for supplying and giving me permission to publish figures 2.100, 3.4 and 4.40b showing respectively a solar cell element, a processed silicon wafer and an F100-L microprocessor chip. I also appreciate the assistance of Ed Evans of Mullard Limited and John Edwards of Siliconix Limited in giving me permission to include published data sheets of selected devices thus enabling the theoretical work to be related to typical performance characteristics and parameter values.

The electron micrographs of a bipolar transistor chip (figure 2.11a) and a portion of a UV EPROM chip (figure 4.38c) were produced by Bill Edwards of the School of Chemical and Physical Sciences at Kingston Polytechnic using the School's Cambridge Scientific Instruments S600 scanning electron microscope. Other photographs were taken by Jerry Humphreys and David Gore of the Polytechnic's Audio-Visual Services (AVS) unit, while the figure outlines were drawn by Sue Hepworth also of the AVS unit at Kingston.

My special thanks must be to my wife Judith for her patience and encouragement during the preparation of the text and for typing the manuscript — thank you.

MALCOLM GOODGE

1 Semiconductor Physics

The composition of matter on a microscopic scale is a mystery. The most fundamental deduction in this context has been the link between mass and energy (Einstein, 1905). First-order definitions describe mass as being a measure of the *quantity of matter* and energy as the *capacity to do work*. On a macroscopic scale the concepts of a mass possessing kinetic energy due to movement and potential energy due to position can be readily appreciated. The energy state of the mass is a measure of the work the mass can perform in transferring its energy to another system, that is, by being brought to rest or by changing position due to gravitational attraction. The concept of matter possessing energy by virtue of its composition is not so easy to visualise. Qualitative evidence is provided by certain chemical reactions which result in the emission of energy in the form of heat causing the temperature of the surroundings to rise. In such cases the energy state of the product of the reaction is lower than the total energy of the original constituents and the difference is emitted. Energy is termed mechanical, chemical, thermal, electrical or nuclear depending on the form in which it is manifested. The energy of a mass due to movement or position is described as mechanical energy while that involved in chemical reactions is termed chemical energy and the emission of heat is referred to as a release of thermal energy. In general, energy may be considered as the fundamental constituent of the universe; without energy there would be nothing.

The operation of semiconductor devices depends on the energy properties of materials on a microscopic scale. Since the true nature of energy and hence the composition of matter, is unknown, various *models* have been devised representing observed phenomena. Using models of the properties of materials on a microscopic scale it is possible to develop an appreciation of the operation of semiconductor devices.

1.1 DEVELOPMENT OF ELECTRICAL MODELS OF MATERIALS

The earliest observed electrical phenomenon in materials was the force created between materials which had been 'excited by friction'. The material was said to be 'electrified' or 'charged with electricity', the term *electric* being derived from the Greek word for amber, a naturally occurring resin in which the effect was first observed. Subsequently the word *electricity* was used to describe the *cause* of the attractive and repulsive forces between materials and the term *charge* became used to describe the *quantity* of electricity (electrical energy) in the material. Based on these fundamental concepts, early researchers modelled the force between materials in terms of the excess or deficit of electric charge in the materials. If the material had an excess of charge it was said to be *positively* charged while a deficit was termed *negative* charge. Thus a model developed whereby electrification by friction was considered to be due to the addition or removal of charge by the mechanical process of rubbing and the tendency to restore an equilibrium distribution of charge created the force between the materials. Where one material had a charge excess (positive charge) and the other a charge deficit (negative charge) an attractive force was said to exist due to the tendency for the excess charge from one material to transfer to the other to restore equilibrium. If both materials had a charge excess, or both a charge deficit, the tendency for charge transfer did not exist and there was a force of repulsion between the materials. Hence there is the fundamental rule in what is now termed *electrostatics* (stationary charge properties): like charges repel, unlike charges attract. A concentration of charge was said to establish an electrical *potential* and the force tending to move the charge (or the material if the charge would not leave the material) could then be described in terms of the difference in potential, termed the *voltage*, between the materials. The interaction across space between materials having different charge conditions was modelled in terms of an electric *field* between the materials.

In the early nineteenth century the discovery of the voltaic cell which provides a voltage between its plates (electrodes) was an important link between electrical phenomena and the chemical properties of materials. The chemical interaction of the different plate materials with the electrolyte (conducting solution) is a source of electrical charge and an external metal connection between the two plates causes a continuous flow of charge (*current*) to occur. By this time the concept had been developed that all substances were composed of fundamental units, the units being termed *atoms* from the Greek *atomos* meaning 'indivisible'. It was not, however, until the end of the nineteenth century and the early years of this century that studies of electrical discharges, in particular the conductivity of gases,[1] and radioactivity led to useful models of atomic structure being developed.

It was found that continuous conduction through a gas, as distinct from spark discharges, could be obtained if the applied gas pressure was sufficiently low. In a famous demonstration[2] in 1869, using an obstacle in the path of a low pressure gas discharge, Hittorf showed that the charge flow originated from the electrode having the lower or (relatively) negative potential. This electrode was termed the *cathode*, from the Greek *kathodos* meaning 'way down', while the other electrode having the higher or positive potential was named the *anode*, from the Greek *anodos* meaning 'way up'. Goldstein called the charge flow *cathode rays*, and using earlier observations of the interaction between an electric current and a magnetic field (Oersted, 1820), the rays were shown to constitute a current since they could be deflected by a magnetic field.

Towards the end of the nineteenth century the charge flow was modelled as a stream of charged 'particles' called *electrons* (a 'particle of electricity') which, from the direction of flow in the gas discharge, could be said to be negatively charged. Experiments by Thompson and Millikan enabled the charge and mass of an electron to be determined, currently accepted values being -1.602×10^{-19} C and 9.109×10^{-31} kg respectively.

It was found that emission with the same charge–mass ratio was obtained for discharges in different gases and, in addition, the same ratio was obtained for the emissions from a heated metal (thermionic emission) and from a metal irradiated with ultraviolet light (photoemission). From these observations it was concluded that electrons must be a fundamental constituent of materials.

Closer examination of the low pressure gas discharge showed that in addition to the stream of electrons (cathode rays) there was also a movement of positive charge towards the cathode (Goldstein, 1896). The positive charge did not come however from the anode and so it was concluded that it was made up of gas atoms that had lost electrons, termed gas *ions*, the gas being said to be *ionised*. These discoveries led to the proposal in about 1900, of an atomic model consisting of a region of positive charge containing a number of electrons. It was considered that, under equilibrium conditions, the balance between positive and negative charge caused the atom to be neutral but that if electrons were removed, for example, by mechanical (rubbing), electrical (discharge) or chemical means, the remaining portion of the atom would be positively charged—an ion. From earlier measurements of atomic masses it seemed evident that the positive charge made up most of the mass of the atom and consequently the space occupied by the positive charge was considered to determine the size of the atom.

Meanwhile, the legendary work by Becquerel (1896) and the Curies (1898) in the investigation of radioactivity, the natural disintegration of certain elements, showed that three types of emission occurred which were termed α, β and γ-rays. Investigation of the deflection of these emissions in a magnetic field showed α-rays to be positively charged, β-rays negatively charged and γ-rays uncharged. Measurement of the charge–mass ratio of β-rays indicated that they were streams of electrons. It therefore seemed reasonable to regard an atom as a combination of positive, negative and uncharged 'particles'.

In 1911, during investigation of α-ray scattering, Rutherford deduced that the majority of the mass of an atom must be concentrated into a region far smaller than the volume occupied by the complete atom on the basis of atomic diameters proposed by the original model. He therefore suggested a modification to the earlier model, considering almost all the mass of the atom to be concentrated in a central core with encircling negatively charged electrons forming a planetary system which may be represented in two dimensions as shown in figure 1.1a. The core was named the *nucleus,* from the Latin word meaning 'inner part', which from scatter measurements was shown to be of the order of 10^{-14} m diameter, compared with an atomic diameter of 10^{-10} m. Investigation of the lightest element, hydrogen, which was therefore assumed to have the simplest atomic structure, led to the conclusion that its nucleus consists of an elementary positively charged 'particle' termed a *proton*. The nuclei of other atoms were considered to consist of a number of protons but in order to explain the atomic mass of these atoms it became evident that the nuclei of atoms other than hydrogen also contained uncharged 'particles' called *neutrons*. Research showed that the α-particle emitted during radioactive decay was in fact a helium nucleus comprising two protons and two neutrons and since then other elementary nuclear 'particles' have been detected.

Rutherford's particle atomic model formed the basis of the classical model of the structure of material on a microscopic scale. Stability of the electron orbits was considered to be provided by the electrostatic attraction between the electron and the nucleus, the attraction providing the centripetal force necessary to keep the electron in its orbit. Considering the simplest case of a hydrogen atom having a nucleus of charge $+e$ and a single circular electron orbit of radius r, the force of attraction between the electron and the nucleus (from Coulomb's law[3]) is $e^2/(4\pi\epsilon_0 r^2)$, where ϵ_0 is the permittivity of the region between the electron and the nucleus which is assumed to be a vacuum, that is, devoid of material. This force provides the centripetal force mu^2/r acting on the electron, where m is the electron mass and u is its tangential velocity. Thus the velocity of the electron in a circular orbit of radius r is

$$u = \sqrt{\left(\frac{e^2}{4\pi\epsilon_0 mr}\right)} \tag{1.1}$$

Electron energy is made up of two components, 'potential' energy due to the attraction and 'kinetic' energy due to the motion. The potential energy component is the work done in moving the electron from infinity (where the attractive force due to the nucleus is zero) to the orbit position distance r from the nucleus, that is

$$\int_{\infty}^{r} \frac{e^2}{4\pi\epsilon_0 r^2} \, dr = -\frac{e^2}{4\pi\epsilon_0 r}$$

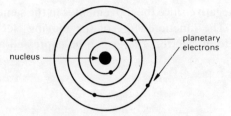

(a) Rutherford's particle atomic model

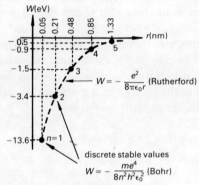

(b) Variation of electron energy with orbit radius for a circular orbit (hydrogen atom)

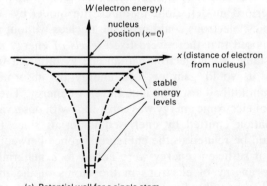

(c) Potential well for a single atom

Figure 1.1   Atomic models

This component is negative since the movement is in the same direction as the force of attraction. The kinetic energy of the moving electron is $mu^2/2 = e^2/8\pi\epsilon_0 r$ and so the total electron energy can be expressed as

$$W = -\frac{e^2}{8\pi\epsilon_0 r} \tag{1.2}$$

showing that the energy of the electron increases (that is, is less negative) as the orbit radius increases (figure 1.1b).

Rutherford's atomic model has a major inconsistency in that the curved motion of the electron means that it is continually being accelerated towards the nucleus. Electromagnetic theory shows that accelerating charge radiates energy, the phenomenon exploited in a transmitting aerial, and so the orbiting electron would lose energy continuously and the system would not be stable.

In parallel with investigations into the structure of materials, research in the radiation of energy had been taking place and in 1901 Planck postulated that energy is not radiated continuously but in discrete pulses called *quanta*. He proposed that the energy of a quantum of radiation is directly related to the frequency of the radiation such that

$$W = hf = \frac{hc}{\lambda} \tag{1.3}$$

where h (Planck's constant) is 6.62×10^{-34} J s and c is the velocity of propagation of electromagnetic radiation in free space. In terms of visible radiation (light), a quantum of energy became known as a *photon*—a 'light particle'.

The significance of Planck's quantum theory in the development of the atomic model was that in 1913 Bohr used the theory to remove the instability of the Rutherford model. Bohr modified the model by suggesting that at certain energies, electrons could orbit the nucleus without radiating energy. He thus proposed that there were fixed *levels* of energy at which electron orbits were stable. If an electron had an energy different from one of these stable levels it would gain or lose energy by absorption or radiation respectively until it had a stable value (level) of energy. The concept of fixed stable levels of electronic energy was consistent with observations of the fixed spectral radiation emitted by energised materials since the frequency of radiation could be related to the energy difference between the fixed stable levels. Thus if sufficient energy were added to a material by heating, for example, the energy of electrons in the atoms of the material could be increased to higher stable levels. Subsequently, when the electrons returned to their original energy state, the difference in energies would be released causing the material to radiate. If the stable levels were fixed, the difference between levels would also be fixed giving rise to a characteristic emitted frequency (equation 1.3) for the material as observed in spectroscopy.[4] Quantitatively, based on the simplest atomic structure, that of hydrogen, Bohr's proposal was that the stable electron orbits were those for which the product of electron momentum and orbit circumference is an integral multiple of Planck's constant, that is

$$(mu)(2\pi r) = nh \tag{1.4}$$

where n is an integer.

If the orbiting electron is considered as a uniform ring of mass m and radius r, its moment of inertia J is mr^2 and its angular momentum $\sigma_m = J\omega = mr^2\omega = mr^2 u/r = mru$, where ω is the angular velocity and u the tangential velocity of the electron. Bohr's proposal (equation 1.4) can be written

$$mru = n \times \frac{h}{2\pi} \tag{1.5}$$

and stated alternatively as *stable orbits are those for which the angular momentum of the electron is an integral multiple of $h/2\pi$*. Equation 1.5 provides a set of relationships between electron velocity u and orbit radius r which, when combined with equation 1.1 defining the orbit from force considerations, enables an expression for the radii of stable orbits to be derived

$$u = \sqrt{\left(\frac{e^2}{4\pi\epsilon_0 mr}\right)} = \frac{nh}{2\pi mr}$$

from which (1.6)

$$r = \frac{n^2 h^2 \epsilon_0}{\pi m e^2}$$

Substituting this expression for r in equation 1.2 shows that the stable electron orbits are for energies given by

$$W = -\frac{me^4}{8n^2 h^2 \epsilon_0^2} \tag{1.7}$$

The integer n is defined as the *principal quantum number* describing the orbit, the $n = 1$ level being the lowest energy state of the electron and is known as the *ground* state.

Figure 1.1b shows that the continuous energy–radius relationship of the Rutherford model becomes a set of discrete relationships for the Bohr model. The diagram includes numerical values for the five lowest stable energy levels calculated from equations 1.6 and 1.7. The latter equation gives W in joules but, due to the order of magnitude involved, it is usual to quote atomic energies in *electronvolts* (eV) as shown in figure 1.1b. An electronvolt is the change in energy of an electron as it passes through a potential difference of one volt. Since the electron carries a charge e coulombs

$$
\begin{aligned}
1 \text{ electronvolt} &= e \text{ coulomb} \times 1 \text{ volt} \\
&= e \text{ joule} \\
&= 1.6 \times 10^{-19} \text{ joule}
\end{aligned}
\tag{1.8}
$$

The ground state ($n = 1$) has the lowest energy and so it requires a greater amount of energy to remove an electron from the atom if it is in the ground state than if it is in one of the higher states. Electrons in the lower energy states are therefore said to be tightly *bound* to the nucleus while electrons occupying higher levels, that is, having more energy and hence higher radius orbits, are less tightly bound to the nucleus and therefore more easily removed from the atom. A model used to represent the tightness of binding between the electrons at various energy levels and the nucleus is termed a *potential well* which is a plot of electron energy against distance from the nucleus (figure 1.1c). Electrons having energy corresponding to the lower stable levels are said to be 'deep within the well' and therefore cannot easily escape the influence of the nucleus while those at higher levels are 'less deeply bound' and can escape more easily.

It should be noted that at the time Bohr postulated stable energy levels (1913), the mathematical description of the levels (equation 1.4) had no justification except that the differences in energies between the levels calculated according to this relationship corresponded to the wavelengths of radiation obtained from spectrographic observations using Planck's quantum relationship (equation 1.3). Thus Bohr devised the mathematical description of stable levels to agree with observed phenomena.

The establishment of *relativity* between mass and movement led to Einstein's classic relationship between mass and energy

$$W = mc^2 \tag{1.9}$$

and subsequently de Broglie (1924) extended the concept of wave–particle duality to materials. Combining Einstein's mass–energy equation and Planck's energy–wavelength relation (equations 1.9 and 1.3), de Broglie derived an expression relating the momentum of an energy quantum to the wavelength of the equivalent wave

$$mc^2 = \frac{hc}{\lambda}$$

from which

$$\lambda = \frac{h}{mc} \tag{1.10}$$

where mc is the momentum of the quantum. De Broglie extended this relation to apply to any mass moving with velocity u such that the equivalent wavelength λ is

$$\lambda = \frac{h}{mu} \tag{1.11}$$

Combination of de Broglie's relation (equation 1.11) and Bohr's postulate for a stable electron orbit (equation 1.4) shows that, in wave terms, the stable circular orbits for the hydrogen atom are those having a circumference equal to an integral number of wavelengths, that is, $2\pi r = n\lambda$. The wave nature of the electrical charge within a material (electrons) was first demonstrated practically in 1927 when an electron beam was diffracted by the regular atomic array provided by a crystal.

Formulation of a mathematical representation of the electron was provided in 1926 by Schrödinger.[5] He used the de Broglie relation (equation 1.11) together with a differential equation describing the properties of a travelling wave to represent the electron as a function of position and time. The function describes the probability of an electron being present at a particular point in space at a particular time. This mathematical representation of the behaviour of electronic charge developed rapidly during the late 1920s and today forms a comprehensive description of the electronic aspects of materials. The branch of physics that uses these concepts became known as *quantum mechanics* or *wave mechanics* to distinguish it from the classical or Newtonian physics that describe the properties of materials on a macroscopic scale. The quantum and Newtonian mechanical descriptions of material do not conflict; the quantum description is the more general which reduces to the Newtonian model in the special case of material on a macroscopic scale.

Schrödinger's model of electronic behaviour is a more general representation than the particle atomic model of Rutherford and Bohr. However, provided the limitations of the particle model are acknowledged, it remains a useful model of material structure since it is conceptually simple and can be used as the basis for discussion of many of the effects involved in semiconductor devices. It must always be remembered however that *the model is a representation of the properties of the material and is not reality*; thus if a particular model does not explain a phenomenon observed in practice it is due to the inadequacy of the model and a more detailed model or a different form of model must be used.

In terms of the particle model the various elements have different numbers of electrons and protons, the latter being termed the *atomic number Z* of the element. Hydrogen has a single electron and the nucleus comprises a single proton; thus $Z = 1$. Helium has two electrons and a nucleus consisting of two protons and two neutrons; hence $Z = 2$ for helium. Silicon has 14 protons ($Z = 14$) and germanium 32 ($Z = 32$). The electron orbits are grouped into *shells* corresponding to the quantum number n. The ground state or K shell consists of a single circular orbit having a maximum capacity of two electrons. Higher states (L,M,N... shells corresponding to $n = 2,3,4...$) comprise multiple orbits, each shell having a maximum electron capacity of $2n^2$. Electrons in the outer shell of an atom are responsible for the chemical properties of the element. They are termed *valence* electrons since the valency of the element and hence the combinational properties with other elements is dependent on the number of electrons in the outer shell. Silicon for example has its K and L shells full, with two and eight electrons respectively, and has four electrons in the M shell; the valency of silicon is thus four. Oxidation of silicon produces silicon dioxide SiO_2, each molecule comprising one atom of silicon (valency 4) and two atoms of oxygen (valency 2). Arrangement of the 105 known elements into groups according to shell occupation results in the *periodic table* of elements. Groups I to VII contain elements having one to seven valence electrons respectively while a separate group comprises those elements known as inert gases.

For a detailed consideration of the electronic structure of atoms based on this model, the reader is referred to texts concerned primarily with atomic physics or physical electronics such as references 6, 7 and 8. A comprehensive table of the electronic configuration of the elements is given in reference 9.

When atoms are close together, the atomic charges cause interatomic forces which bind the atoms together. Since atoms comprise positive and negative charges, both attractive and repulsive forces exist and the equilibrium spacing between atoms in a solid depends on the balance between these interatomic forces. The forces between atoms holding them together are called *binding forces* or *bonds*. Depending on the electronic structure of the atoms involved, different *bonding models* have been developed, the most important of which are termed ionic, covalent and metallic.[10] *Ionic* bonds are those between different elemental atoms where the electrons from the incomplete outer shell of one atom are 'transferred' to fill the outer shell of the other atom. The transfer of electrons from one atom to the other forms two ions, one positive, the other negatively charged, resulting in ionic attraction. *Covalent* bonding occurs where the electrons in the outer shells of the constituent atoms are 'shared' between the various atoms thereby filling the outer shell of each atom for part of the time. In terms of the particle model this type of bonding results from distortion of the outer electron orbits of each atom to encompass other atoms. In metals, the outer electrons are only loosely attached to parent nuclei and only a small energy addition, for example, thermally from the surroundings, is necessary to cause these electrons to break free. In a metal crystal such as copper, the outer electrons form a cloud of 'free' electrons which results in its high electrical conductivity. The type of bonding formed by the electron cloud, termed a metallic bond, is stable because the total energy of the positive metal ions and the cloud is lower than the sum of the energies of the individual atoms.

Solids are therefore formed by the bonds between atoms. If the bonding results in a regular structure or lattice, the solid is said to be *crystalline* and to form a *crystal*. Where a regular structure is created but it is not perfect due to faults, the structure is termed *polycrystalline*. Substances in which there is no atomic ordering, atoms being bound in a random array, are described as *amorphous*. For semiconductor devices, the crystalline structures of group IV elements silicon (Si) and germanium (Ge) and the group III–V compounds such as gallium arsenide (GaAs), gallium phosphide (GaP) and indium antimonide (InSb) are of most interest. Silicon and germanium have a tetrahedral crystal structure, atoms being equispaced at the centre and vertices of a regular three-dimensional four-sided figure (a tetrahedron), which is created from interpenetrating face-centred cubic structures.[10] Each atom in such a structure has four close neighbours to which it is covalently bonded. The group III–V compounds form slightly more complex tetrahedral crystals.

When atoms are in close proximity, as in a crystal, the energy-level model is modified from that for a single atom. Pauli's exclusion principle (1925) states that no two electrons can be in the same place at the same time. In terms of the energy-level model this means that electrons in close proximity cannot have the same energy. For two closely spaced identical atoms, the single levels of the individual atoms become dual levels with a minute energy difference. In a crystal with a large number of closely spaced atoms, typically of the order of $10^{29}/m^3$, the energy level structure becomes *bands* of levels (figure 1.2a), the difference between the individual levels within a band being so small ($\approx 10^{-8}$ eV) that the band is often considered as a continuous range of energies. The stable orbits of individual atoms in the particle model, which are represented by stable levels on an energy diagram, become *allowed bands* of electron energies for a crystal. Ranges of energies between these allowed bands corresponding to unstable orbits in the Bohr–Rutherford particle model, are termed *forbidden bands*. The difference in energy between the highest energy ('top') of an allowed band and the lowest energy ('bottom') of the next highest allowed band, that is, the energy span of a forbidden band is described as an *energy gap*. As the levels within the gap are unstable, an electron is restricted to remain in a lower allowed band unless it can acquire sufficient additional energy to be in a higher allowed energy band; often termed 'jumping the energy gap'.

In terms of the potential-well model, figure 1.2b shows how the single atom model must be modified to represent the situation in a crystal. The spatial energy distributions of individual atoms (figure 1.1c) become overlapped when the interatomic spacing is small as in a crystal; typically 0.2–0.3 nm.

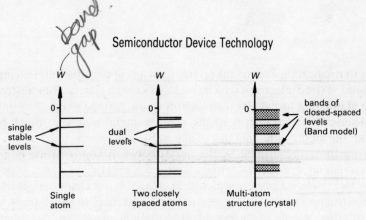

(a) Energy-level models for atoms in close proximity

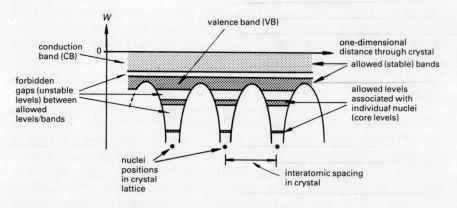

(b) Potential-well model for closely spaced atoms in a crystal

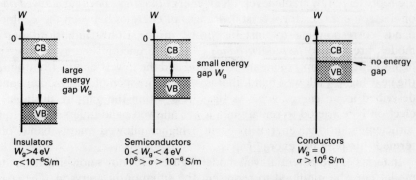

(c) Energy-band model of insulators, semiconductors and conductors

Figure 1.2 Electron energy modelling for crystals

Electrons having energies corresponding to the lower stable or allowed levels, that is, the inner orbits in the particle model, are largely unaffected by neighbouring atoms and these electrons remain tightly bound to their parent nuclei. These levels are thus often referred to as *core levels*. Higher energy electrons, those occupying the 'outer orbits', are those involved in the bonding process. These electrons form the covalent bonds which bind the atoms together in a crystal and the band of energies they have is termed the *valence band* (VB). Thus in figure 1.2b, the allowed band labelled the 'valence band' is shown linking the atoms in the array. If electrons have sufficient energy to break free from their parent atoms, or groups of atoms in the case of covalently bonded crystals, they become free to move throughout the crystal, that is, to take part in electrical conduction through the crystal. The allowed band of energies higher than ('above') the valence band is thus termed the *conduction band* (CB) and the electrons having energies in this band are those that take part in conduction. It must be emphasised that the energy bands do not have physical significance in the sense that description of an electron as being 'in the CB' does not describe a region within the crystal but a 'range of energies', thus CB electrons are more energetic than VB electrons.

In general there are forbidden bands (energy gaps) between the allowed bands. As regards the electrical conduction properties of crystals, however, it is the gap between the VB and CB which is most significant; *the energy gap* or *band gap* W_g of the material. In the absence of any energy supply to a material, all outer-shell electrons are taking part in atomic bonding and none are 'free' to move through the material. Thus, at absolute zero temperature ($T = 0$ K) all levels in the VB are 'occupied' and no levels in the CB are occupied, that is, the VB is said to be 'full' and the CB 'empty'. For temperatures above 0 K the supply of thermal energy to the structure enables some electrons in the VB to gain sufficient energy to metaphorically 'jump the energy gap' and to occupy levels in the CB. Those electrons having CB energies are responsible for the conduction properties of the material, described in terms of *conductivity*.

The magnitude of the energy gap W_g is significant in determining the conductivity of a material. A large W_g causes a relatively low conductivity since at a particular level of excitation only few VB electrons receive sufficient energy to 'reach' the CB compared with a small W_g case. Figure 1.2c shows energy-band models corresponding to insulators, conductors and semiconductors. Materials classed as *insulators* have a large energy gap (>4 eV) resulting in a conductivity <10^{-6} S/m, the conductivity of widely used insulators such as alumina ceramic, mica, polyvinylchloride (PVC), polytetrafluoroethylene (PTFE), polystyrene, polyamides (nylon) and acrylics (for example, Perspex) being <10^{-9} S/m and as low as 10^{-19} S/m in some cases. *Conductors*, the metallic elements, have no energy gap between the VB and CB representing the properties of the metallic bond described above whereby even at low temperature, that is, low excitation, there is a significant number of electrons having CB energies (previously described as 'free' electrons) providing a high conductivity, typically $>10^6$ S/m and up to 10^8 S/m for silver, copper and aluminium. A few pure crystals (silicon, germanium, selenium and tellurium) with energy gaps in the range $0 < W_g < 4$ eV act as *poor conductors* at normal temperatures having electrical conductivities between 10^{-6} and 10^6 S/m; they are called *semiconductors*.

To summarise this consideration of materials on a microscopic scale it can be said that the actual composition of matter is unknown but the observation of physical phenomena has enabled matter to be represented in terms of *energy*, that abstract constituent of the universe. Research has shown that matter comprises regular patterns of energy distribution, the fundamental pattern for an element being termed an *atom* of that element. *Models* developed to represent the energy pattern at their simplest describe the spatial energy distribution in terms of *particles*. These atomic particles have such energy properties in terms of mass, movement and charge that the results of analytical calculations using the model agree with observed properties of materials. The resulting *particle atomic model* and the *energy band model* developed from it provide a basis for the description of the majority of phenomena involved in the operation of semiconductor devices and are thus used extensively in the following sections. More sophisticated models represent the energy content of matter in terms of *waves* which, although conceptually more difficult to handle, are required to describe certain phenomena where the particle model is inadequate. It must be remembered that the work that follows describes the properties of materials and devices in terms of models.

1.2 SEMICONDUCTORS

Pure elements and compounds having electrical conductivities in the semiconducting range of $10^{-6} < \sigma < 10^6$ S/m at normal temperatures are said to be *intrinsic semiconductors*, that is, the degree of electrical conduction is an *inherent* property of the material. The most important semiconducting elements and compounds are listed in table 1.2. At temperatures above absolute zero (0 K, $-273\,°C$) the energy imparted to such a material from the surroundings causes some electrons to break free from their parent atoms and to move through the crystal, that is, to take part in conduction. In terms of the band model this means that some electrons taking part in bonding the atoms together (valence band, VB, electrons) attain energies corresponding to levels in the conduction band, CB. Thus, depending on the degree of excitation, a number of electrons are said to 'jump the energy gap W_g' to reach the CB.

When an electron (charge $-e$) leaves its associated nucleus, the remaining atomic structure acquires a charge of $+e$. The vacancy in the atomic structure left by the electron is termed a *hole* and the excitation of electrons from VB to CB thus creates holes in the VB (figure 1.3a). As the excitation of each electron from CB to VB leaves a hole in the VB, the number of holes in the VB equals the number of electrons in the CB. The process therefore results in the creation (generation) of *electron–hole pairs*. Having an effective charge $+e$, a hole attracts a neighbouring valence electron which completes the atomic structure that the original electron left but leaves a hole elsewhere and so holes as well as electrons contribute to the movement of charge within the material. If a voltage is applied across the material, electrons are attracted towards the positive potential and holes towards the negative potential (figure 1.3b). Conduction is therefore via electrons in the CB *and* holes in the VB. Holes do not flow out into the conductor linking the material with the negative terminal of the voltage source due to the overwhelming number of 'free' electrons in the conductor. Holes tending to flow out of the semiconductor are 'neutralised' at the conductor–semiconductor junction, drawing electrons from the conductor, resulting in an electron current in the circuit. The neutralisation process whereby electrons and holes combine causing annihilation is called *recombination* (section 1.3.1). Conduction in intrinsic semiconductors is via electron–hole pairs and is described as *intrinsic conduction*. The densities of free electrons and holes in an intrinsic semiconductor are given the symbols n_i and p_i respectively, n and p being derived from *n*egative and *p*ositive charge carriers, and the suffix i indicating *i*ntrinsic semiconductor. Since these electron and hole densities result from the generation of electron–hole pairs they must be equal at any temperature, thus

$$n_i = p_i \qquad (1.12)$$

At $T = 300$ K, the values of n_i (the *intrinsic carrier density*) for silicon and germanium are 1.5×10^{16} and 2.5×10^{19} carriers/m^3 respectively.

1.2.1 Doping

The number of CB electrons or VB holes in a semiconductor, and hence its electrical conductivity, is very sensitive to the presence of traces of certain impurities. If atoms of phosphorus (P) that have five VB electrons are introduced into pure silicon (Si) that has only four VB electrons, the phosphorus atoms take up positions in the crystal structure, four of the five electrons forming bonds with neighbouring silicon atoms. The fifth electron from each P atom is surplus to the bonding requirements and being an electron in the outer shell it is only loosely bound to its parent atom. The binding energy of this electron is only about 0.05 eV and so even at low temperature it becomes 'free' and takes part in conduction, increasing the conductivity of the crystal. The introduction of foreign atoms into the intrinsic semiconductor is called *doping*, the element introduced being termed *dopant*, and the resulting doped crystal is referred to as an *extrinsic semiconductor*. In the above example of silicon doped with phosphorus, each phosphorus atom *donates* one CB electron to the crystal and phosphorus is thus described as *donor-dopant* in silicon. The release of the surplus electron from a donor-dopant atom is termed *ionisation* and the ionised donor then becomes a *positive ion* fixed in the lattice structure of the crystal. Although the dopant atoms are ionised, the positive and negative charge within the material is still balanced and so electrical neutrality is maintained. As the donated electron is surplus to bonding requirements, its excitation to the CB does not result in a deficiency in the bonding structure and hence does not result in a hole in the VB. The 'free' electron density in such a crystal is thus greater than the hole density and due to the predominance of electrons (*n*egative charge) this type of extrinsic semiconductor is termed *n-type*. The hole and electron densities in an *n*-type semiconductor are given the symbols p_n and n_n, the suffix indicating the *type* of semiconductor. As $n_n > p_n$, electrons are referred to as *majority* (charge) *carriers* and holes as *minority carriers* in an *n*-type semiconductor.

The element chosen as donor-dopant depends not only on the number of VB electrons but also on the size of the atom. For the dopant atom to take the place of a semiconductor atom in the crystal structure without causing flaws in the crystal, it must have a similar atomic size as the atom it replaces. For the intrinsic semiconductors silicon and germanium, suitable donor-dopants are phosphorus, arsenic and antimony each having five VB electrons compared with four for silicon and germanium. Table 1.1 shows the relevant section of the periodic table regarding semiconducting elements and compounds and dopants. Flaws in the structure reduce the conductivity of the crystal because they upset the uniform periodic potential distribution (electric field) in the crystal causing free carriers to be 'trapped' and annihilated (section 1.3.1). Considerable care has to be taken during the preparation of semiconductor material for device applications to avoid flaws (sections 3.1.1 and 3.1.2).

The value of binding energy for the 'surplus' electron can be calculated by modifying the electron energy expression for an isolated atom (equation 1.7). The donor atom in the semiconductor structure with its single loosely bound electron can be approximated to a hydrogen atom which has a single electron. In the crystal case, however, the effective mass of the electron differs from the value m due to the effect of the electric field within the crystal. It has been shown that the effective mass m^* is of the order of 0.8 m for silicon and 0.3 m for germanium. Also, the close spacing of atomic nuclei in the crystal affects the force on the electron which can be included by replacing the free-space value of permittivity ϵ_0 in equation 1.7 by the permittivity of the semiconductor $\epsilon = \epsilon_0 \epsilon_r$, where the relative permittivities (ϵ_r) for silicon and germanium are 12 and 16 respectively. Using these modifications, the binding energy of the 'surplus' electron is shown to be approximately 0.05 eV for silicon and 0.01 eV for germanium.

Table 1.1 Relevant section of the periodic table regarding semiconducting elements and compounds

Period	Outer shell	IIB 2	IIIB 3	IVB 4	VB 5	VIB 6	Group Electrons in outer shell
2	L		Boron(B)	Carbon(C)	Nitrogen(N)	Oxygen(O)	
3	M		Aluminium(Al)	Silicon(Si)	Phosphorus(P)	Sulphur(S)	
4	N	Zinc(Zn)	Gallium(Ga)	Germanium(Ge)	Arsenic(As)	Selenium(Se)	
5	O	Cadmium(Cd)	Indium(In)	Tin(Sn)	Antimony(Sb)	Tellurium(Te)	
6	P	Mercury(Hg)		Lead(Pb)			

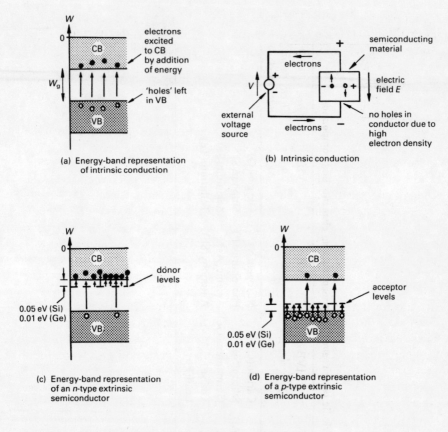

Figure 1.3 Intrinsic and extrinsic conduction in semiconductors

Figure 1.3c shows the energy-band representation of an *n*-type semiconductor. Each dopant atom effectively introduces a single energy level just 'below' the CB, described as *donor levels*. The energy difference between the donor levels and the lowest CB energy is the binding energy of the surplus electron. In terms of the energy-band diagram, the donor levels can be considered as allowed levels within the energy gap of the semiconductor. The *population* of electrons in the CB thus comprises electrons from ionised donor atoms together with a few thermally generated electrons from the VB. As the lower CB levels are occupied by electrons supplied by the donor atoms, fewer electron–hole pairs are generated than in the intrinsic case at the same level of excitation (for example, temperature) since more energy is required for electrons from the VB to 'reach' the unoccupied 'higher' levels in the CB. At normal operating temperatures almost all the donor-dopant atoms are ionised, each contributing one electron to the CB. In addition, at normal temperatures and typical dopant densities, the number of thermally generated electrons is negligible compared with dopant-derived electrons and therefore, if a semiconductor contains only donor-dopant of density N_d atoms/m³, the density of CB electrons is given by

$$n_n \approx N_d \tag{1.13}$$

If silicon or germanium is doped with a group III element that has only three valence electrons, such as boron, aluminium, gallium or indium (table 1.1), the bonding requirements of the adjacent semiconductor atoms in the structure cannot be satisfied; there is a deficiency of one electron to the bonding requirement. Thus the use of group III dopant introduces *holes* into the VB. This vacancy in the bonding structure can *accept* a VB electron leaving a hole elsewhere in the structure and so VB holes can move through the crystal contributing to conduction. Group III dopant elements are thus termed *acceptor-dopants* in silicon. This type of dopant therefore increases the hole density in the crystal without increasing the electron density and due to the majority of holes (*p*ositive charge), the resulting extrinsic semiconductor is termed *p*-type. The hole and electron densities in a *p*-type semiconductor are represented by p_p and n_p respectively and, since $p_p > n_p$, holes are the *majority carriers* and electrons are the *minority carriers*. In terms of the energy band diagram for *p*-type material (figure 1.3d), each acceptor-dopant atom introduces an acceptor level just 'above' the VB which accepts an electron from the VB resulting in a hole in the VB, described as ionisation of the acceptor-dopant atoms. These ionised acceptor atoms are fixed in the crystal lattice and, as each receives an electron, they are *negative* ions. The energy that a VB electron must acquire to fill the vacancy introduced by the acceptor-dopant corresponds to the binding energy calculated for the 'surplus' electron in the *n*-type case. Almost all the acceptor-dopant atoms are ionised at normal operating temperatures, each contributing a hole to the VB. As the number of thermally generated holes is negligible under normal conditions, the density of VB holes for a semiconductor containing only acceptor-dopant of density N_a atoms/m³ is

$$p_p \approx N_a \tag{1.14}$$

Pure silicon contains about 5×10^{28} atoms/m³ and the typical dopant densities used in the fabrication of semiconductor devices range from 10^{20} to 10^{28} atoms/m³, corresponding to the replacement of one silicon atom in 5×10^{20} to as much as one in five by a dopant atom. The lower end of this range is described as *light* doping and the upper end as *heavy* doping, the symbols p^-, p^+ and n^-, n^+ being used to indicate lightly doped and heavily doped *p*-type and *n*-type semiconductor respectively.

Although there appear to be a number of suitable donor and acceptor-dopants in terms of number of VB electrons (table 1.1) and atomic size, in practice the choice of dopant is somewhat limited. With particular regard to silicon, the acceptors boron, aluminium, gallium and indium all appear to be suitable acceptor-dopants. However, gallium diffuses appreciably in silicon dioxide so that the use of silicon dioxide as a masking medium in the fabrication process (sections 3.1.5 and 3.3) would not be possible. Aluminium reacts with any oxygen in the crystal and indium has a relatively high acceptor level causing lower acceptor ionisation. In practice, therefore, boron is the most popular choice as an acceptor-dopant in silicon. It has a moderate diffusion coefficient which is convenient in controlling the fabrication process. The choice of donor-dopant for silicon is more flexible, phosphorus, arsenic and antimony all being practical alternatives. Phosphorus has a higher diffusion coefficient than arsenic or antimony leading to shorter diffusion times. The range of diffusion rates provides a degree of flexibility in that arsenic or antimony can be used for early stages in the process, antimony being preferred because of its lower toxicity and phosphorus for later stages. The shorter diffusion time necessary for phosphorus means that regions doped with arsenic or antimony earlier in the process are not significantly disturbed during later stages; an important aspect in fabrication (section 3.3). The greater choice of donor-dopant compared with acceptor-dopant has implications in the choice of device structure from a fabrication point of view, particularly in the choice of the *npn* structure for a BJT in preference to the *pnp* arrangement (section 2.2.1). Another important factor is that the solid solubility of donor-dopants tends to be higher than for acceptor-dopants allowing heavier donor-doping (n^+ regions) than acceptor-doping (p^+ regions).

1.2.2 Quantitative Description of Carrier Densities in Semiconductors

The conductivity of a semiconductor depends on the densities of CB electrons and VB holes, and is thus dependent on dopant density and temperature. These free carrier densities depend on the availability of levels in the CB and VB and on the probability of electrons acquiring sufficient energy to occupy the available levels. The higher the level of energy considered, the less likely it is that an electron will have that energy, that is, the probability of occupation of an energy level W by an electron decreases as W increases.

Early work by Boltzmann modelled energy distribution by a probability function $P(W)$ where

$$P(W) = \exp\left(-\frac{W}{kT}\right) \tag{1.15}$$

The constant k (Boltzmann's constant, 1.38×10^{-23} J/K) provides the link between energy and temperature, where the thermal energy corresponding to a temperature T is kT joules. At 300 K (27 °C), $kT = 4.14 \times 10^{-21}$ J $= 0.0258$ eV. The probability of an energy level W being occupied $P(W)$ can take values between one and zero, $P(W) = 1$ states that level W is always occupied while $P(W) = 0$ states that level W is always unoccupied. Equation 1.15 implies that at absolute zero temperature ($T = 0$ K) only the $W = 0$ state can be occupied. However, this is not consistent with the exclusion principle applicable to closely spaced electrons as in a solid, since all electrons in the solid would then occupy the same state. The concept was modified by Fermi and Dirac who proposed that in a solid at $T = 0$ K all the levels up to a certain value of energy W_F are occupied and all the levels above W_F are unoccupied. In mathematical terms this probability function describing energy distribution is written

$$P(W) = \frac{1}{1 + \exp\left(\frac{W-W_F}{kT}\right)} \qquad (1.16)$$

and is known as the *Fermi–Dirac function*. The energy value W_F is termed the *Fermi level* or *Fermi energy*.

At $T = 0$ K, $P(W) = 1$ for $W < W_F$ and $P(W) = 0$ for $W > W_F$. The probability function, that is, the variation of the probability of occupation with energy, for a continuous range of allowed levels is shown in figure 1.4a. As the temperature increases from absolute zero, the energy imparted to the structure (kT) causes some electrons having energy just below W_F to attain energies just above W_F causing the probability of occupation below W_F to be reduced and that above W_F to be increased. A value of probability between one and zero implies that the level is occupied for part of the time. For any temperature the level of energy corresponding to W_F has a probability of occupation of 0.5, since from equation 1.16, $P(W) = 0.5$ for $W = W_F$ whatever the value of T. Thus, in terms of the model, the Fermi level W_F is defined as that *level of energy in a continuous range of allowed levels that is occupied by an electron for half of the time*. It should be noted that the statement 'energy level W is occupied' should not be considered as 'physical' occupation but that 'within the structure there exists an electron of energy W'.

The above discussion of the Fermi–Dirac distribution applies to a continuous range of allowed electron energies as exists for a conductor. In a semiconductor this is not the case since values of energy corresponding to levels within the energy gap are not stable and therefore not allowed in the band model at equilibrium. Figure 1.4b shows the corresponding distribution for an intrinsic semiconductor. The same function applies but the portion corresponding to energies within the gap is meaningless and therefore shown dashed. At $T = 0$ K, all VB levels are 'full' ($P(W) = 1$) and all CB levels are 'empty' ($P(W) = 0$). For $T > 0$ K some VB electrons gain sufficient energy to occupy CB levels and the density of electrons in the CB (n_i) equals the density of holes left in the VB (p_i).

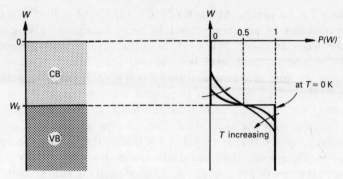

(a) Probability of occupation of a continuous range of energy levels in a conductor

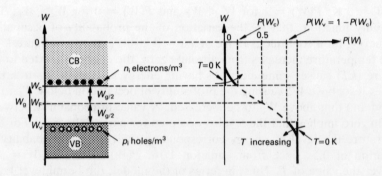

(b) Probability of occupation of VB and CB levels for an intrinsic semiconductor

Figure 1.4 Fermi–Dirac energy distribution statistics

An expression for the free electron density n may be developed from the probability of levels at the bottom of the CB (energy W_c, figure 1.4b) being occupied ($P(W_c)$) and the number of available levels in the CB. From equation 1.16

$$P(W_c) = \frac{1}{1 + \exp\left(\dfrac{W_c - W_F}{kT}\right)} \approx \exp\left(\frac{W_F - W_c}{kT}\right) \qquad (1.17)$$

the approximation being valid for most intrinsic and lightly doped semiconductors as $W_c - W_F \gg kT$ making the exponential term dominant. If the density of available levels in the CB is N_c

$$n = N_c P(W_c) \approx N_c \exp\left(\frac{W_F - W_c}{kT}\right) \tag{1.18}$$

The hole density p may be considered in a similar way. If the probability of the level at the top of the VB (energy W_v figure 1.4b) being occupied by an electron is $P(W_v)$, the probability of it being occupied by a hole (that is, not being occupied by an electron) is $1 - P(W_v)$. From equation 1.16

$$1 - P(W_v) = 1 - \frac{1}{1 + \exp\left(\frac{W_v - W_F}{kT}\right)} \approx \exp\left(\frac{W_v - W_F}{kT}\right), \tag{1.19}$$

noting that $W_F - W_v \gg kT$ for most intrinsic and lightly doped semiconductors. If the density of available levels in the VB is N_v

$$p = N_v(1 - P(W_v)) \approx N_v \exp\left(\frac{W_v - W_F}{kT}\right) \tag{1.20}$$

For intrinsic semiconductors, $n = n_i$, $p = p_i$ and from equation 1.12, n_i and p_i are equal. Thus from equations 1.18 and 1.20

$$N_c \exp\left(\frac{W_F - W_c}{kT}\right) = N_v \exp\left(\frac{W_v - W_F}{kT}\right)$$

from which

$$W_F = \frac{W_c + W_v}{2} \tag{1.21}$$

assuming $N_c = N_v$. The Fermi energy is thus the average of W_c and W_v, that is the mid-gap energy for an intrinsic semiconductor. From equations 1.18 and 1.20 the product of the free carrier densities is

$$pn = N_c N_v \exp\left(\frac{W_v - W_c}{kT}\right)$$

$$= N_c N_v \exp\left(-\frac{W_g}{kT}\right) \tag{1.22}$$

as $W_c - W_v = W_g$. As the terms on the RHS of equation 1.22 are constant for a particular semiconductor at fixed temperature, this leads to the important result that *the product of the free carrier densities is constant*. The constant being dependent on the energy gap of the semiconductor and the temperature.

In the case of an intrinsic semiconductor, $n = n_i$, $p = p_i$ and $n_i = p_i$ thus

$$pn = n_i^2 \qquad (1.23)$$

and combining equations 1.22 and 1.23, the intrinsic carrier density is given by

$$n_i = \sqrt{(N_c N_v)} \exp\left(-\frac{W_g}{2kT}\right) \qquad (1.24)$$

Detailed consideration [11] reveals that $N_c N_v$ is a function of T^3. The values of n_i for silicon and germanium at $T = 300$ K are 1.5×10^{16} and 2.5×10^{19} carriers/m³ respectively.

Figure 1.5 shows the energy band models and corresponding Fermi–Dirac energy distribution statistics for extrinsic semiconductors. As the degree of doping determines the majority carrier density, the probability of occupation of CB levels in n-type material and the probability of holes at VB levels in p-type material are increased compared with the intrinsic case and so the Fermi energy is a function of dopant density.

For temperatures of the order of 300 K, the number of thermally generated majority carriers is insignificant at practical doping levels. Thus the free electron density in an n-type semiconductor n_n is given approximately by the donor density N_d (equation 1.13). Corresponding to equation 1.18, n_n can therefore be written

$$n_n \approx N_c \exp\left(\frac{W_{Fn} - W_c}{kT}\right) \approx N_d \qquad (1.25)$$

where W_{Fn} is the Fermi energy for the n-type material. As $n_n > n_i$, W_{Fn} is higher than the intrinsic value W_F. The energy difference between the Fermi level and the bottom of the CB is thus

$$W_c - W_{Fn} \approx kT \ln \frac{N_c}{N_d} \qquad (1.26)$$

showing that an increase of donor density N_d causes the Fermi energy W_{Fn} to increase. The increased probability of CB level occupation and the reduced probability of holes in the VB due to reduced thermal generation for an n-type semiconductor is shown in figure 1.5a. The corresponding situation for

Semiconductor Physics

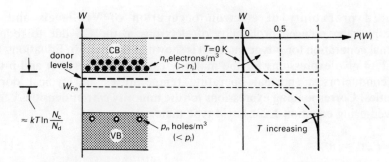

(a) Probability of occupation of VB and CB levels for an n-type semiconductor

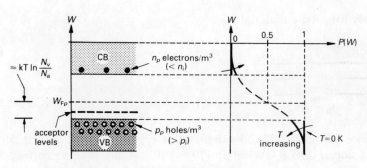

(b) Probability of occupation of VB and CB levels for a p-type semiconductor

Figure 1.5   Fermi–Dirac energy distribution statistics for extrinsic semiconductors

a p-type semiconductor in which $p_p \approx N_a$ can be deduced by modifying equation 1.20 for the p-type case, that is

$$p_p \approx N_v \exp\left(\frac{W_v - W_{Fp}}{kT}\right) \approx N_a \qquad (1.27)$$

from which the Fermi energy for the p-type material W_{Fp} is

$$W_{Fp} - W_v \approx kT \ln \frac{N_v}{N_a} \qquad (1.28)$$

above the top of the VB. Comparison of equations 1.27 and 1.20 shows that as $p_p > p_i$, W_{Fp} is lower than the intrinsic value W_F. From equation 1.28, an increase in acceptor-dopant N_a causes the Fermi energy W_{Fp} to decrease. The increased hole density in the VB due to the acceptor-dopant, that is, the reduced probability of electron occupation of VB levels and the corresponding reduced probability of electrons in the CB due to reduced thermal generation for *p*-type material is shown in figure 1.5b. Equations 1.13 and 1.14 give expressions for the majority carrier densities in *n* and *p*-type semiconductors at normal operating temperatures and typical dopant densities. Corresponding expressions for the minority carrier densities can be derived using equation 1.23. For *n*-type material

$$P_n n_n = n_i^2 \tag{1.29}$$

thus

$$p_n = \frac{n_i^2}{n_n} \approx \frac{n_i^2}{N_d} \tag{1.30}$$

Similarly for *p*-type material

$$p_p n_p = n_i^2 \tag{1.31}$$

and so

$$n_p = \frac{n_i^2}{p_p} \approx \frac{n_i^2}{N_a} \tag{1.32}$$

Equations 1.30 and 1.32 show that for the typical range of dopant densities used in silicon device fabrication ($10^{20} - 10^{28}$ atoms/m^3), the corresponding minority carrier density range is 2.25×10^{12} to 2.25×10^4 carriers /m^3 respectively at $T = 300$ K ($n_i = 1.5 \times 10^{16}$ carriers /m^3). The majority to minority carrier density ratio is thus in the range 10^8 to 10^{24}.

1.2.3 Degenerate Semiconductors

As the dopant density in an extrinsic semiconductor is increased, the Fermi energy increases for *n*-type material approaching CB energies and decreases for *p*-type material approaching VB energies. In the case of an *n*-type semiconductor, if the doping is sufficiently heavy, the Fermi energy W_{Fn} has a value within the range of CB energies (figure 1.6a). Under these conditions the lower levels in the CB are occupied for most of the time, resulting in high conductivity. Such heavily doped semiconductor is described as *degenerate*, the conduction properties having degenerated to those of a conductor. For degenerate *p*-type material the Fermi energy W_{Fp} has a value within the VB range (figure 1.6b), the upper VB levels being unoccupied, that is, containing holes, for most of time, again resulting in high conductivity.

In quantitative terms, the Fermi energy is at the bottom of the CB for *n*-type material when $W_c - W_{Fn} = 0$ and at the top of the VB for *p*-type material when $W_{Fp} - W_v = 0$. From equations 1.26 and 1.28 these conditions correspond to doping levels of $N_d = N_c$ and $N_a = N_v$ respectively, which are

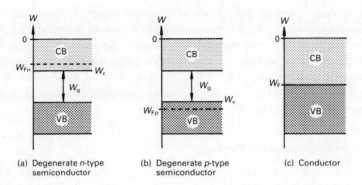

Figure 1.6 Energy band models of degenerate semiconductors in comparison with that of a conductor

of the order of 10^{25} atoms/m^3. Thus a semiconductor becomes degenerate for doping levels greater than about 10^{25} atoms/m^3. As a semiconductor approaches degeneracy, that is, for dopant densities greater than about 10^{23} atoms/m^3, the simplifying approximations made in equations 1.17 and 1.19 become invalid. For example, at a doping level of 10^{24} atoms/m^3, the ratios N_c/N_d and N_v/N_a are of the order of ten so that ignoring the unity in equations 1.17 and 1.19 results in a ten per cent error. Physically, the use of the approximate probability expression for low dopant densities, which is termed the Maxwell–Boltzmann model, corresponds to the case of low carrier density within the crystal where the exclusion principle of level occupancy is less significant. At higher carrier densities the exclusion principle has major effect and the Fermi–Dirac model must be used.

1.2.4 Temperature Effects

The product of the majority and minority carrier densities has been shown to be fixed at constant temperature (equations 1.22, 1.23, 1.29 and 1.31). The ratio of majority to minority density is typically in the range 10^8–10^{24} at 300 K, showing that in a homogeneous extrinsic semiconductor the contribution of minority carriers to current flow is insignificant. The intrinsic carrier density (n_i), however, increases with temperature (equation 1.24), and so both the majority and minority densities increase with temperature. Since thermal generation produces electron–hole pairs, the ratio of majority to minority density reduces as the temperature increases until at the *transition temperature* the density of thermally generated carriers is the same as the density of dopant-derived carriers, that is, $n_i = N_d$ (*n*-type) or $p_i = N_a$ (*p*-type). At and above this temperature $n_n \neq N_d$ and $p_p \neq N_a$ due to the significant proportion of thermally generated carriers and the semiconductor acquires *intrinsic* properties as the hole and electron densities are then of similar order. Thermal generation at elevated temperature is important in certain semiconductor devices as the increased current flow resulting from the rise in carrier densities causes the power dissipation and hence the semiconductor temperature to rise, further increasing the carrier densities. Beyond a certain limit, determined by the flow of heat away from the semiconductor, the effect escalates resulting in device failure. This positive feedback effect is termed *thermal runaway*.

1.2.5 Semiconductors in Practice

The most important semiconducting materials have been the elements germanium and silicon. Semiconductor devices throughout the 1950s were based almost entirely on germanium which has an intrinsic conductivity (equation 1.47) of 2.3 S/m at 300 K. Since the early 1960s the majority of discrete semiconductor devices and all commercial integrated circuits have used silicon doped with boron, phosphorus, antimony or arsenic. The larger energy gap of silicon (1.1 eV compared to 0.7 eV for germanium) together with the reduced carrier mobilities (table 1.2) mean that silicon has a lower conductivity than germanium, the intrinsic conductivity of silicon at 300 K being 4.5×10^{-4} S/m. Although of lower conductivity, the lower cost, fabrication advantages (section 3.2) and improved device performance (section 2.1.1), particularly at elevated temperature, offered by silicon have led to its popularity. Table 1.2 lists the most important semiconductors and their physical properties. The compound semiconductor gallium arsenide doped with tellurium, sulphur, zinc and oxygen or zinc and nitrogen is currently of particular interest for high performance devices and ICs due to its high electron mobility and low intrinsic conductivity although there are fabrication difficulties. The wide range of energy gaps available in compound semiconductors, which include the ternary (three element), quaternary (four element) and even quinternary (five element) compounds with various proportions of each element, are particularly important for optical devices (section 2.6).

1.3 FREE CARRIERS IN SEMICONDUCTORS

As far as the conduction properties of semiconductors and semiconductor devices are concerned, it is the electrons in the outer shells of the atoms in the crystal that are of interest. These are the electrons which, if given sufficient energy either by thermal or electrical means, are able to break free from their parent nuclei and take part in conduction through the crystal, that is, they become CB electrons. These electrons and the holes left in VB levels are termed *free carriers* since they are able to move through the crystal and it is the numbers and energies of these carriers that determine the electrical

Table 1.2 Physical properties of some semiconductors at 300 K

Semiconductor	Energy gap W_g (eV)	Electron mobility $\mu_n (m^2/V\,s)$	Hole mobility $\mu_p (m^2/V\,s)$
Elements			
Silicon (Si)	1.1	0.14	0.05
Germanium (Ge)	0.7	0.39	0.19
Selenium (Se)	2.6	–	10^{-4}
Tellurium (Te)	0.3	0.12	0.06
Group IV compound			
Silicon carbide (SiC)	3.0	0.01	0.001
Group III–V compounds			
Gallium arsenide (GaAs)	1.4	0.85	0.05
Gallium phosphide (GaP)	2.3	0.05	0.02
Gallium antimonide (GaSb)	0.7	0.40	0.07
Indium arsenide (InAs)	0.4	2.30	0.01
Indium phosphide (InP)	1.3	0.45	0.05
Indium antimonide (InSb)	0.2	7.50	0.12
Aluminium arsenide (AlAs)	2.2	0.02	–
Aluminium phosphide (AlP)	2.5	0.008	–
Aluminium antimonide (AlSb)	1.8	0.02	0.03
Group II–VI compounds			
Zinc oxide (ZnO)	3.2	0.02	–
Zinc sulphide (ZnS)	3.8	0.01	–
Zinc selenide (ZnSe)	2.8	0.06	–
Zinc telluride (ZnTe)	2.4	–	0.010
Cadmium sulphide (CdS)	2.4	0.02	0.002
Cadmium selenide (CdSe)	1.8	0.07	–
Cadmium telluride (CdTe)	1.6	0.10	0.010
Mercury telluride (HgTe)	0.2	2.50	0.040
Group IV–VI compounds			
Lead sulphide (PbS)	0.4	0.08	0.02
Lead selenide (PbSe)	0.3	0.12	0.12
Lead telluride (PbTe)	0.3	0.12	0.05
Ternary compounds			
Gallium arsenide-phosphide ($GaAs_{1-x}P_x$)	1.4–2.3	–	–
Gallium aluminium arsenide ($Ga_xAl_{1-x}As$)	1.4–2.2	–	–
Quaternary compounds			
Gallium indium arsenide-phosphide ($Ga_xIn_{1-x}As_yP_{1-y}$)	0.5–2.1	–	–

properties of devices. The lower energy electrons associated with each atom, those in the core levels (figure 1.2b), do not take part in conduction; they are part of the ions left fixed in the crystal lattice.

1.3.1 Generation and Recombination

At a fixed level of excitation, electron–hole pairs are continually being created or *generated* as the supplied energy enables some electrons having VB levels of energy to gain sufficient energy to break free from their parent nuclei and to take part in conduction through the crystal. The excitation of VB electrons to CB energy levels leaves holes in the VB which also contribute to conduction. At the same time that free carriers are being generated, the reverse process of *recombination* is also taking place whereby free electrons and holes meet, annihilating each other. In physical terms, recombination is the process whereby a free electron travelling through the crystal is *captured* by an atom having an incomplete bonding structure. The captured electron completes the structure and is then no longer free to take part in conduction. In the energy band model this mechanism is described by an electron returning to a VB energy level from a CB level with a corresponding loss of energy. In most cases the energy lost by the electron in returning to the lower energy state is transferred to the rest of the crystal contributing to its thermal energy state although in certain cases the released energy can be emitted from the surface of the semiconductor which is the phenomenon used in semiconductor optical sources (section 2.6). If the level of crystal excitation is constant, a dynamic equilibrium condition is reached whereby the rates of generation and recombination are equal and the free electron and hole densities for this condition are termed the *equilibrium densities*.

Practical crystals contain regions of non-uniform structure due to missing or displaced atoms in the lattice and the presence of impurities. Such regions are collectively termed crystal *flaws* or *defects*. Non-uniformity also occurs at an interface between two crystals and at the surface due to the change or termination of the regular array of atoms. Such non-uniformities upset the periodic nature of the electric field in the crystal causing regions which tend to capture or *trap* free carriers. In the energy band model these disturbances in the electric field create levels in the energy gap range which have a high probability of capturing an electron, a hole or both types of carrier (figure 1.7). Such levels corresponding to energies approaching CB energies are termed *electron traps* as they have a high probability of capturing free electrons, while those having energies near VB levels are called *hole traps*, having a high probability of capturing a hole. During its *life* as a free carrier, an electron or hole is trapped by such levels many times, being 'freed' to the adjacent band by energy absorption. While trapped, electrons and holes cannot take part in conduction and may be considered as being out of circulation for a brief time. Trapping levels corresponding to energies near

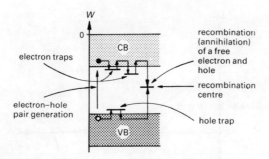

Figure 1.7 Band model representation of generation and recombination in a semiconductor

the centre of the energy gap have a high probability of capturing both an electron and a hole resulting in annihilation (recombination) of the trapped carriers; such traps are termed *recombination centres*.

The average time that an electron or hole remains free after generation is known as its *lifetime* (τ).

If energy is supplied to a material, for example, by irradiating the crystal, generation causes the densities of free carriers to increase above the equilibrium values corresponding to the crystal temperature. The additional free carriers created are termed *excess carriers* and the total density at any instant is the sum of the equilibrium density and the excess density. If the increased rate of generation is maintained by continuing irradiation of the crystal, the rate of recombination increases and a new equilibrium is reached. However, if the irradiation is suddenly terminated, the generation rate falls to the previous equilibrium value and recombination of the excess carriers causes the free carrier densities to fall to their previous equilibrium values. As a first-order model, the rate of recombination of excess carriers is taken as proportional to the excess carrier density present. Thus, considering the recombination of excess electrons in a semiconductor

$$-\frac{d(\Delta n)}{dt} \propto \Delta n$$

from which

$$\frac{d(\Delta n)}{dt} = -\left(\frac{1}{\tau}\right)\Delta n \tag{1.33}$$

and

$$\Delta n(t) = \Delta n(0) \exp\left(-\frac{t}{\tau}\right) \tag{1.34}$$

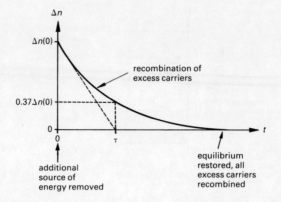

Figure 1.8 Lifetime of excess electrons

where $\Delta n(t)$ is the excess electron density at time t, $\Delta n(0)$ is the value of Δn at $t = 0$ (the instant at which the source of energy causing the generation of excess carriers is removed) and τ is the electron lifetime. Figure 1.8 shows the decay of the excess electron density due to recombination as represented by equation 1.34.

At $t = \tau$ the excess density $\Delta n(\tau)$ is given by equation 1.34 to be

$$\Delta n(\tau) = \Delta n(0) \exp(-1) \approx 0.37 \Delta n(0) \tag{1.35}$$

The excess carrier lifetime can thus be defined quantitatively as

(1) the time taken for the excess density to fall to 37 per cent of its initial value, or
(2) the hypothetical time taken for the excess density to fall to zero if it continued to fall at its initial rate (figure 1.8)

A carrier requires one of opposite type with which to recombine and so in an extrinsic semiconductor, which has relatively few of one type of carrier, it is the lifetime of minority carriers that is of interest. Due to the different energy states and the different trapping probabilities for the two types of carrier, the lifetimes of minority electrons (τ_n) and holes (τ_p) are generally different. In practice, semiconductors have minority carrier lifetimes from less than a nanosecond to several milliseconds. The rate of recombination tends to be greater at the surface of a crystal than in the interior (bulk) owing to the high density of traps at the interface. Carrier lifetime is therefore longer in the bulk of the material than at the surface and the surface value therefore dominates. In device fabrication, particular attention has to be given to the semiconductor surface to avoid surface effects masking the internal device properties.

1.3.2 Drift

In a perfect (defect free) crystal the periodic electric field enables electrons and holes to move freely as if in a vacuum. In this case the wave model of the electron is more appropriate than the particle model. The defects in real crystals cause the periodic electric field to be disturbed and, for $T > 0$ K, thermal energy can be considered to cause the lattice ions to vibrate. These imperfections cause scattering of the free carriers resulting in finite conductivity.

By considering the free electrons in a semiconductor to be analogous to the molecules in a low pressure gas, the kinetic energy of a free electron due to its thermal energy has been shown [12] to be $1.5\ kT$ from which its mean (thermal) velocity can be estimated to be of the order of 10^5 m/s. In the absence of any externally applied electric field, the random motion of free carriers within a crystal does not result in a net transfer of charge since charge movement in any direction is balanced by charge movement in other directions. When a voltage is applied across the material, each carrier experiences a force attracting it to one end of the material, electrons being attracted to the positive potential due to their negative charge and holes to the negative potential. This net movement in charge, termed *drift*, is superimposed on the random 'thermal' movement and results in current flow through the crystal.

The electric field (strength) E is defined as the force on a unit charge and so the force on an electron within the material due to the field created by the applied voltage is eE newtons since the electron carries a charge of magnitude e. Examination of units shows that force/unit charge is equal to the potential gradient (volt/metre) of the field at that point

$$\frac{\text{force}}{\text{charge}} = \frac{\text{force} \times \text{distance}}{\text{charge} \times \text{distance}} = \frac{\text{energy/charge}}{\text{distance}} = \frac{\text{voltage}}{\text{distance}} \tag{1.36}$$

As E is defined in terms of the force on a *positive* charge

$$E = -\frac{dV}{dx} \tag{1.37}$$

as illustrated in figure 1.9a. In a first-order model of the situation, an electron can be assumed to be brought to rest by a collision and be accelerated by the force eE (acceleration = force/mass = eE/m) to a velocity eEm/t_c at the instant before the next collison, where t_c is the mean time between collisions. The mean velocity of electrons through the crystal due to the field (drift velocity u_d) is thus

$$u_d = \frac{eEt_c}{2m} \tag{1.38}$$

from which

$$u_d \propto E$$
$$u_d = \mu E \tag{1.39}$$

where the parameter μ is the electron *mobility* within the crystal. This linear model of electron drift is only valid for drift velocities that are small compared with the thermal velocity, that is, for E less than about 10^5 V/m. For higher values of E, u_d reaches a limiting value approaching 10^5 m/s.

Figure 1.9b shows a voltage applied across a conductor of uniform cross-section. The uniform electric field established in the conductor ($|E| = V/l$ from equation 1.37) causes electrons to drift along the conductor with velocity u_d. The resulting current is given by the rate of charge flow passed across a plane perpendicular to the movement such as X in figure 1.9b. Assuming uniform charge distribution

I = (charge/unit length of conductor) × (drift velocity)
  = (charge density) × (area of plane X) × (drift velocity)
  = $neAu_d$

Thus current density

$$J = \frac{I}{A} = neu_d = ne\mu E \tag{1.40}$$

since $u_d = \mu E$ from equation 1.39.

By comparison with general conduction field equation $J = \sigma E$, the conductivity of the conductor σ (S/m) is given by

$$\sigma = ne\mu \tag{1.41}$$

or in terms of resistivity $\rho(\Omega m)$

$$\rho = \frac{1}{\sigma} = \frac{1}{ne\mu} \tag{1.42}$$

Typical values of electron mobility in conductors at 300 K are 0.0066, 0.0032, 0.0047 and 0.0012 m²/V s for silver, copper, gold and aluminium respectively giving conductivities of 6.1×10^7, 5.9×10^7, 4.4×10^7 and 3.6×10^7 S/m respectively.

If the material in figure 1.9b were a semiconductor, the current flow would be due to electron and hole movement. Corresponding to equation 1.40 the current densities due to electron drift and hole drift are

$$J_n \text{ (drift)} = ne\mu_n E \tag{1.43}$$

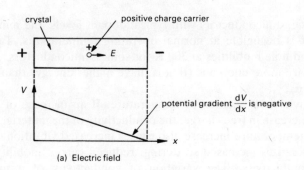

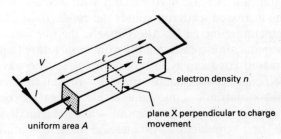

(b) Carrier drift in an applied electric field

Figure 1.9 Carrier drift

$$J_p \text{ (drift)} = pe\mu_p E \tag{1.44}$$

respectively, where μ_n and μ_p are the mobilities of electrons and holes in the semiconductor. Although these charge movements are in opposite directions, the electron movement is negative charge movement while the hole component is positive charge movement and so the total drift current is the *sum* of the two components, thus

$$J \text{ (drift)} = J_n \text{ (drift)} + J_p \text{ (drift)}$$
$$= ne\mu_n E + pe\mu_p E \tag{1.45}$$

Comparison with $J = \sigma E$ shows that the conductivity of a semiconductor is

$$\sigma = ne\mu_n + pe\mu_p \tag{1.46}$$

For an intrinsic semiconductor $n = p = n_i$ and so the intrinsic conductivity is given by

$$\sigma_i = n_i e(\mu_n + \mu_p) \tag{1.47}$$

For extrinsic semiconductors with typical dopant levels, the minority carrier component is negligible at normal operating temperatures. Table 1.2 lists electron and hole mobilities at 300 K for some semiconductors. As electrons in the CB are more energetic (that is, have higher energy) than holes in the VB, $\mu_n > \mu_p$.

Conductivity is a function of temperature. If an increase of temperature causes an increase in free charge, the conductivity of the material is increased. However, temperature increase also causes increased vibration of the lattice ions which causes increased scattering, reduced carrier mobility and hence reduced conductivity. The variation of conductivity of a material with temperature depends on which of these effects dominates. For a conductor there is not a significant increase in free charge with increase of temperature and therefore the increased scattering causes the conductivity of a conductor to fall as the temperature increases. Alternatively, the free carrier density of an *intrinsic* semiconductor increases rapidly with temperature (equation 1.24) resulting in increased conductivity. A conductor is said to have a *positive temperature coefficient* of resistance since its resistance increases with temperature while an intrinsic semiconductor has a negative coefficient. At normal operating temperatures an *extrinsic* semiconductor has properties similar to those of a conductor. An increase of temperature does not significantly increase the free carrier density since most of the dopant atoms are already ionised; thus increased scattering causes conductivity to fall. However, as the transition temperature is approached, the generation of electron–hole pairs does cause the free carrier density to increase significantly and the material takes on intrinsic properties. Thus an extrinsic semiconductor has a positive temperature coefficient of resistance at normal temperatures changing to a negative coefficient near the transition temperature.

1.3.3 Diffusion

The free electron density in a homogeneous semiconductor in the absence of an applied electric field is uniform at equilibrium due to a combination of the random 'thermal' motion of the electrons and the repulsive force between them caused by their like charge. If excess carriers are introduced locally within such a semiconductor, either by causing carrier generation by heating or incident radiation or by *injecting* carriers into the material via a surface contact (section 1.6), a non-uniform distribution is created. Ignoring the process of recombination which is an additional phenomenon, the excess carriers move from the region of higher density to regions of lower density tending to produce a uniform distribution. This transport mechanism is called *diffusion* and it takes place in addition to *drift* caused by any applied electric field.

Fick's law states that the rate at which carriers diffuse is proportional to the

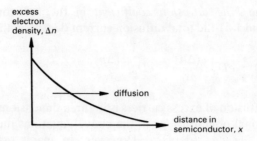

Figure 1.10 Carrier diffusion

density gradient and that the movement is in the direction of negative gradient (figure 1.10) thus, in one space dimension, the diffusion of electrons is described by

$$\frac{\partial(\Delta n)}{\partial t} \propto -\frac{\partial(\Delta n)}{\partial x}$$

$$= -D_n \frac{\partial(\Delta n)}{\partial x} \quad (1.48)$$

where D_n is the *diffusion coefficient* for electrons in the semiconductor concerned. Partial derivatives are used in equation 1.48 as Δn is a function of both time and distance. This movement constitutes an *electron diffusion current density* and since conventional current is the rate of movement of *positive* charge

$$J_n \text{ (diffusion)} = -(e)\left(\frac{\partial(\Delta n)}{\partial t}\right)$$

$$= +eD_n \frac{\partial(\Delta n)}{\partial x} \quad (1.49)$$

If an excess hole concentration is created in the same region, hole diffusion takes place in the same direction at a rate

$$\frac{\partial(\Delta p)}{\partial t} = -D_p \frac{\partial(\Delta p)}{\partial x} \quad (1.50)$$

resulting in a *hole diffusion current density*

$$J_p \text{ (diffusion)} = +(e)\left(\frac{\partial(\Delta p)}{\partial t}\right)$$

$$= -eD_p \frac{\partial(\Delta p)}{\partial x} \quad (1.51)$$

where D_p is the *hole diffusion coefficient* in the semiconductor. From equations 1.49 and 1.51 the total diffusion current density in the x direction is

$$J(\text{diffusion}) = eD_n \frac{\partial(\Delta n)}{\partial x} - eD_p \frac{\partial(\Delta p)}{\partial x} \tag{1.52}$$

In general the diffusion of excess carriers is in three dimensions, in which case the one-dimensional operator $\partial/\partial x$ in the above equations must be replaced by $\partial/\partial x + \partial/\partial y + \partial/\partial z$ (del ∇). However, in many practical cases a one-dimensional description is adequate due to the device geometry.

1.3.4 Einstein's Relation

Although the carrier transport mechanisms drift and diffusion have been considered independently, they are both dependent on the scattering processes responsible for hindering the flow of carriers in a semiconductor. Einstein showed that the parameters describing the two processes, mobility μ and diffusion coefficient D respectively, are directly related. At equilibrium with no applied electric field, the free electron distribution is uniform and there is no net current flow. Any tendency to disturb the state of equilibrium which would lead to a diffusion current, creates an internal electric field and a drift current balancing the diffusion component. Under equilibrium conditions, therefore, the drift and diffusion currents due to an excess density of electrons Δn are equal, and from equations 1.43 and 1.49

$$\Delta n e \mu_n E = eD_n \frac{\partial(\Delta n)}{\partial x} \tag{1.53}$$

The force F on the excess carriers restoring equilibrium is given by the product of excess charge and electric field (from the definition of E), thus from equation 1.53

$$F = \Delta n e E = e \frac{D_n}{\mu_n} \frac{\partial(\Delta n)}{\partial x} \tag{1.54}$$

This force also depends on the thermal energy of the excess carriers. By making an analogy between the excess carriers in a semiconductor and gas molecules in a low pressure gas, the force corresponding to the pressure gradient is equal to $kT \, \partial(\Delta n)/\partial x$ as, from the kinetic theory of gases, gas pressure is equal to nkT where n is the molecule concentration. Comparing this force term with equation 1.54 gives

$$\frac{D_n}{\mu_n} = \frac{kT}{e} \tag{1.55}$$

similarly for holes

$$\frac{D_p}{\mu_p} = \frac{kT}{e} \quad (1.56)$$

thus

$$\frac{D_n}{D_p} = \frac{\mu_n}{\mu_p} \quad (1.57)$$

The relation between diffusion coefficient and mobility of a charge carrier is termed Einstein's relation. Use of the analogy with a *low pressure* gas corresponds to conditions of low carrier densities in the semiconductor, that is, it applies to non-degenerate semiconductors. A rigorous derivation of this relation is given in reference 13.

At $T = 300$ K, $kT/e \approx 26$ mV. Using this relationship the electron and hole diffusion coefficients for various semiconductors can be calculated from the mobility values given in table 1.2. The values for silicon are $D_n = 3.6 \times 10^{-3}$ m^2/s and $D_p = 1.3 \times 10^{-3}$ m^2/s while those for germanium are $D_n = 0.01$ m^2/s and $D_p = 5 \times 10^{-3}$ m^2/s.

1.3.5 General Excess Carrier Movement: the Continuity Equations

The behaviour of excess carriers in a semiconductor is due to the combined effects of generation, recombination, drift and diffusion. The mathematical relations combining these effects are termed the electron and hole *continuity equations* which can be derived by considering the combined time rate of change of the excess charge density within an elemental volume (figure 1.11a). The number of excess electrons in such a volume is $\Delta n A \delta x$ and the rate of increase of this number is $\partial(\Delta n)/\partial t \, A \delta x$. This rate of change is due to three effects: the net rate of electron recombination and movement due to drift and diffusion.

Net Recombination Rate

Equation 1.33 is a first-order representation of the net rate of increase of the number of excess electrons in the volume due to the difference between the rates of recombination and generation, thus

$$\left.\frac{\partial(\Delta n)}{\partial t}\right|_{\text{gen. and recomb.}} = -\frac{\Delta n}{\tau_n} \quad (1.58)$$

where τ_n is the lifetime of excess electrons.

Rate of Increase due to Drift

Due to the applied electric field E, electrons drift into the volume at plane 2 and out at plane 1 (figure 1.11b). The excess electron density at plane 1 is Δn

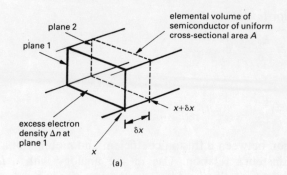

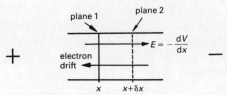

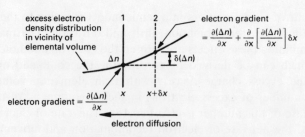

Figure 1.11 General excess carrier movement

and so from equation 1.43 the current density at this plane due to excess electron drift is $\Delta n e \mu_n E$ and the current leaving the volume is therefore $\Delta n e \mu_n E A$ where A is the cross-sectional area of the element. At plane 2 the excess density is $\Delta n + \delta(\Delta n)$ as shown in figure 1.11c and so the current entering the volume is $(\Delta n + \delta(\Delta n)) e \mu_n E A$. The net rate of increase of the number of excess electrons in the volume due to drift is thus

$$\frac{(\Delta n + \delta(\Delta n))e\mu_n EA - \Delta n e \mu_n EA}{e} = \delta(\Delta n) \mu_n EA$$

but

$$\delta(\Delta n) = \frac{\partial(\Delta n)}{\partial x} \times \delta x$$

therefore

$$\left.\frac{\partial(\Delta n)}{\partial t}\right|_{\text{drift}} = \frac{\partial(\Delta n)}{\partial x}\mu_n E A \delta x \qquad (1.59)$$

Rate of Increase due to Diffusion
The change of excess electron gradient across the element (figure 1.11c) causes a net increase of electrons in the volume since the rate of diffusion into the volume at plane 2 is greater than the rate of diffusion out of the volume at plane 1. Equation 1.48 relates the rate of diffusion to the excess electron gradient and, as diffusion is in the negative x direction in this case, the respective rates of diffusion into and out of the volume are

$$D_n\left[\frac{\partial(\Delta n)}{\partial x} + \frac{\partial}{\partial x}\left[\frac{\partial(\Delta n)}{\partial x}\right]\delta x\right]A$$

and

$$D_n \frac{\partial(\Delta n)}{\partial x} A$$

The net rate of increase of electrons in the volume is therefore

$$D_n \frac{\partial}{\partial x}\left[\frac{\partial(\Delta n)}{\partial x}\right]\delta x A$$

thus

$$\left.\frac{\partial(\Delta n)}{\partial t}\right|_{\text{diffusion}} = D_n \frac{\partial^2(\Delta n)}{\partial x^2} A \delta x \qquad (1.60)$$

The total rate of increase of excess electrons in the volume is given by the sum of the increases given by equations 1.58, 1.59 and 1.60 resulting in

$$\frac{\partial(\Delta n)}{\partial t} = -\frac{\Delta n}{\tau_n} + \mu_n E \frac{\partial(\Delta n)}{\partial x} + D_n \frac{\partial^2(\Delta n)}{\partial x^2} \qquad (1.61)$$

which is the *continuity equation for excess electrons*. For excess holes the corresponding equation is

$$\frac{\partial(\Delta p)}{\partial t} = -\frac{\Delta p}{\tau_p} - \mu_p E \frac{\partial(\Delta p)}{\partial x} + D_p \frac{\partial^2(\Delta p)}{\partial x^2} \qquad (1.62)$$

The sign difference between equations 1.61 and 1.62 is due to the different directions of drift of electrons and holes in an applied field.

These equations describe the general behaviour of excess minority electrons and holes in a semiconductor and are the basis of analytical work on device performance at a physical level. The equations shown are strictly only valid for *low levels of minority carrier injection,* that is, $\Delta n \ll n_p$ or $\Delta p \ll p_n$, as no account has been taken of the effect of the excess carrier densities on the internal electric field.

From equations 1.43 and 1.49 the combined current density due to excess electron drift and diffusion is

$$J_n = \Delta n e \mu_n E + e D_n \frac{\partial(\Delta n)}{\partial x} \tag{1.63}$$

and for excess holes, combining equations 1.44 and 1.51

$$J_p = \Delta p e \mu_p E - e D_p \frac{\partial(\Delta p)}{\partial x} \tag{1.64}$$

Using these expressions for J_n and J_p the continuity equations 1.61 and 1.62 can be written

$$\frac{\partial(\Delta n)}{\partial t} = -\frac{\Delta n}{\tau_n} + \frac{1}{e} \frac{\partial J_n}{\partial x} \tag{1.65}$$

and

$$\frac{\partial(\Delta p)}{\partial t} = -\frac{\Delta p}{\tau_p} - \frac{1}{e} \frac{\partial J_p}{\partial x} \tag{1.66}$$

1.3.6 Diffusion Length

If excess carriers are injected into a semiconductor they diffuse away from the point of injection as described in section 1.3.3. As they diffuse they recombine at a rate described by their lifetime and within a certain distance the excess carrier density falls to zero. As some semiconductor devices rely on the injection of minority carriers into a semiconductor and their subsequent collection, the decay of the excess carrier density with distance is of particular importance. This decay is described in terms of the *carrier diffusion length* for the semiconductor.

Figure 1.12a represents the injection of excess electrons into a *p*-type

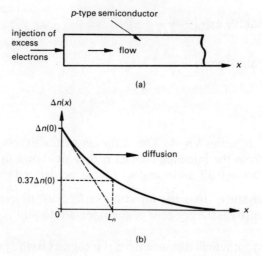

Figure 1.12 Electron diffusion length

semiconductor In several practical cases the area of the injecting surface and the geometry of the semiconductor are such that the resultant carrier flow can be assumed to be in one dimension. The injected electrons diffuse in the x-direction recombining as they do so, the density at the injecting contact $\Delta n(0)$ being maintained constant by the carrier source. The resultant electron distribution is given by the solution $\Delta n(x,t)$ of the continuity equation for excess electrons (equation 1.61). At steady-state a continuous flow is established and the time rate of change of the electron density in the semiconductor $\partial(\Delta n)/\partial t$ is zero. Under these conditions the variation of the excess electron density with distance $\Delta n(x)$ is given by the solution of

$$D_n \frac{d^2(\Delta n)}{dx^2} + \mu_n E \frac{d(\Delta n)}{dx} - \frac{\Delta n}{\tau_n} = 0 \tag{1.67}$$

If the semiconductor can be assumed to be neutral, that is, there is no applied electric field and the level of injection is low, drift effects can be neglected and equation 1.67 reduces to

$$\frac{d^2(\Delta n)}{dx^2} - \frac{\Delta n}{D_n \tau_n} = 0 \tag{1.68}$$

Using the boundary conditions $\Delta n(0)$ and $\Delta n(\infty) = 0$, the solution of equation 1.68 is

$$\Delta n(x) = \Delta n(0) \exp\left(-\frac{x}{\sqrt{(D_n \tau_n)}}\right) \tag{1.69}$$

$$= \Delta n(0) \exp\left(-\frac{x}{L_n}\right) \tag{1.70}$$

where

$$L_n = \sqrt{(D_n \tau_n)} \tag{1.71}$$

is the *electron diffusion length*. The variation of the excess electron density with distance from the injecting contact is then as shown in figure 1.12b.

At $x = L_n$, $\Delta n \approx 0.37\, \Delta n(0)$ and so L_n can be defined either as

(1) the distance from the injecting surface to the point at which the excess density is 37 per cent of the injected density

or

(2) the hypothetical distance from the contact to the point at which the excess density would be zero if the initial gradient were maintained.

Similarly for excess holes in an n-type semiconductor, the *hole diffusion length* L_p is given by

$$L_p = \sqrt{(D_p \tau_p)} \tag{1.72}$$

For silicon $D_n = 3.6 \times 10^{-3}$ m²/s at 300 K and so for a typical electron lifetime of 1 μs, the electron diffusion length is 60 μm.

The diffusion length of injected carriers is important in determining certain dimensions in semiconductor devices. In a bipolar junction transistor for example, excess carriers injected from the emitter into the base are collected at the collector–base junction. For efficient operation, the width of the base region must be small compared with the diffusion length of the injected carriers otherwise a large proportion of them would be lost by recombination before reaching the collector–base junction (section 2.2.10). Solution of equation 1.68 to find the carrier distribution in such a restricted region yields a result different from that given in equation 1.69 due to the change of boundary conditions.

1.3.7 Haynes–Shockley Measurements

Haynes and Shockley developed a laboratory method for the measurement of the mobility, diffusion coefficient and lifetime of minority carriers in a semiconductor. The technique involves the local generation of excess carriers, by incident radiation on a thin semiconductor sample, which are caused to drift toward a sense contact (a reverse-biased *pn* junction) by a constant applied field (figure 1.13a). Measurement of the delay time t_d between the instant the radiation is applied and the time of arrival of the peak

Semiconductor Physics

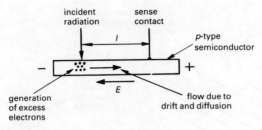

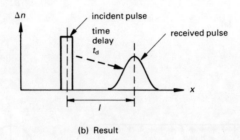

Figure 1.13   Haynes–Shockley method for measurement of μ, D and τ

of the pulse at the sense contact using an oscilloscope, enables the drift velocity and hence the electron mobility to be calculated.

As well as lateral movement due to drift, pulse-widening occurs due to carrier diffusion and comparison of the widths of the received and incident pulses enables the diffusion coefficient to be determined.

The number of excess carriers generated and the number remaining after time t_d are proportional to the respective pulse areas, comparison of which enables the electron lifetime to be determined. Quantitative consideration of the results of this experiment to determine μ, D and τ can be found in reference 14.

1.3.8 Hall Effect

Free carrier movement in a semiconductor is deflected by a magnetic field causing accumulations of holes along one surface and electrons along the opposite surface. This is called the *Hall effect* and the voltage established across the semiconductor due to the charge accumulation is the *Hall voltage* V_H. Figure 1.14a shows a uniform semiconductor bar conducting a constant current I due to an external voltage across its ends. If $l \gg w$ the current flow can be assumed to be uniform. A magnetic field of flux density B is applied perpendicular to the current and the resultant force on the moving charge

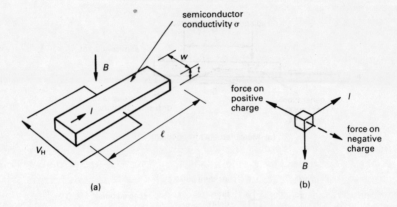

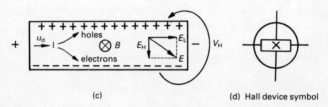

(d) Hall device symbol

Figure 1.14 Hall effect

(Fleming's left-hand rule, figure 1.14b) causes accumulation of holes and electrons along opposite sides. The electric field established by the charge accumulation E_H (figure 1.14c) opposes deflection and an equilibrium condition is reached with a constant voltage V_H across the semiconductor. At equilibrium the force due to movement in the magnetic field Beu_d is balanced by the force due to the transverse field eE_H, thus

$$E_H = Bu_d \tag{1.73}$$

where u_d is the drift velocity of carriers in the direction of current. From equation 1.40 the drift current density in the material due to the applied voltage is

$$neu_d = \frac{I}{wt} \tag{1.74}$$

where n is the free carrier density and wt is the cross-sectional area of the sample. Combining equations 1.73 and 1.74

$$E_H = \frac{BI}{newt} = K_H \frac{BI}{wt} \quad (1.75)$$

the parameter $K_H = 1/ne$ is the *Hall coefficient* of the material. Assuming linearity, the Hall voltage V_H is then

$$V_H = E_H w = \frac{BI}{net} = K_H \frac{BI}{t} \quad (1.76)$$

The polarity of V_H depends on the sign of the majority carriers in the semiconductor and the phenomenon can therefore be used to determine the type of an extrinsic semiconductor, the free carrier density being approximately p_p for p-type and n_n for n-type. If V_H is measured for known B, I and t, the majority carrier density can be determined using equation 1.76. The conductivity of an extrinsic semiconductor is approximately $\sigma = ne\mu$ where n is the majority carrier density and μ the majority carrier mobility (from equation 1.46). Measurement of the conductivity and Hall coefficient of the semiconductor thus enables the carrier mobility, termed the *Hall mobility* μ_H, to be determined

$$\mu_H = \frac{\sigma}{ne} = K_H \sigma \quad (1.77)$$

Due to non-linearities particularly at low B and/or low doping levels, K_H is only approximately equal to $1/ne$, causing μ_H to be slightly different from μ_p or μ_n.

Hall-effect devices have been used as analog multipliers making use of the relation $V_H \propto BI$ from equation 1.76. They can also be used to measure magnetic flux density since $V_H \propto B$ for constant I, the device being called a Hall probe. The circuit symbol for a Hall-effect device is given in figure 1.14d.

1.4 SPACE CHARGE

A homogeneous semiconductor is electrically *neutral* at equilibrium. For intrinsic material, free carriers are generated as electron–hole pairs and as the numbers of electrons and holes are equal, neutrality is maintained. In the extrinsic case, the ionisation of dopant atoms providing an increased number of free carriers of one sign leaves charged ions of opposite sign fixed in the semiconductor crystal lattice and, as the number of free carriers equals the number of ions, the semiconductor remains neutral.

When excess carriers of one type are injected into a semiconductor they instantaneously form an excess of one sign of charge termed *space charge*. This charge imbalance causes an electric field which attracts charge carriers of opposite sign to the region neutralising the space charge initially created. The electrical conductivity of semiconductors is usually sufficiently high that such charge movements restoring neutrality are so fast that they can be considered to be instantaneous and the semiconductor considered neutral throughout. A consequence of this effect is that the movements of excess electrons and holes in semiconductors are related via the electric field as shown by the continuity equations 1.61 and 1.62. Any minute charge imbalance (space charge) may be considered to create an electric field which redistributes charge to restore equilibrium.

An important consequence of charge neutrality in a homogeneous semiconductor is that the *minority* charge distribution controls the situation. Due to the relative sparsity of the minority carriers, charge neutrality would not be maintained if there were a sudden change in majority carrier distribution. In practice, therefore, the distribution of majority carriers is controlled by the minority carrier distribution via the mechanism of charge neutrality. It is for this reason that discussion of free carrier movement earlier in this chapter, particularly in terms of recombination, lifetime, diffusion length and in development of the continuity equations, has been concerned with minority carriers.

The above consideration of space charge has been concerned with homogeneous material. In cases where the crystal is not homogeneous, for example, a junction between p and n-type semiconductors, a net space charge can exist due to charge of opposite sign being prevented from moving into a region to restore neutrality. In such a situation the electric field established by the space charge is fundamental in maintaining equilibrium as discussed in section 1.6.2.

1.5 SURFACE CONDITIONS

At the crystal surface the periodic array of atoms forming the crystal ceases abruptly causing a departure from the regular potential distribution (figure 1.15a). Also any surface contaminant causes surface charge which affects the potential distribution in the surface layer of the semiconductor. In terms of the particle atomic model, this change in potential modifies the outer electron orbits of atoms in the surface layer in such a way that stable electron orbits are created corresponding to electron energies which would be unstable in the bulk material. The additional stable levels are termed *surface states* and some are at levels of energy corresponding to the energy gap for the bulk material (figures 1.15b and c). These surface states affect the electrical properties as the excess charge they contain is compensated to maintain neutrality by drawing additional charge of opposite sign to the surface resulting in non-uniform free carrier distributions.

A freshly formed silicon surface in air rapidly oxidises which reduces the number of surface states by completing the bonding structure at the surface, albeit in a slightly different form from that present within the semiconductor.

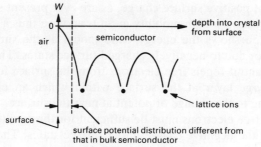

(a) Energy distribution near the surface

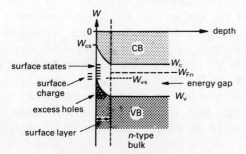

(b) Energy-band model of an n-type semiconductor surface

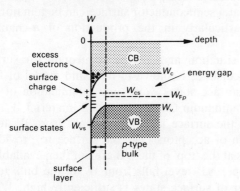

(c) Energy-band model of a p-type semiconductor surface

Figure 1.15 Surface states

In terms of the particle model this continuation of the crystal structure into a silicon dioxide (SiO_2) layer reduces the deformation of electron orbits at the interface thus reducing the number of surface states. However, any foreign atoms trapped at the Si–SiO_2 interface or buried in the oxide layer introduce additional states. In particular, oxygen vacancies in an imperfect SiO_2 layer are a cause of positive surface charge, as are ever present sodium ions which unfortunately have a high mobility in SiO_2 and are thus a major problem.

Figure 1.15b shows the energy band model at the surface of an n-type semiconductor due to negatively charged surface states. This negative charge is immobile and it repels free electrons from the surface forming a *depletion* or *space–charge* layer at the surface within which an electric field exists corresponding to the change of potential near the surface. The degree of this depletion of free electrons must be sufficient for the resulting ionised donors to neutralise the negative charge in the surface states. The repulsion of free electrons from the surface region means that energy levels corresponding to lower CB energies in the bulk semiconductor are not allowed, that is, they are not stable levels, near the surface. As a result the minimum CB energy increases from W_c in the bulk to W_{cs} at the surface, termed *band bending*. Being at lower energies than the CB, the surface states enable additional holes to be created in the VB resulting in the maximum VB energy to increase from the bulk value W_v to W_{vs} at the surface. This accumulation of holes causes the surface layer to be less n-type than the bulk and if the density of surface states is sufficiently high, the excess hole density can be large enough to cause the surface layer of the n-type semiconductor to *invert* to p-type; an *inversion layer* then exists. In terms of the band model (figure 1.15b), inversion corresponds to the Fermi level W_{Fn} being nearer the 'top' of the VB than the 'bottom' of the CB, that is, $W_{Fn} - W_{vs} < W_{cs} - W_{Fn}$. This inherent tendency of inversion at a semiconductor surface can be a major difficulty in device fabrication, particularly in the production of n-channel MOSTs (section 2.3.4).

The corresponding situation at a p-type semiconductor surface due to positive charged surface states is shown in the band model of figure 1.15c. The presence of the surface states causes an accumulation of electrons at energies less than the minimum CB levels in bulk material W_c, causing the minimum CB level in the surface layer to decrease from W_c to W_{cs} at the surface. The ionisation of acceptors near the surface to provide neutrality means that some levels at the 'top' of the VB are not then available to provide free holes and the upper VB level falls from W_v in the bulk to W_{vs} at the surface. If the number of surface states is sufficiently high, $W_{Fp} - W_{vs} > W_{cs} - W_{Fp}$, and an n-type inversion layer is formed at the surface of the p-type semiconductor.

If special processing were not used to reduce surface states in semiconductors, they would dominate device performance masking the useful bulk effects. The following methods are used to reduce surface effects.

(1) Surface polishing to reduce crystal defects.
(2) Extreme cleanliness to reduce surface contamination.
(3) Use of surface insulating layers such as silicon dioxide and silicon nitride, which is more dense than the oxide, at each stage to protect the

surface. The use of a nitride layer considerably reduces troublesome sodium ion migration through the protecting layer which occurs via oxygen vacancies in an oxide layer.

(4) Formation of a 'phosphorus glass' layer on top of the oxide by exposing the oxide layer to phosphorus pentoxide (P_2O_5). This has two major effects: the abundance of oxygen helps fill oxygen vacancies in the oxide layer and, as sodium ions are more soluble in the phosphorus glass layer than in the silicon dioxide, the surface layer tends to collect the sodium ions present and hold them away from the surface (termed 'gettering').

The protection of the semiconductor surface to reduce contamination and the resultant density of surface states is called *passivation*.

Surface conditions are important in the leakage and breakdown properties of a junction and so particular care has to be taken in the region where a *pn* junction meets the surface. The presence of mobile ions in the protecting surface layer and free charge attracted to the surface by fixed surface states, greatly increases the reverse leakage of a junction degrading its blocking properties. The accumulation of charge at the surface due both to surface states and charge within the oxide layer, causes the electric field in the junction depletion layer to be different at the surface from that in the bulk semiconductor. Thus the breakdown voltage at the junction edge is different from that in the bulk and premature breakdown at the edge can occur. This problem is avoided in practice by either including a diffused region at the surface of lower density than the bulk, often termed a *guard ring*, or by physically altering the angle the junction makes with the surface. Both methods modify the depletion layer width at the surface thereby increasing the breakdown voltage relative to that of the junction in the bulk. The guard-ring method achieves this by the use of lower doping (section 2.1, figure 2.3b) while physical alteration of the junction–surface angle is achieved by etching the semiconductor surface forming a bevelled *mesa* ('table-shaped hill') structure. Surface etching is normally only used for devices required to have a very high breakdown voltage, that is, > 1 kV.

In conductors, surface states are of little consequence as the overwhelming number of free electrons can neutralise the surface charge without disturbing the free electron distribution significantly.

1.6 JUNCTIONS

Depending on the thermal energy state of a material, binding (outer orbit) electrons are continually breaking free from their parent nuclei, contributing to the material conductivity and returning to their binding role (recombining). In terms of the band model, such electrons leave the VB for

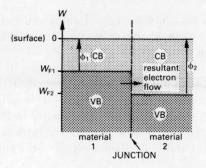

(a) Energy model at instant of formation

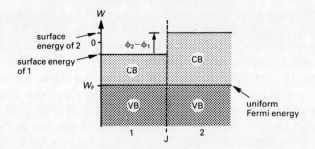

(b) Energy model at equilibrium

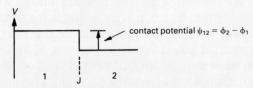

(c) Potential distribution across junction at equilibrium

Figure 1.16 Energy conditions at a contact

the CB, returning to the VB some time later. The mean energy of these electrons is the Fermi level of energy for the material and the difference between the surface energy of the material (zero) and the Fermi level is the mean energy required to cause emission from the surface of the material, this energy difference being termed the *work function* φ of the material.

If a *contact* or *junction* is formed between two *different* materials, initially there is a flow of electrons from the material with the higher Fermi energy (material 1) to that of lower Fermi energy (material 2) as some VB electrons

Semiconductor Physics

in material 1 have sufficient energy to occupy CB levels in material 2 (figure 1.16a). As a result of the flow, W_{F1} is reduced and W_{F2} is increased and the transfer of charge takes place until equilibrium is reached when the Fermi energy is the same on each side of the junction (figure 1.16b). Since material 1 has lost electrons and matrial 2 has gained electrons, material 1 has become positively charged relative to 2 and a potential difference then exists across the junction ψ_{12} which is called the *contact potential* (figure 1.16c). This potential difference opposes further electron movement from 1 to 2 and is therefore termed a *potential barrier*, electrons having 'to climb or cross the barrier' (energywise) for further flow to take place. An alternative name for the potential difference ψ_{12} is thus the *barrier potential*. Work function has units of electronvolts and so the surface energy difference between electrons on side 1 and those on side 2, which creates the contact potential, is $e(\phi_1 - \phi_2)$ joules. As voltage is defined as the change in energy of *unit-positive* charge between the points concerned

$$\psi_{12} = \frac{-e(\phi_1 - \phi_2)}{e} = \phi_2 - \phi_1 \tag{1.78}$$

1.6.1 Metal–Semiconductor Junctions

The conduction properties of a metal–semiconductor junction depend on the type (p or n) of the semiconductor and the relative magnitudes of the work functions ϕ_m and ϕ_s of the metal and semiconductor. If $\phi_m > \phi_s$ in the metal/n-type semiconductor case or $\phi_m < \phi_s$ in the metal/p-type semiconductor case, the redistribution of charge at formation creates a region depleted of free charge and the resulting electric field produces non-linear conduction properties whereby the junction conducts well in one direction but poorly in the opposite direction. Such junctions are termed *rectifying* and form the basis of the Schottky-barrier diode used in fast switching circuits (table 2.1 and section 2.4.2). Alternatively, if $\phi_m < \phi_s$ (n-type) or $\phi_m > \phi_s$ (p-type), the charge redistribution results in similar CB energies on either side of the junction so that free (CB) electrons can flow across the junction in either direction. Such junctions are described as *ohmic* and are used to form the contacts between the semiconductor chip and the external connecting leads in semiconductor devices and ICs. For simplicity, only metal/n-type semiconductor junctions are considered below but a corresponding explanation applies to the p-type case taking account of the different Fermi energy involved.

Rectifying Metal/n-type Semiconductor Junctions ($\phi_m > \phi_s$)
If a junction is formed between a metal and an n-type semiconductor such that $\phi_m > \phi_s$, the mean energy of electrons in the semiconductor is higher

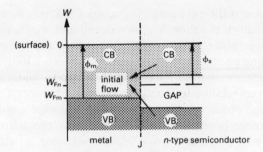

(a) Energy model at instant of contact

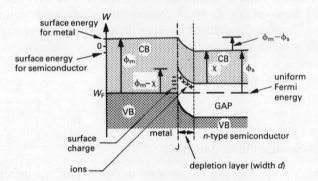

(b) Energy model at equilibrium

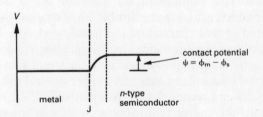

(c) Potential distribution across junction at equilibrium

Figure 1.17 Energy conditions at a rectifying metal/n-type semiconductor junction ($\phi_m > \phi_s$).

than for the metal ($W_{Fn} > W_{Fm}$) and a transfer of electrons therefore takes place from the semiconductor to the metal (figure 1.17a). Due to the electron movement, the Fermi energy of the metal rises and that of the semiconductor falls until at equilibrium the Fermi energy is the same on either side of the junction and there is then no net charge transfer (figure 1.17b). The flow of electrons from the layer of semiconductor adjacent to the junction causes this region to be depleted of free carriers; termed a *depletion layer*. The width of the depletion layer, that is, the distance the depleted region spreads into the semiconductor from the junction, depends on the degree of initial electron movement to create equilibrium. The depletion layer width therefore depends on the difference in work functions ($\phi_m - \phi_s$) and hence on the dopant density of the semiconductor.

Being depleted of free carriers, the depletion layer contains positive ions (space charge) and correspondingly the electron movement into the metal forms a net negative surface charge. An electric field therefore exists in the depletion layer corresponding to the contact potential ψ ($=\phi_m - \phi_s$) across the region at equilibrium (figure 1.17c). Once equilibrium has been established, further electron movement across the junction is controlled by the potential 'barriers' created by the original movement. For an electron to cross from semiconductor to metal it must have energy ψ greater than the mean electron energy in the semiconductor so as to overcome the potential barrier $\phi_m - \phi_s$, the metal being negative relative to the semiconductor (figure 1.17c). To cross the junction in the opposite direction, an electron must have an energy $\phi_m - \chi$ greater than the mean electron energy in the metal to overcome the barrier created by the electric field. The energy parameter χ is called the *affinity* of the semiconductor and is the additional energy required by an electron with minimum CB energy (W_c) to be emitted from the surface of the semiconductor. Equilibrium corresponds to the case where the electron flows in either direction against the two barriers are equal resulting in zero net flow.

Conduction Properties of a Rectifying Metal/n-type Semiconductor Junction under Applied Bias Conditions

Due to the absence of free carriers, the resistance of the depletion layer is considerably greater than that of the bulk metal and bulk semiconductor regions and so an applied voltage across the extremities of the metal and semiconductor regions falls almost entirely across the depletion layer (figure 1.18a). Such an externally applied voltage alters the relative free electron energies on either side of the junction causing an *energy bias* which alters the net flow across the junction. Conventionally the applied bias V is defined as the potential applied to the metal side of the junction relative to the *n*-type semiconductor; it is the opposite for a metal/*p*-type semiconductor junction.

If V is positive (figure 1.18b), the semiconductor is more negative relative to the metal compared with the unbiased case. The mean energy of free electrons in the semiconductor is thus increased and more electrons then have sufficient energy to cross the junction to the metal. This increase in semiconductor electron energy is shown in figure 1.18b by $W_{Fn} > W_{Fm}$, and the difference in surface energies, that is, the barrier to flow from semiconductor to metal, is reduced from the unbiased value ψ to $\psi - V$. Under these conditions the flow of electrons from semiconductor to metal is

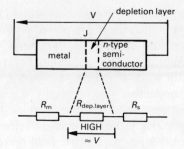

(a) Applied voltage falls almost entirely across depletion layer

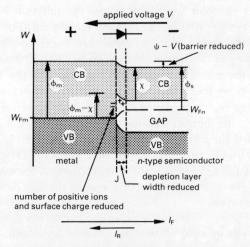

(b) Energy model for forward bias

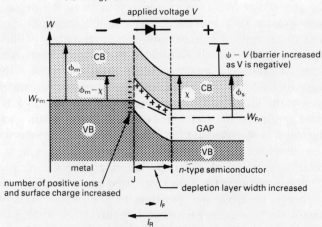

(c) Energy model for reverse bias

Figure 1.18 Rectifying metal/n-type semiconductor junction under applied bias conditions

increased resulting in an increased 'conventional' current flow (from positive to negative potential) from metal to semiconductor which is termed the *forward current* I_F, and correspondingly the application of positive V is called *forward bias*. The barrier to flow of electrons from metal to semiconductor causing a *reverse current* I_R is not affected by the applied voltage and so this component remains unchanged. The reduction of voltage across the depletion region, from ψ in the unbiased case to $\psi - V$ (V positive) for forward bias, causes a reduction in the electric field in this region with a corresponding reduction in space charge. Since the distribution of space charge is fixed by the dopant density, it follows that a reduction corresponds to a decrease in the depletion layer width compared with the unbiased case (figure 1.18b).

With V negative (figure 1.18c) the semiconductor is more positive relative to the metal than in the unbiased case and so the mean energy of free electrons in the semiconductor is reduced, $W_{Fn} < W_{Fm}$ (figure 1.18c). Compared with the unbiased case, fewer electrons in the semiconductor then have sufficient energy to cross the junction to the metal causing the forward current I_F to be reduced. In terms of the energy barrier to electron flow from semiconductor to metal given by $\psi - V$, the reduced flow corresponds to the increased barrier since V is negative. This condition is termed *reverse bias*. As the barrier to electron flow from metal to semiconductor $\phi_m - \chi$ is not dependent on the applied bias, the reverse current I_R is unchanged. The voltage across the depletion layer is increased from the unbiased value ψ to $\psi - V$ for reverse bias, V being negative. This corresponds to an increase in the electric field and hence an increase of space charge in the region. As the space charge density is fixed, the effect of reverse bias is therefore to increase the depletion layer width compared with the case of zero bias (figure 1.18c).

The net current flow across the junction in the *forward* direction I is given by $I_F - I_R$. For moderate forward bias (V positive) $I_F \gg I_R$ and $I \approx I_F$ while for moderate reverse bias (V negative) $I_F \to 0$ and $I \approx I_R$ (figure 1.19a). The conduction properties of this type of junction are therefore dependent on the *polarity* of the applied bias. It is called a *rectifying* junction and forms the basis of rectifiers such as copper–copper oxide and metal alloy–selenium rectifiers used before the introduction of semiconductor junction diodes to convert an alternating signal to a unidirectional signal. It should be noted that copper oxide and selenium are in fact *p*-type semiconductors. More recently, aluminium–silicon Schottky-barrier diodes have been used as high speed switches.

I–V Characteristic of a Rectifying Junction
A quantitative relation between the current flow across a rectifying junction and the applied bias can be obtained from the probability of free carriers acquiring sufficient energy to cross the potential barriers. The Fermi–Dirac function (equation 1.16) gives the probability of an electron acquiring energy W. For intrinsic and lightly doped semiconductors, the energy difference

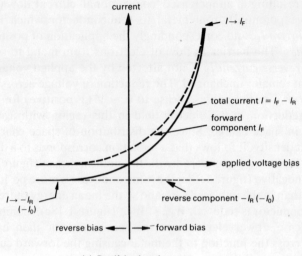

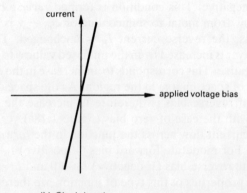

Figure 1.19 Theoretical I–V characteristics for rectifying and ohmic junctions

between CB electrons and the Fermi energy $W - W_F$ is $\gg kT$ and the exponential term is therefore much greater than unity. In this case the probability function simplifies to

$$P(W) = \exp\left(\frac{W_F - W}{kT}\right) \tag{1.79}$$

The forward current I_F due to electrons flowing from the semiconductor to the metal is thus proportional to $\exp(-W/kT)$. With zero-applied bias the forward and reverse current components are equal and the energy barrier to flow in the forward direction is $e\psi$, therefore

$$I_F = I_R = K \exp\left(-\frac{e\psi}{kT}\right) \tag{1.80}$$

where K is a constant.

Under applied-bias conditions the energy barrier is $e(\psi - V)$, where V is positive for forward bias and negative for reverse bias, in which case

$$\begin{aligned} I_F &= K \exp\left[-\frac{e(\psi - V)}{kT}\right] \\ &= K \exp\left(-\frac{e\psi}{kT}\right) \times \exp\left(\frac{eV}{kT}\right) \\ &= I_R \exp\left(\frac{eV}{kT}\right) \end{aligned} \tag{1.81}$$

The total current flow in the forward direction is then

$$\begin{aligned} I &= I_F - I_R \\ &= I_R \left[\exp\left(\frac{eV}{kT}\right) - 1\right] \end{aligned} \tag{1.82}$$

For moderate reverse bias (V negative) the exponential term becomes insignificant and I approaches the constant value I_R. The fact that electron flow from metal to semiconductor, which is responsible for I_R, is constant, is due to the non-dependence of the barrier $\phi_m - \chi$ on the applied reverse bias. Under forward bias (V positive), with V greater than about 100 mV, the exponential term dominates and $I \to I_R \exp(eV/kT)$ as shown in figure 1.19a. The reverse current I_R is often termed the *reverse saturation* or *leakage current* and given the symbol I_0.

Ohmic Metal/n-type Semiconductor Junction ($\phi_m < \phi_s$)
If the work function of the metal is less than that of the semiconductor for a metal/n-type semiconductor junction, the mean energy of electrons in the metal is initially higher than for the semiconductor ($W_{Fm} > W_{Fn}$) resulting in electron transfer from the metal to the semiconductor (figure 1.20a). The charge redistribution causes the Fermi energy of the metal to fall and that of the semiconductor to rise, equilibrium being reached when the Fermi energy is the same on each side of the junction when there is no net charge flow. The energy band model at equilibrium is given in figure 1.20b. As the

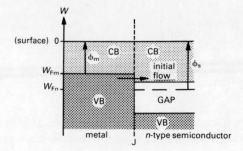

(a) Energy model at instant of formation

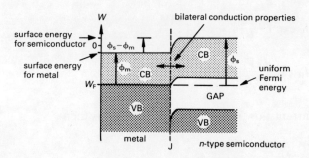

(b) Energy model at equilibrium

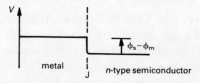

(c) Potential distribution across junction at equilibrium

Figure 1.20   Energy conditions at an ohmic metal/n-type semiconductor junction ($\phi_m < \phi_s$)

semiconductor *gains* electrons, in contrast to losing electrons in the rectifying junction case, no depletion layer is formed and therefore there is no barrier to flow from the semiconductor to the metal as there is for a rectifying junction due to the depletion region and its associated field. Electron flow in the opposite direction, from metal to semiconductor, is against the contact potential of the junction (figure 1.20c). However, this presents only a low resistance to flow and due to the high concentration of CB electrons in the metal the potential difference across the junction is unaffected by an applied bias. The essential lack of barriers means that this type of junction conducts well in both directions and it is therefore termed an *ohmic* junction. In practice a small ohmic voltage drop occurs across the bulk metal and semiconductor regions resulting in the *I–V* characteristic as shown in figure 1.19b. Since this type of junction conducts well in both directions, it is used to form contact pads on the semiconductor chip surface allowing connection to the package leads.

In semiconductor device fabrication, aluminium is widely used to form surface contacts, not only for economic reasons and its good conductivity but also because it adheres well to both silicon and silicon dioxide which is commonly used as a surface insulating/protecting layer. Aluminium forms an ohmic contact with *p*-type silicon but a rectifying contact with *n*-type silicon as it acts as an acceptor-dopant in the surface layer of the silicon forming *pn* junction properties. If an ohmic contact is required between aluminium and *n*-type silicon it is necessary to increase the donor-dopant in the silicon to swamp the effect of the acceptor-dopant. Thus shallow n^+ regions are formed at the surface of *n*-type silicon regions allowing ohmic contacts with aluminium.

1.6.2 pn Junction

A rectifying junction is produced in a single semiconductor crystal if the net dopant changes from acceptor (*p*-type-dopant) to donor (*n*-type-dopant) with distance through the crystal; termed a *pn junction*. As with metal–semiconductor junctions, the electrical properties depend on the close proximity of the two materials and such junctions cannot therefore be formed by physically putting the two types of material in contact since, on a microscopic scale, surface irregularities would cause a 'wide' spacing of the materials and the effects described below would not occur. In practice, *pn* junction devices are formed by doping an *n*-type crystal with acceptor-dopant (or a *p*-type crystal with donor-dopant) until the added dopant density in the surface region exceeds the dopant density of opposite type already present; thus converting the surface region from *n* to *p*-type or vice versa. The junction is the plane within the crystal where the acceptor density N_a is equal to the donor density N_d. More detailed consideration of the formation of *pn* junctions can be found in sections 3.1.5 and 3.3. For purposes of explanation the net acceptor and donor densities on either side of the junction will be assumed constant and the change assumed to occur over zero distance in the crystal, termed an *abrupt* or *step* junction (figure 1.21b). The 'sharpness' of the change affects the electrical properties of the junction; in practical terms an abrupt junction is taken to be one in which the change takes place in less than 10 nm, that is, a distance corresponding to less than about 40 lattice atoms.

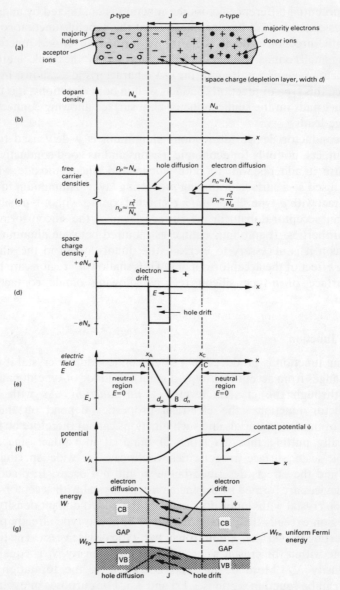

Figure 1.21 Equilibrium condition at a *pn* junction for zero applied bias

pn Junction at Equilibrium

To appreciate the equilibrium condition of a *pn* junction consider the situation immediately following the hypothetical creation of a junction from separate *p* and *n*-type crystals. The *p*-type material has a majority of holes while the *n*-type material has a majority of electrons and so at contact holes *diffuse* from p to n-type material and electrons *diffuse* in the reverse direction due to the density differentials across the junction (figure 1.21c). As the charge movement takes place, the regions on either side of the junction become depleted of free carriers leaving space charge; negative ions on the p-side which have lost holes and positive ions on the n-side which have lost electrons (figure 1.21a). The space charge causes an electric field in the *depletion region* (also termed the *transition region* or *layer* as it occurs at the junction where the transition in semiconductor type takes place) which tends to cause holes to *drift* towards the p-region and electrons to *drift* towards the n-region (figure 1.21d). These drift components are in the opposite directions to the diffusion components for the two types of carrier and equilibrium is established when the tendency to diffuse is balanced by the tendency to drift in the opposite direction.

For the typical dopant concentrations used in devices (10^{20}–10^{28} atoms/m^3) the density of electron–hole pairs is insignificant and the majority carrier densities on either side of the depletion layer are $p_p \approx N_a$ in the p-region and $n_n \approx N_d$ in the n-region (from equations 1.13 and 1.14). Since the product of majority and minority carrier densities is fixed for a particular semiconductor at constant temperature (equations 1.29 and 1.31), the minority carrier densities in each region are given by $n_p \approx n_i^2/N_a$ and $p_n \approx n_i^2/N_d$ (from equations 1.30 and 1.32), where n_i is the intrinsic carrier density for the semiconductor at the temperature concerned.

Spatial Variation of Electric Field at Equilibrium
Outside the depletion layer charge neutrality exists, the charge of the dopant ions being balanced by the majority carriers, and the electric field in these regions is zero; they are termed *neutral regions*. The variation of electric field within the depletion layer due to the space charge either side of the junction, sometimes called a *dipole layer*, is shown in figure 1.21e. It can be seen that the field is a maximum at the junction (E_J).

$$D = \epsilon E \tag{1.83}$$

Equation 1.83 is the fundamental electric field equation where D is the electric flux density established in a material of permittivity ϵ by an electric field (strength) E. Considering the variation of electric field in one dimension perpendicular to the junction (x direction, figure 1.21e), from equation 1.83

$$\frac{dE}{dx} = \frac{1}{\epsilon} \frac{dD}{dx} \tag{1.84}$$

The flux density in a semiconductor depends directly on the net positive charge density which is given generally by

$$\rho = e(p - n + N_d - N_a) \tag{1.85}$$

where p and n are the hole and electron densities, N_d is the donor (positive ion) density and N_a is the acceptor (negative ion) density. Each free carrier and ion has a charge magnitude of e coulombs. The charge contained in a region of length x is thus ρx/unit cross-sectional area and since, by definition, one unit of electric flux originates and terminates on one unit of charge, the flux density in the region is

$$D = \rho x$$
$$= e(p - n + N_d - N_a)x \tag{1.86}$$

Thus from equation 1.84

$$\frac{dE}{dx} = \frac{e}{\epsilon}(p - n + N_d - N_a) \tag{1.87}$$

Applying this relation to the various regions of the pn junction shows that outside the depletion layer the term $(p - n + N_d - N_a)$ is zero, leading to $dE/dx = 0$. Within the depletion layer the free carrier densities p and n are zero and so in the p-type portion of the layer

$$\frac{dE}{dx} = \frac{e}{\epsilon}(-N_a) \tag{1.88}$$

while in the n-type portion

$$\frac{dE}{dx} = \frac{e}{\epsilon}(N_d) \tag{1.89}$$

Equations 1.88 and 1.89 describe the field gradients of AB and BC in figure 1.21e.

Spatial Variation of Potential at Equilibrium
From equation 1.37 the electric field in a certain direction is equal to the negative potential gradient in that direction. Combining this relation with equation 1.87 gives

$$\frac{d^2V}{dx^2} = -\frac{\rho}{\epsilon} = -\frac{e}{\epsilon}(p - n + N_d - N_a) \tag{1.90}$$

which is a one-dimensional statement of Poisson's equation.

As there is no electric field in the neutral regions ($p - n + N_d - N_a = 0$), there is no change of potential in these regions. In the p-type section of the depletion layer, the space charge density is $-N_a$ and so the variation of potential is given by

$$\frac{d^2V}{dx^2} = \frac{e}{\epsilon}N_a \tag{1.91}$$

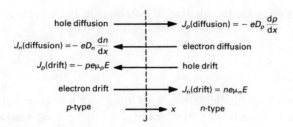

Figure 1.22 Current components across a *pn* junction

while in the *n*-type section

$$\frac{d^2V}{dx^2} = -\frac{e}{\epsilon} N_d \tag{1.92}$$

the corresponding variation of potential across the depletion layer being as shown in figure 1.21f.

With no externally applied bias, the potential difference across the depletion layer, known as the *contact potential* ψ, can be determined by considering the balance between the diffusion and drift tendencies of carriers across the junction. Figure 1.22 shows the four current components at the junction. The arrows show the direction of carrier movement while the current density expressions, developed from equations 1.43, 1.44, 1.49 and 1.50, are in terms of conventional current. Note that electron diffusion and hole drift are in the negative *x* direction and since d*n*/d*x* and *E* refer to the positive *x* direction, the signs of J_n (diffusion) and J_p (drift) have to be altered accordingly. At equilibrium therefore

$$ne\mu_n E = -eD_n \frac{dn}{dx} \tag{1.93}$$

and

$$-pe\mu_p E = -eD_p \frac{dp}{dx} \tag{1.94}$$

thus

$$E\,dx = \frac{-D_n}{\mu} \frac{dn}{n} = \frac{+D_p}{\mu_p} \frac{dp}{p} \tag{1.95}$$

Using Einstein's relations for the ratio of diffusion coefficient to mobility (equations 1.55 and 1.56), equation 1.95 becomes

$$E\,dx = -\frac{kT}{e} \frac{dn}{n} = +\frac{kT}{e} \frac{dp}{p} \tag{1.96}$$

From equation 1.37, the change of potential in the direction of the electric field is

$$V = -\int_x E\, dx \tag{1.97}$$

The boundaries of the depletion layer are x_A and x_C (figure 1.21e) at which the hole and electron densities are p_p, n_p and p_n, n_n respectively (figure 1.21c). Thus the potential difference across the depletion layer, the contact potential, is

$$\psi = -\int_{x_A}^{x_C} E\, dx$$

$$= \frac{kT}{e} \int_{n_p}^{n_n} \frac{dn}{n} = \frac{kT}{e} \ln \frac{n_n}{n_p} \tag{1.98}$$

$$= -\frac{kT}{e} \int_{p_p}^{p_n} \frac{dp}{p} = \frac{kT}{e} \ln \frac{p_p}{p_n} \tag{1.99}$$

But

$$\frac{n_n}{n_p} \approx \frac{N_d}{n_i^2/N_a} = \frac{N_a N_d}{n_i^2} \tag{1.100}$$

and

$$\frac{p_p}{p_n} \approx \frac{N_a}{n_i^2/N_d} = \frac{N_a N_d}{n_i^2} \tag{1.101}$$

thus the contact potential for an abrupt *pn* junction is

$$\psi = \frac{kT}{e} \ln \frac{N_a N_d}{n_i^2} \quad \text{built in potential} \tag{1.102}$$

At $T = 300$ K the intrinsic carrier densities for silicon and germanium are 1.5×10^{16} and 2.5×10^{19} carriers/m³ respectively and the term $kT/e \approx 26$ mV. A step junction with $N_a = 10^{23}$ atoms/m³ and $N_d = 10^{21}$ atoms/m³ would therefore have a contact potential of 0.70 V for silicon and 0.31 V for germanium at $T = 300$ K.

Equations 1.98 and 1.99 enable the minority and majority carrier densities on either side of the depletion layer to be related in terms of the contact potential. From equation 1.98

$$n_p = n_n \exp\left(-\frac{e\psi}{kT}\right) \tag{1.103}$$

while from equation 1.99

$$p_n = p_p \exp\left(-\frac{e\psi}{kT}\right) \tag{1.104}$$

These relationships are shown on the idealised free carrier distributions in figure 1.25a.

As in all junctions, the contact potential can be examined by reference to the energy band model (figure 1.21g). At equilibrium, with zero-applied bias, the distribution of charge is such that the Fermi energy is the same on either side of the junction. The contact potential established thus depends on the Fermi energies of each material (W_{Fp} and W_{Fn}) and therefore on the dopant densities as shown by the quantitative relation (equation 1.102).

Depletion Layer Width at Equilibrium
With no external applied bias, the width of the depletion layer is determined by the electric field necessary for carrier drift to balance carrier diffusion. The magnitude of the electric field depends on the space charge and hence on the degree of depletion.

From charge balance considerations, the magnitude of space charge is the same on either side of the junction. If the depths of depletion on either side are d_p and d_n (figure 1.21e), the magnitude of the negative space charge on the *p*-side is $eN_a d_p$/unit junction area while that of positive space charge on the *n*-side is $eN_d d_n$/unit junction area, thus

$$eN_a d_p = eN_d d_n$$

or

$$\frac{d_p}{d_n} = \frac{N_d}{N_a} \tag{1.105}$$

The depth of depletion on one side of the junction is therefore inversely proportional to the dopant density on that side of the junction. If the dopant densities on the *p* and *n*-type sides are equal ($N_a = N_d$), $d_p = d_n$ and the depletion layer is thus equispaced about the junction. Normally, however, one side of the junction is more heavily doped than the other in which case *the depletion layer spreads mainly into the more lightly doped (lower conductivity) side.*

The variations of potential with x in the *p* and *n*-type portions of the depletion layer are given by equations 1.91 and 1.92. Using the junction as a reference ($V = 0$, $x = 0$; figure 1.23), the variation of potential in the *p*-type portion of the layer can be obtained by integrating equation 1.91 twice using the boundary conditions $V = 0$ at $x = 0$ and $dV/dx = 0$ at $x = -d_p$, from which

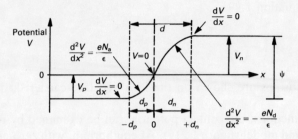

Figure 1.23 Determination of depletion layer width at equilibrium

$$V \text{ (in } p\text{-type layer)} = \frac{eN_a x}{\epsilon}\left(\frac{x}{2} + d_p\right) \tag{1.106}$$

Thus the voltage across the p-type portion is

$$V_p = \frac{eN_a d_p^2}{2\epsilon} \tag{1.107}$$

Similarly the variation of potential in the n-type portion of the layer can be obtained from equation 1.92 using the boundary conditions $V = 0$ at $x = 0$ and $dV/dx = 0$ at $x = d_n$. Thus

$$V(\text{in } n\text{-type layer}) = \frac{eN_d x}{\epsilon}\left(d_n - \frac{x}{2}\right) \tag{1.108}$$

from which the voltage across the n-type portion is

$$V_n = \frac{eN_d d_n^2}{2\epsilon} \tag{1.109}$$

The junction contact potential ψ is then given by

$$\psi = V_p + V_n = \frac{e}{2\epsilon}(N_a d_p^2 + N_d d_n^2) \tag{1.110}$$

Using equation 1.105, the width of the depletion layer d is

$$d = d_p + d_n = d_p + \frac{N_a}{N_d}d_p = \left(\frac{N_a + N_d}{N_d}\right)d_p$$

thus

$$d_p = \left(\frac{N_d}{N_a + N_d}\right)d \tag{1.111}$$

and similarly

$$d_n = \left(\frac{N_a}{N_a + N_d}\right)d \tag{1.112}$$

Substituting for d_p and d_n in equation 1.110

$$\psi = \frac{ed^2}{2\epsilon}\left(\frac{N_a N_d}{N_a + N_d}\right) \tag{1.113}$$

from which the depletion layer width for zero-applied bias is

$$d = \sqrt{\left[\frac{2\epsilon}{e}\left(\frac{N_a + N_d}{N_a N_d}\right)\psi\right]} \tag{1.114}$$

Previous calculation of ψ from equation 1.102 for a step silicon junction with $N_a = 10^{23}$ atoms/m^3 and $N_d = 10^{21}$ atoms/m^3 showed that $\psi = 0.70$ V at $T = 300$ K. Absolute permittivity $\epsilon = \epsilon_0 \epsilon_r$, where ϵ_0 is the free-space value (8.85 × 10^{-12} F/m) and ϵ_r is the relative permittivity of the material; $\epsilon_r = 12$ for silicon. Equation 1.114 shows that the depletion layer width for the junction with zero-applied bias is approximately 1 μm. From equation 1.105, $d_n = 100\, d_p$ in this case, showing that the 100:1 doping ratio causes approximately 99 per cent of the depletion layer width to be in the n-type (lower conductivity) region.

For p^+n and pn^+ junctions in which $N_a \gg N_d$ and $N_d \gg N_a$ respectively, the term $(N_a + N_d)/N_a N_d$ in equation 1.114 reduces to $1/N_d$ and $1/N_a$ respectively, so that for a step p^+n junction

$$d = \sqrt{\left(\frac{2\epsilon\psi}{eN_d}\right)} \tag{1.115}$$

while for a step pn^+ junction

$$d = \sqrt{\left(\frac{2\epsilon\psi}{eN_a}\right)} \tag{1.116}$$

Junction Field at Equilibrium
The electric field at the junction E_J can be calculated from either equation 1.88 or 1.89. As $E = 0$ at the boundaries of the depletion layer, the junction field is simply the product of the field gradient on either side of the junction and the width of that portion of the layer, thus

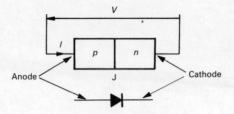

(a) Bias convention

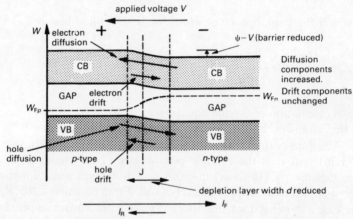

(b) Energy model for forward bias

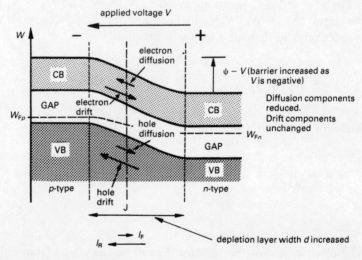

(c) Energy model for reverse bias

Figure 1.24 *pn* junction under applied bias conditions

Semiconductor Physics

$$E_J = -\frac{eN_a d_p}{\epsilon} = -\frac{eN_d d_n}{\epsilon} \tag{1.117}$$

Substituting for d_p or d_n from equations 1.111 or 1.112 and using the total depletion layer width equation (1.114) gives

$$E_J = -\sqrt{\left[\frac{2e}{\epsilon}\left(\frac{N_a N_d}{N_a + N_d}\right)\psi\right]} \tag{1.118}$$

which simplifies to $-\sqrt{(2eN_d\psi/\epsilon)}$ for a step p^+n junction ($N_a \gg N_d$) and to $-\sqrt{(2e\,N_a\psi/\epsilon)}$ for a step pn^+ junction ($N_d \gg N_a$). For the previous example with $N_a = 10^{23}$ atoms/m^3 and $N_d = 10^{21}$ atoms/m^3, $E_J = -1.4 \times 10^6$ V/m.

pn Junction under Applied Bias

As in the case of a rectifying metal–semiconductor junction (section 1.6.1) an applied voltage bias falls almost entirely across the depletion layer of the *pn* junction due to its high resistance relative to that of the neutral regions. The bias V is defined conventionally as the potential applied to the *p*-type region (the anode) relative to the *n*-type region (the cathode), figure 1.24a.

If V is positive (*forward bias*) the mean energy of electrons on the *n*-side is increased causing the tendency for electrons to diffuse across the junction to the *p*-side to increase. In terms of the energy-band model (figure 1.24b), the increase of electron energy on the *n*-side causes W_{Fn} to increase relative to the unbiased case (figure 1.21g) and the difference in surface energies between the *p* and *n* regions, that is, the potential barrier opposing the diffusion of electrons from *n* to *p*, is reduced from the unbiased value ψ to $\psi - V$. The positive bias also increases the energy of holes on the *p*-side of the junction (W_{Fp} is reduced compared with the unbiased case) so that there is an increased tendency for holes to diffuse to the *n*-side of the junction. The forward bias thus increases electron diffusion from *n* to *p* and increases hole diffusion from *p* to *n* resulting in an increased 'conventional' forward current I_F from *p* to *n*. Due to the reduced voltage across the depletion layer compared with the unbiased case, the electric field in this region is reduced and correspondingly the space charge and depletion layer width are reduced. To a first-order approximation, the electron and hole drift components in the depletion layer due to the electric field, which together form the reverse current I_R, are unaffected by the applied bias since, although the electric field is reduced, all the thermally generated minority carriers in the depletion layer are still attracted out into the neutral regions by the field. The combined effect of the applied forward bias is therefore to alter the balance of the diffusion and drift components causing a net current flow $I (= I_F - I_R)$ in the forward direction.

The situation for *reverse bias* (V negative) is the converse of that for forward bias. Under reverse bias conditions the mean energies of electrons on the n-side and holes on the p-side are reduced so that W_{Fn} is reduced and W_{Fp} is increased (figure 1.24c) compared with the unbiased case. The reduction in free carrier energies means that the tendency to diffuse across the junction is reduced and the 'conventional' forward current I_F correspondingly falls. In terms of the energy-band model, the reduced forward current can be attributed to the increased differential in surface energies between the p and n-sides of the junction. Under bias conditions this difference, that is, the barrier to carrier diffusion, is $\psi - V$ and since V is negative for reverse bias, the magnitude of the barrier is increased. The increased voltage across the depletion layer under these conditions corresponds to an increased electric field, increased space charge and a wider depletion layer. As in the forward bias case, to a first-order approximation, the drift components due to thermally generated carriers within the depletion layer are unaffected by the reverse bias since all minority carriers in and near the depletion layer are attracted across the junction by the influence of the electric field regardless of its magnitude. The over-all effect of the reverse bias is therefore to reduce the forward diffusion component leaving the net current flow $I \ (= I_F - I_R)$ comprising mainly of the reverse drift component.

The degree of conduction across a *pn* junction is thus dependent on the polarity of the applied bias; it is therefore a *rectifying* junction which forms the basis of junction diodes used in a variety of applications (table 2.1).

Quantitative Consideration of pn Junction Conduction
For zero-applied bias the equilibrium minority carrier densities in the p and n-neutral regions can be written in terms of the contact potential (equations 1.103 and 1.104) and the free carrier distributions in these regions are as shown in figure 1.25a.

Under forward bias conditions the diffusion of holes from p to n and of electrons from n to p causes the free carrier densities at the edges of the depletion layer to increase (figure 1.25b). Due to the large ratio of majority to minority carrier densities, the relatively large increase of minority charge just outside the depletion layer is neutralised by minute percentage increases in the majority carrier densities maintaining charge neutrality. To a first-order approximation the majority carrier distributions can be considered to be unaffected by the bias. The minority carrier densities at the edges of the layer n_{po} and p_{no} are very sensitive to the applied forward voltage V, the quantitative relationship being obtainable by the method previously used to obtain the contact potential ψ (equation 1.102), that is, by integrating the electric field across the depletion layer, but in this case the voltage across the layer is $\psi - V$ and the boundary conditions are p_p, n_{po} and n_n, p_{no}. In this way, corresponding to equations 1.98 and 1.99

Semiconductor Physics

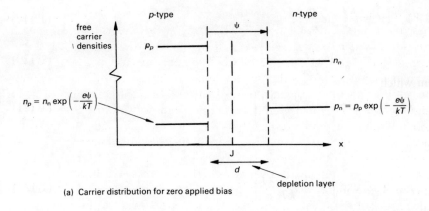

(a) Carrier distribution for zero applied bias

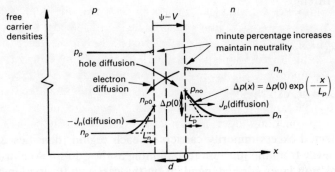

(b) Carrier distribution for forward bias

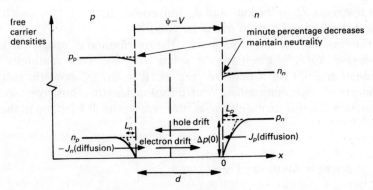

(c) Carrier distribution for reverse bias

Figure 1.25   Free carrier distributions near a *pn* junction

$$\psi - V = \frac{kT}{e} \ln \frac{n_n}{n_{po}} \qquad (1.119)$$

$$= \frac{kT}{e} \ln \frac{p_p}{p_{no}} \qquad (1.120)$$

from which

$$n_{po} = n_n \exp\left[-\frac{e(\psi - V)}{kT}\right] \qquad (1.121)$$

and

$$p_{no} = p_p \exp\left[-\frac{e(\psi - V)}{kT}\right] \qquad (1.122)$$

Combining these relations with equations 1.103 and 1.104 enables n_{po} and p_{no} to be written as

$$n_{po} = n_p \exp\left(\frac{eV}{kT}\right) \qquad (1.123)$$

and

$$p_{no} = p_n \exp\left(\frac{eV}{kT}\right) \qquad (1.124)$$

The injected excess minority carriers in each region (densities Δp and Δn) diffuse away from the junction recombining as they do so. At a distance from the depletion layer edge, dependent on the carrier diffusion length (section 1.3.6), all the excess carriers have recombined and the minority densities reduce to the equilibrium values p_n and n_p dependent on the dopant densities. These movements of minority carriers away from the junction are the current components J_p (diffusion) and J_n (diffusion), figure 1.25b, which together form the forward current of the junction.

Considering the hole component, the distribution of excess carriers in the n-region $\Delta p(x)$ is given by the solution of the hole continuity equation (equation 1.62) for a field-free region ($E = 0$), as shown in section 1.3.6 concerned with determination of diffusion length. Thus, corresponding to equation 1.70 for excess electrons, the excess hole distribution in the n-region is

$$\Delta p(x) = \Delta p(0) \exp\left(-\frac{x}{L_p}\right) \qquad (1.125)$$

where $x = 0$ is taken as the position of the depletion layer edge on the n-side. Using equation 1.51, J_p (diffusion) is given by

Semiconductor Physics

$$J_p \text{ (diffusion)} = -eD_p \frac{d}{dx}\left[\Delta p(0) \exp\left(-\frac{x}{L_p}\right)\right]$$

$$= \frac{eD_p}{L_p}\Delta p(0) \exp\left(-\frac{x}{L_p}\right) \qquad (1.126)$$

From figure 1.25b, using equation 1.24

$$\Delta p(0) = p_{no} - p_n$$

$$= p_n\left[\exp\left(\frac{eV}{kT}\right) - 1\right] \qquad (1.127)$$

Thus

$$J_p \text{ (diffusion)} = \frac{eD_p}{L_p}p_n\left[\exp\left(\frac{eV}{kT}\right)-1\right]\exp\left(-\frac{x}{L_p}\right)$$

and so at the edge of the depletion layer ($x = 0$)

$$J_p \text{ (diffusion)} = \frac{eD_p}{L_p}p_n\left[\exp\left(\frac{eV}{kT}\right)-1\right] \qquad (1.128)$$

Similar consideration of excess electrons injected into the p-region shows that the diffusion current density at the edge of the depletion region on the p-side is

$$J_n \text{ (diffusion)} = \frac{eD_n}{L_n}n_p\left[\exp\left(\frac{eV}{kT}\right)-1\right] \qquad (1.129)$$

If A is the junction area, the total current in the forward direction is the sum of these diffusion currents, that is

$$I = A[J_p \text{ (diffusion)} + J_n \text{ (diffusion)}]$$

$$= eA\left(\frac{D_p p_n}{L_p} + \frac{D_n n_p}{L_n}\right)\left[\exp\left(\frac{eV}{kT}\right)-1\right] \qquad (1.130)$$

From equations 1.30 and 1.32 the equilibrium minority carrier densities p_n and n_p are given approximately by n_i^2/N_d and n_i^2/N_a respectively, therefore

$$I = eAn_i^2\left(\frac{D_p}{L_p N_d} + \frac{D_n}{L_n N_a}\right)\left[\exp\left(\frac{eV}{kT}\right) - 1\right] \qquad (1.131)$$

which may be written

$$I = I_0\left[\exp\left(\frac{eV}{kT}\right) - 1\right] \quad (1.132)$$

where

$$I_0 = eAn_i^2\left(\frac{D_p}{L_p N_d} + \frac{D_n}{L_n N_a}\right) \quad (1.133)$$

Equation 1.132 is the *theoretical diode equation* and can be seen to be the same as equation 1.82 for a rectifying metal–semiconductor junction developed directly from the Fermi–Dirac energy probability function.

At $T=300$ K, $kT/e \approx 26$ mV and so for forward biases greater than 60 mV the exponential term is greater than ten at this temperature. Theoretically, therefore, the exponential term is dominant at practical levels of forward bias and the net forward current is then

$$I \approx I_0 \exp\left(\frac{eV}{kT}\right) \quad (1.134)$$

The exponential term in equation 1.132 is less than 0.1 for reverse biases greater than 60 mV at $T=300$ K and so for practical levels of reverse bias the exponential term is negligible. Thus the reverse current is theoretically given by the value I_0 which is constant for a particular junction at fixed temperature (equation 1.133) and is termed the *reverse saturation* or *leakage current*. The theoretical I–V relationship is shown in figure 1.19a.

The derivation of the theoretical diode equation (1.132) assumes that diffusion currents are dominant and that the junction current can be written in terms of those components (equation 1.130). It is found, however, that this is valid only for semiconductors with an energy gap less than about 0.8 eV, such as germanium (table 1.2). In the case of $W_g > 0.8$ eV, such as for silicon and gallium arsenide, generation and recombination in the depletion layer is important at low forward bias, in which case the junction current is given [15] by

$$I \approx \frac{eAdn_i}{\tau}\left[\exp\left(\frac{eV}{2kT}\right) - 1\right] \quad (1.135)$$

where τ is the minority carrier lifetime in the depletion layer. From this equation, the reverse leakage current I_0 is given by

$$I_0 = \frac{eAdn_i}{\tau} \quad (1.136)$$

in which case, as d increases with reverse bias, so does I_0. Under forward bias conditions the lower index of the exponential term $eV/2kT$ (equation 1.135) compared with eV/kT (equation 1.132), means that the rate of rise of forward current with voltage is lower for junctions in semiconductors with $W_g > 0.8$ eV than for lower energy gap semiconductors. At moderate to high forward bias the diffusion term dominates whatever the value of energy gap and equation 1.132 then applies.

The expressions derived above for carrier densities at the depletion layer edges (equations 1.121, 1.122, 1.123 and 1.124) also apply to the case of reverse bias, although as V is then negative, n_{po} and p_{no} approach zero for practical values of reverse bias (figure 1.25c). Under these conditions current flow across the junction is due to thermal generation within the depletion layer together with minority carriers drawn in from the edge of the neutral region. Charge neutrality is maintained outside the depletion layer by minute percentage decreases in the majority carrier densities which, as in the forward bias case, are so small that the majority carrier densities can be regarded as unaffected by the bias.

For junctions in low energy gap semiconductors ($W_g < 0.8$ eV) where thermal generation in the depletion layer is negligible, the hole diffusion component J_p (diffusion) is given by equation 1.126, where $\Delta p(0)$ is p_n for reverse bias (figure 1.25c). Summation of this term at the depletion layer edge ($x = 0$) with the corresponding J_n (diffusion) component gives the reverse current component I_o as given by equation 1.133. For junctions in semiconductors with $W_g > 0.8$ eV, thermal generation in the depletion layer dominates and the leakage current is given by equation 1.136.

Variation of Electric Field, Potential and Depletion Layer Width of a pn Junction with Applied Bias

The voltage across the depletion layer is $\psi - V$ for an applied bias V, where ψ is the junction contact potential and V is positive for forward bias and negative for reverse bias. The variation of voltage across the layer causes a change in the space charge, which corresponds to a change in the width of the depleted region, and hence a change in the electric field.

Equation 1.114 gives the depletion layer width for zero-applied bias. Under bias conditions the same analysis is valid except that the voltage across the layer is $\psi - V$ and so the expression for depletion layer width becomes

$$d = \sqrt{\left[\frac{2\epsilon}{e}\left(\frac{N_a + N_d}{N_a N_d}\right)(\psi - V)\right]} \qquad (1.137)$$

Note that $d \to 0$ as $V \to \psi$ indicating that the depletion layer disappears as the forward bias approaches the contact potential for the junction.

Figure 1.26a shows how the electric field is affected by bias. The field gradients in the p and n regions of the layer are determined by the dopant densities (equations 1.88 and 1.89) and the field at the junction, given by equation 1.118 for $V = 0$, becomes

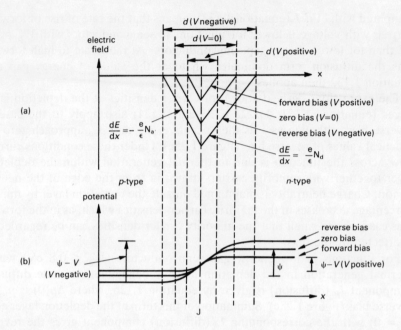

Figure 1.26 Variation of electric field, potential and depletion layer width of a *pn* junction with applied bias

$$E_J = -\sqrt{\left[\frac{2e}{\epsilon}\left(\frac{N_a N_d}{N_a + N_d}\right)(\psi - V)\right]} \qquad (1.138)$$

for an applied bias V. The corresponding variations of potential across the layer are given in figure 1.26b, where the junction has been used as reference.

Temperature Dependence of Reverse Leakage Current and Forward Voltage for a pn Junction

The electrical properties of a *pn* junction are a function of temperature due mainly to the dependence of intrinsic carrier density n_i on temperature. To a first-order approximation the value of n_i for a semiconductor of energy gap W_g is given by equation 1.24 and consideration of the term $N_c N_v$ shows [11] that

$$n_i = K_1 T^{3/2} \exp\left(-\frac{W_g}{2kT}\right) \qquad (1.139)$$

where K_1 is a constant.

For practical purposes, the temperature variations of reverse leakage current I_0 and forward voltage V are of most interest. For a *pn* junction in a low energy gap semiconductor ($W_g < 0.8$ eV) such as germanium, reverse leakage is given theoretically by equation 1.133, which, substituting for n_i from equation 1.139 becomes

$$I_0 = K_1^2 \, eAT^3 \left(\frac{D_p}{L_p N_d} + \frac{D_n}{L_n N_a} \right) \exp\left(-\frac{W_g}{kT} \right) \tag{1.140}$$

Ignoring any temperature dependence of D_p, D_n, L_p and L_n, the ratio of the reverse leakage currents at temperatures T_1 and T_2 is therefore

$$\frac{I_0(T_2)}{I_0(T_1)} = \left(\frac{T_2}{T_1}\right)^3 \exp\left[\frac{W_g}{k} \left(\frac{1}{T_1} - \frac{1}{T_2} \right) \right] \tag{1.141}$$

For germanium $W_g/k = 8.1 \times 10^3$ K.

In the case of semiconductors having a high energy gap ($W_g > 0.8$ eV) such as silicon and gallium arsenide, reverse leakage is given by equation 1.136, having dependence on n_i rather than n_i^2 for $W_g < 0.8$ eV junctions (equation 1.133). Thus

$$I_0 = \frac{K_1 eAd}{\tau} T^{3/2} \exp\left(-\frac{W_g}{2kT} \right) \tag{1.142}$$

and the ratio of leakage currents at temperatures T_1 and T_2 is

$$\frac{I_0(T_2)}{I_0(T_1)} = \left(\frac{T_2}{T_1}\right)^{3/2} \exp\left[\frac{W_g}{2k} \left(\frac{1}{T_1} - \frac{1}{T_2} \right) \right] \tag{1.143}$$

$W_g/2k = 6.3 \times 10^3$ K for silicon.

For practical levels of forward bias, the net forward current of the junction is given by equation 1.134 from which the junction voltage is

$$V \approx \frac{kT}{e} \ln\left(\frac{I}{I_0} \right) \tag{1.144}$$

Under such conditions (moderate forward bias) I_0 is given by equation 1.140 as diffusion components are dominant, thus

$$I_0 = K_2 T^3 \exp\left(-\frac{W_g}{kT} \right) \tag{1.145}$$

For constant forward current I therefore

$$\frac{I}{I_0} = K_3 T^{-3} \exp\left(\frac{W_g}{kT} \right)$$

where K_2 and K_3 are constants, hence, from equation 1.144

$$V \approx \frac{kT}{e} \ln\left[K_3 T^{-3} \exp\left(\frac{W_g}{kT}\right)\right] \tag{1.146}$$

The sensitivity of forward junction voltage to change of temperature at constant forward current I is then

$$\left.\frac{\partial V}{\partial T}\right|_{I \text{ constant}} \approx \frac{\partial}{\partial T}\left\{\frac{kT}{e} \ln\left[K_3 T^{-3} \exp\left(\frac{W_g}{kT}\right)\right]\right\}$$

$$= \frac{V - W_g/e}{T} - \frac{3k}{e} \tag{1.147}$$

Capacitance of a pn Junction

There are two components of *pn* junction capacitance: *depletion (transition) layer capacitance* C_T due to the space charge in the depletion layer (figure 1.21d) and *diffusion capacitance* C_D due to the accumulation of free charge on either side of the junction under forward bias conditions (figure 1.25b).

The dipole layer on either side of the junction formed by acceptor ions on the *p*-side and donor ions on the *n*-side is a store of charge with a voltage $\psi - V$ across it, where V is the applied bias. If V is altered, the depletion layer width changes (equation 1.137) so that the quantity of charge stored is a function of bias. The change of charge does not occur instantaneously thus affecting the speed of response of junction conduction to changing bias. This speed limitation is described in terms of the depletion layer capacitance of the junction, the time lag in response being the time taken to charge or discharge C_T corresponding to increased or decreased depletion layer width. From figure 1.27a the magnitudes of space charge on either side of the junction Q_p and Q_n are

$$Q_p = eN_a d_p A \tag{1.148}$$

and

$$Q_n = eN_d d_n A \tag{1.149}$$

where A is the junction area.

From charge balance considerations $Q_p = Q_n = Q$ and using equations 1.111, 1.112 and 1.137

$$Q = \frac{AeN_a N_d}{N_a + N_d} \sqrt{\left[\frac{2\epsilon}{e}\left(\frac{N_a + N_d}{N_a N_d}\right)(\psi - V)\right]}$$

Semiconductor Physics

$$= A\sqrt{\left[2\epsilon e \left(\frac{N_a N_d}{N_a + N_d}\right)(\psi - V)\right]} \quad (1.150)$$

Under static conditions, capacitance is the ratio of charge stored to voltage across the medium containing the electric field due to the charge, that is, $C = Q/V$. In the dynamic case, where the voltage varies with time, a small increase in voltage δV causes a small increase in charge δQ provided by a charging current $\delta Q/\delta t$. Since $\delta Q/\delta t = (\delta Q/\delta V) \delta V/\delta t$, the effective capacitance as $\delta V \to 0$ is therefore dQ/dV. If Q is a *linear* function of V then dQ/dV is numerically equal to Q/V but, as equation 1.150 shows, this is not the case for space charge in the depletion layer of a *pn* junction. Further, as an *increase* in V causes a *decrease* in Q, the depletion layer capacitance of the junction is defined as

$$C_T = -\frac{dQ}{dV}$$

$$= A\sqrt{\left[\frac{\epsilon e}{2}\left(\frac{N_a N_d}{N_a + N_d}\right)\frac{1}{\psi - V}\right]} \quad (1.151)$$

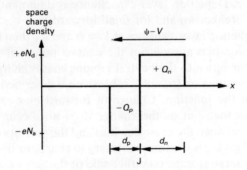

(a) Space charge in depletion layer

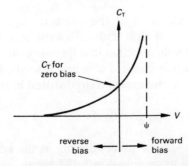

(b) Variation of depletion layer capacitance with bias

Figure 1.27 Depletion layer capacitance

using the Q/V relation of equation 1.150.

Combining this result with that for the depletion layer width d (equation 1.137) gives

$$C_T = \frac{\epsilon A}{d} \tag{1.152}$$

and so the depletion layer of an abrupt pn junction may be considered as a parallel-plate capacitor having plate area equal to the junction area and plate separation equal to the depletion layer width. For an abrupt silicon junction ($\epsilon = 12\epsilon_0$) with $N_a = 10^{23}$ atoms/m³ and $N_d = 10^{21}$ atoms/m³, $\psi = 0.70$ V at $T = 300$ K (equation 1.102), giving $C_T = 100 pF/mm^2$ junction area at zero bias. For the same junction, C_T is 145 pF/mm² at a forward bias of 0.3 V and 28 pF/mm² at a reverse bias of 10 V. From equation 1.151

$$C_T \propto (\psi - V)^{-1/2} \tag{1.153}$$

for a step junction (figure 1.27b). A similar calculation for a junction with a linearly graded dopant profile shows that C_T is proportional to $(\psi - V)^{-1/3}$, while for a practical diffused junction the proportionality is to $(\psi - V)^{-n}$ where $1/3 < n < 1/2$. Depletion layer capacitance is dominant for reverse bias conditions below breakdown and for small forward bias $V < \psi$. It should be noted that the depletion layer vanishes at $V = \psi$ and C_T then has no meaning.

When the junction bias approaches the contact potential ψ, the increased injection of free carriers into the neutral regions causes charge accumulation on either side of the junction (figure 1.25b) which is described by the *diffusion capacitance* C_D of the junction. Change of forward bias causes the charge accumulation to change but as the change does not occur instantaneously there is a time lag between the change in bias and the corresponding change in forward current. This lag is the time necessary to charge or discharge C_D. The degree of excess charge storage on either side of the junction, and hence the value of C_D, depends on the lifetime of minority carriers in the neutral regions.

As a first-order consideration of the value of C_D, consider an abrupt junction with $N_a > N_d$ such that under forward bias the majority of the forward current is due to hole injection into the n-region. The resulting excess hole distribution is shown in figure 1.25b and described mathematically by equation 1.125 from which the excess charge stored in the n-region is

$$Q = \int_0^\infty A e \Delta p(x) dx = A e L_p \Delta p(0) \tag{1.154}$$

If the forward bias changes, the excess density at the edge of the depletion layer $\Delta p(0)$ varies according to equation 1.127 giving

$$Q = A e L_p p_n \left[\exp\left(\frac{eV}{kT}\right) - 1 \right] \tag{1.155}$$

Since Q is not a linear function of V, and Q increases as V increases, the effective capacitance C_D is given by

$$C_D = +\frac{dQ}{dV} = \frac{e^2}{kT} AL_p p_n \exp\left(\frac{eV}{kT}\right) \quad (1.156)$$

$$= \left(\frac{e}{kT}\frac{L_p^2}{D_p}\right)\frac{eAD_p p_n}{L_p} \exp\left(\frac{eV}{kT}\right)$$

$$= \frac{e}{kT} \tau_p I \quad (1.157)$$

since $L_p^2 = D_p \tau_p$ (equation 1.72) and, from equation 1.130, I is the forward hole current for moderate forward bias. Diffusion capacitance is thus dependent on current flow and minority carrier lifetime. For a hole lifetime of 10 ns, equation 1.157 gives C_D as about 400 pF/mA forward current at $T=300$ K. A more general analysis based on determination of the junction admittance for a sinusoidal variation of forward bias is presented in reference 16 leading to the result

$$C_D = \frac{e^2}{2kT} A (L_p p_n + L_n n_p) \exp\left(\frac{eV}{kT}\right) \quad (1.158)$$

Diffusion capacitance is dominant for forward bias conditions and also at reverse breakdown when a substantial reverse current flows.

1.7 JUNCTION BREAKDOWN

When a rectifying junction is reverse-biased, a small reverse current flows, termed *reverse leakage*, due to carrier drift in the junction field. If the magnitude of the reverse bias is increased sufficiently, the reverse current increases sharply (figure 1.28a), the condition being called *reverse breakdown* of the junction. Two mechanisms are responsible for breakdown; *avalanche multiplication* and *Zener effect*.

Avalanche Multiplication
If the acceleration of a free electron in the junction field is sufficiently high and the time between collisions is long enough, an electron can acquire sufficient energy to ionise a lattice atom by bombardment. Additional free electrons created in this way are accelerated by the field causing further ionisation and so on, as represented diagrammatically in figure 1.28b. The multiplication of the number of free carriers causes the reverse current of the junction to increase rapidly (avalanche). For silicon the minimum field

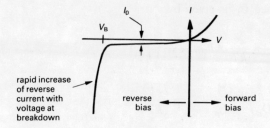

(a) Junction breakdown characteristic

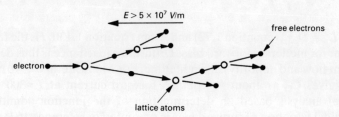

(b) Avalanche multiplication

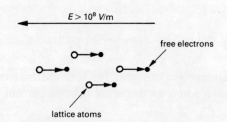

(c) Zener effect

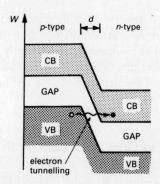

(d) Tunnelling

Figure 1.28 Junction breakdown

necessary to impart to free electrons sufficient energy between collisions for this to occur is about 5×10^7 V/m.

Zener Effect

When the doping is high ($\geqslant 10^{25}$ atoms/m^3) on one side of the junction, the depletion layer is very narrow (equations 1.115 and 1.116) and the resulting electric field causes *field emission* whereby the force on outer orbit electrons due to the field ($\geqslant 10^8$ V/m for silicon) is so high that they are 'pulled away' from their parent nuclei to become free carriers. This ionisation by electrostatic attraction is known as the *Zener effect* (figure 1.28c) and causes an increase in the free carrier density and hence an increase in the reverse current of the junction. Zener breakdown cannot be described in terms of the energy-band model in the same way as normal conduction where carriers have to acquire sufficient energy to cross the energy barrier. Instead, due to the heavy doping, VB electrons on the p-side can have energies corresponding to CB levels on the n-side even for low reverse bias. The VB electrons and the CB levels that they can occupy are separated by the *very narrow* depletion layer and under these conditions electrons can flow across the junction. This conduction mechanism is described as *tunnelling* and electrons are said to tunnel 'under' or 'through' the energy barrier (figure 1.28d) which is a figurative representation of this 'unconventional' process. The particle model of the electron is of little use here and quantitative consideration involves use of the wave model. Tunnelling relies on the narrow separation of the VB electrons and the available CB levels. It therefore relies on the narrow depletion region and heavy doping.

The reverse bias at which the reverse current begins to increase is the breakdown voltage V_B of the junction. Once at breakdown, the increased current flow causes power dissipation at the junction to increase and the semiconductor temperature rises. Providing the junction temperature does not rise sufficiently to cause irreversible physical changes to take place, breakdown is not damaging and the junction returns to normal operation when the reverse bias is reduced below V_B. Knowing the minimum field necessary to cause breakdown, the dopant densities required to provide breakdown at a certain voltage V_B can be found from equation 1.138, the term $N_a N_d/(N_a+N_d)$ simplifying to N_d for a p^+n junction ($N_a \gg N_d$) and to N_a for a pn^+ junction ($N_d \gg N_a$).

Zener breakdown requires a very narrow depletion layer which corresponds to high doping and low reverse bias. In practice, for junctions with a breakdown voltage < 5 V, Zener breakdown dominates while for those having $V_B >$ 8 V the wider depletion layer prevents tunnelling and avalanche multiplication dominates. For those junctions with 5 V < V_B < 8 V the breakdown process is a combination of Zener and avalanche effects.

1.8 JUNCTION MODELLING

A first-order network model of the static performance of a rectifying junction can be produced using a piecewise-linear approximation of the I–V junction characteristic (appendix A.1 and figure 1.29a). If r_f, r_r and r_b are *slope* resistances describing the I–V relationships in the forward, reverse and breakdown regions respectively, the corresponding network model is as given in figure 1.29b, where the unblanked diode symbol is used to represent hypothetical voltage-controlled switches which are either short-circuit (s/c) or open-circuit (o/c). For positive voltages less than V_1 all the switches are o/c and the model current is zero. When V exceeds V_1, S_1 is s/c while S_2 and S_3 are both o/c so that the increase of current with voltage is represented by r_f. Under reverse-bias conditions below breakdown ($0 > V > V_B$), S_1 and S_3 are o/c while S_2 is s/c, enabling the reverse leakage to be represented by r_r. When the reverse bias exceeds the breakdown voltage V_B, r_f and r_r are switched out of the model leaving r_b representing the breakdown characteristic. The approximation provided by such a model can be improved by increasing the number of segments, although the greater the complexity of the model the more cumbersome is its use in circuit analysis.

Under small-signal dynamic conditions where the applied voltage varies due to the superposition of a low amplitude time-varying voltage on the static bias, the junction capacitance must be included in the model (figure 1.29c). For moderate forward bias the capacitance is predominantly diffusion capacitance C_D given by equation 1.158 and the shunt resistance r gives the slope dV/dI of the forward characteristic which from equation 1.134 is

$$r = \frac{dV}{dI} = \frac{kT}{eI} \qquad (1.159)$$

where I is the static bias current.

When the junction is reverse-biased below breakdown, the capacitance in the model of figure 1.29c is the depletion layer capacitance C_T (equation 1.151). Under these conditions r represents the change of reverse leakage with voltage which is usually negligible.

Although network models of junction performance are often useful in basic analytical work, a more fundamental representation based on charge variations near the junction is particularly useful; it is known as the *charge-control model* and is a mathematical model. Consider the junction under forward-bias conditions (figure 1.25b), the hole component of current density J_p (diffusion) being given by equation 1.128 from which the hole current is

$$I_p = \frac{eAD_p}{L_p} p_n \left[\exp\left(\frac{eV}{kT}\right) - 1 \right] \qquad (1.160)$$

Semiconductor Physics

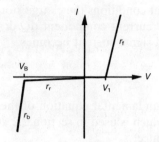

(a) Piecewise linear approximation

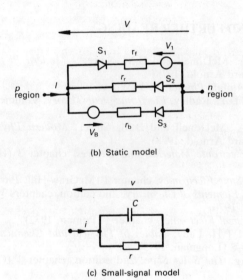

(b) Static model

(c) Small-signal model

Figure 1.29 Rectifying junction modelling

The charge due to excess holes Q_p corresponding to this current is given by equation 1.155, showing that

$$I_p = \frac{Q_p}{\tau_p} \tag{1.161}$$

where τ_p is the hole lifetime ($= L_p^2/D_p$, equation 1.72). From equation 1.161 the total static current flow across the junction may be written

$$I = \frac{Q}{\tau} \tag{1.162}$$

where Q is the total charge stored and τ is a time parameter, termed a charge-control parameter, for the junction.

Under a.c. or transient conditions the charge stored changes with time and an additional 'charging' current component dQ/dt must be included so that the combined static and signal current becomes

$$i = \frac{Q}{\tau} + \frac{dQ}{dt} \qquad (1.163)$$

Equation 1.163 is the fundamental equation of the charge-control model of junction performance which is used in section 2.2.26 to form a mathematical model of BJT performance.

REFERENCES AND FURTHER READING

1. D.E. Caro, J.A. McDonell and B.M. Spicer, *Modern Physics*, 3rd edition, chapter 2 (Edward Arnold, 1978)
2. J.C. Willmott, *Atomic Physics*, p. 9 (Wiley, 1975)
3. J.B. Davies and D.E. Radley, *Electromagnetic Theory*, Volume 1, p. 9 (Oliver & Boyd, 1969)
4. D.E. Caro, J.A. McDonell and B.M. Spicer, *Modern Physics*, 3rd edition, chapter 7 (Edward Arnold, 1978)
5. D.H. Navon, *Electronic Materials and Devices*, chapter 3 (Houghton Mifflin, 1975)
6. S. Wang, *Solid-State Electronics*, chapter 1 (McGraw-Hill, 1966)
7. L.P. Eblin, *The Elements of Chemistry*, 2nd edition, chapters 3 and 4 (Harcourt, Brace & World, 1970)
8. J.Seymour, *Physical Electronics*, chapter 1 (Pitman, 1972)
9. G.W. Kaye and T.H. Laby, *Tables of Physical and Chemical Constants*, 14th edition, pp. 266–8 (Longman, 1973)
10. H.M. Rosenberg, *The Solid State*, 2nd edition, chapter 1 (Oxford University Press, 1978)
11. J. Seymour, *Physical Electronics*, appendix 3 (Pitman, 1972)
12. A. Bar-Lev, *Semiconductors and Electronic Devices*, chapter 2 (Prentice-Hall, 1979)
13. S. Wang, *Solid-State Electronics*, chapter 5 (McGraw-Hill, 1966)
14. A. Bar-Lev, *Semiconductors and Electronic Devices*, chapter 5 (Prentice-Hall, 1979)
15. C.T. Sah, *et al.*, Carrier Generation and Recombination in pn Junctions and pn Junction Characteristics. *Proc. IRE*, **45**, (1957), 1228
16. J. Allison, *Electronic Engineering Materials and Devices*, chapter 9 (McGraw-Hill, 1971)
17. B.G. Streetman, *Solid State Electronic Devices*, 2nd edition, chapters 2–5 (Prentice-Hall, 1980)
18. A. Bar-Lev, *Semiconductors and Electronic Devices*, chapters 2–8 (Prentice-Hall, 1979)
19. J.E. Carroll, *Physical Models for Semiconductor Devices*, chapters 2–4 and 6 (Edward Arnold, 1974)
20. D.A. Fraser, *The Physics of Semiconductor Devices*, chapters 1–4 (Oxford University Press, 1977)
21. D.H. Navon, *Electronic Materials and Devices*, chapters 2–6 (Houghton Mifflin, 1975)

Semiconductor Physics

22. J. Seymour, *Physical Electronics*, chapters 1–3 (Pitman, 1972)
23. J. Allison, *Electronic Engineering Materials and Devices*, chapters 1–6 and 9 (McGraw-Hill, 1971)
24. H.M. Rosenberg, *The Solid State*, 2nd edition, chapters 7–10 (Oxford University Press, 1978)
25. R.S. Muller, and T.I. Kamins, *Device Electronics for Integrated Circuits*, chapters 1–4 (Wiley, 1977)
26. N.M. Morris, *Semiconductor Devices*, chapter 1 (Macmillan, 1978)
27. F.R. Connor, *Electronic Devices*, chapters 2 and 3 (Edward Arnold, 1980)
28. J. Millman, *Microelectronics*, chapters 1 and 2 (McGraw-Hill, 1979)
29. J. Millman, and C.C. Halkias, *Electronic Fundamentals and Applications*, chapters 1 and 2 (McGraw-Hill, 1976)
30. J. Millman and C.C. Halkias, *Integrated Electronics*, chapters 1–3 (McGraw-Hill, 1972)
31. C.A. Holt, *Electronic Circuits*, chapter 1 (Wiley, 1978)
32. J.D. Ryder, *Electronic Fundamentals and Applications*, 5th edition, chapter 1 (Pitman, 1977)
33. M.J. Morant, *Introduction to Semiconductor Devices*, 2nd edition, chapters 1–3 (Harrap, 1970)
34. P. Lynch, and A. Nicolaides, *Worked Examples in Physical Electronics*, sections 3 and 5 (Harrap, 1972)
35. S.M. Sze, *Physics of Semiconductor Devices* (Wiley, 1969)
36. A.S. Grove, *Physics and Technology of Semiconductor Devices* (Wiley, 1967)
37. R.B. Adler, A.C. Smith and R.L. Longini, *Introduction to Semiconductor Physics* (Wiley, 1964)

TUTORIAL QUESTIONS

Values of physical constants and semiconductor parameters are given in appendix C.

1.1 Show that according to the Bohr particle model of the hydrogen atom, the lowest stable electron energy is -13.6 eV and that this corresponds to a circular orbit of radius 0.05 nm. Calculate the energy and orbital radius of the four lowest excited stable levels and hence produce a potential well model for the hydrogen electron.
(Answers: -3.4, 0.21; -1.5, 0.48; -0.9, 0.85; -0.5 eV, 1.33 nm)

1.2 By representing a donor atom in silicon by the Bohr model of a hydrogen atom, calculate the binding energy of the 'surplus' electron. What is the significance of the low value obtained? Mention any simplifying assumptions made in this calculation.
(Answer: 0.09 eV)

1.3 The Fermi level for copper is -7 eV. Assuming that the distribution of electron energies is described by the Fermi–Dirac function, find the probabilities of electrons having energies 0.1 eV and 1 eV above the Fermi level at temperatures of 300 K and 400 K. Comment on the results obtained.
(Answers: 2×10^{-2}, 1.6×10^{-17}; 5×10^{-2}, 2.6×10^{-13})

1.4 Assuming that the densities of states in the conduction and valence bands of silicon are $2.5 \times 10^{25}/\text{m}^3$ and that this density is not a function of temperature, calculate the intrinsic carrier density of silicon at 300 K and 400 K. If silicon is doped with 10^{23} phosphorus atoms/m^3, calculate the majority and minority carrier densities at

these two temperatures, stating any assumptions made. What is the significance of the results?
(Answers: 1.47×10^{16}, 2.98×10^{18}, 10^{23}, 2.16×10^9; 10^{23}, $8.88 \times 10^{13}/\text{m}^3$)

1.5 The intrinsic carrier density for silicon at 300 K is 1.5×10^{16} carriers/m^3. Find the equilibrium majority and minority carrier densities at 300 K when silicon is uniformly doped with 10^{16} phosphorus atoms/m^3. If the resulting extrinsic semiconductor is then uniformly doped with 2×10^{16} boron atoms/m^3, what are the final free carrier densities at 300 K? Assume that charge neutrality exists throughout. Comment on the relative results of the two parts of this question. Why is the assumption used in question 4 not valid here?
(Answers: 2.08×10^{16}, 1.08×10^{16}; 2.08×10^{16}, $1.08 \times 10^{16}/\text{m}^3$)

1.6 If the intrinsic carrier density of a semiconductor is proportional to $T^{3/2}$ exp $(-W_g/2kT)$ and n_i for silicon is 1.5×10^{16} carriers/m^3 at 300 K, calculate n_i at 400 K. Hence repeat question 1.5 for a temperature of 400 K. What is the significance of the result?
(Answer: $4.69 \times 10^{18}/\text{m}^3$)

1.7 An intrinsic chip of silicon is used as a thermistor (temperature dependent resistor). If carrier mobilities vary as $T^{-3/2}$, where T is the temperature in degrees kelvin, calculate the ratio of the conductivity of the chip at 100 °C to the value at 25 °C. The solution requires information from question 1.6.
(Answer: 74)

1.8 Calculate the transition temperature for extrinsic silicon having a uniform dopant concentration of 10^{23} atoms/m^3. The intrinsic carrier density for silicon may be taken as $10^{22} T^{3/2}$ exp $(-6377/T)$ carriers/m^3 for temperature in degrees kelvin.
(Answer: 822 K)

1.9 Copper contains 8.5×10^{28} free electrons/m^3 and their mobility is 3.2×10^{-3} m^2/V s. If 10 V d.c. is applied across a 10 m length of copper wire having a cross-sectional area of 2.5 mm^2, calculate the conductivity of copper, the distance electrons drift per second, the current flowing and the resistance of the conductor.
(Answers: 4.35×10^7 S/m, 3.2 mm, 108.75 A, 92 mΩ)

1.10 The density of free electrons in pure silicon at 300 K is $1.5 \times 10^{16}/\text{m}^3$ and its conductivity is 4.5×10^{-4} S/m. Given that electron mobility in silicon is 0.14 m^2/V s, calculate the hole mobility at 300 K and hence find the conductivities of extrinsic silicon with uniform donor densities of 10^{20}, 10^{23} and 10^{26} atoms/m^3. Assume that carrier mobilities are independent of dopant density. Compare the calculated conductivities with that for copper (question 1.9).
(Answers: 0.0475 m^2/V s, 2.24, 2.24×10^3, 2.24×10^6 S/m)

1.11 At 300 K pure silicon has a conductivity of 4.5×10^{-4} S/m. If the electron and hole mobilities are 0.14 and 0.05 m^2/V s respectively, what is the density of electron–hole pairs at this temperature? Calculate the proportion of current flow carried by electrons in silicon at 300 K when it is uniformly doped with 10^{23} boron atoms/m^3. Comment on this result.
(Answers: $1.48 \times 10^{16}/\text{m}^3$, 6.132×10^{-12} per cent)

1.12 A rectangular strip of uniformly doped n-type silicon 5 mm × 2 mm × 1 mm has an electrical resistance of 100 Ω between its ends. If the intrinsic carrier density of

silicon is 1.5×10^{16} carriers/m^3 and the mobilities of electrons and holes are 0.14 and 0.05 m^2/V s respectively, what is the density of dopant in the silicon?
(Answer: 1.12×10^{21} atoms/m^3)

1.13 A strip of intrinsic silicon 1 mm $\times$ 2 mm $\times$ 10 mm long has 10 V d.c. applied across its ends. Calculate the current flowing at 300 K. The sample is uniformly irradiated and the current increases to 2 μA. Calculate the density of electron–hole pairs generated by the radiation. The radiation ceases abruptly and 1 μs later the current has dropped to 1 μA. Calculate the carrier lifetime.
(Answers: 912 nA, 1.79×10^{16}/m^3, 398 ns)

1.14 A p-type silicon chip has a surface contact 100 μm square. If an electron current of 100 nA is injected into the chip from the contact and the silicon has an electron lifetime of 1 μs, calculate the excess electron density directly below the contact and at depths of 100 μm and 1 mm. Assume one-dimensional flow perpendicular to the contact and ignore field effects. Comment on the relative values obtained.
(Answers: 1.04×10^{18}, 1.96×10^{17}, 6×10^{10}/m^3)

1.15 A Hall probe is formed from a rectangular p-type chip of germanium 5 mm long, 2 mm wide and 0.5 mm thick having a dopant concentration of 10^{20} atoms/m^3. Calculate the magnetic flux density perpendicular to the 5 $\times$ 2 mm face of the chip if a Hall voltage of 10 mV is produced across its width when 2 V d.c. is applied across its length. The intrinsic carrier density for germanium is 2.5×10^{19} carriers/m^3.
(Answer: 65.8 mT)

1.16 A Hall device is used as an analog multiplier such that the Hall voltage $V_H = KBI$ where the current I along the length of a rectangular silicon chip and the magnetic flux density B parallel to the thickness of the chip are the two variables and V_H across the width of the chip is proportional to the product. When 5 V d.c. is applied across the ends of the chip, which is 10 mm long, 4 mm wide and 1 mm thick, a current of 10 mA flows. Calculate the donor density and the constant of proportionality K for the multiplier. The intrinsic carrier density for silicon is 1.5×10^{16} carriers/m^3.
(Answers: 2.23×10^{20} atoms/m^3, 28 m^2/A s)

1.17 An abrupt pn junction of area 0.2 mm^2 is formed in silicon with uniform dopant densities of 10^{22} and 10^{21} atoms/m^3 in the p and n-regions respectively. For zero-applied bias calculate the contact potential, the depletion layer width, the electric field at the junction and the depletion layer capacitance at 300 K. What proportion of the total depletion layer width is in the p-region?
(Answers: 637 mV, 0.96 μm, 1.32×10^6 V/m, 22 pF, 9.1 per cent)

1.18 The depletion layer capacitance of an abrupt silicon p^+n junction with uniform dopant densities is found to be 20 pF at a reverse bias of 1 V and the contact potential is 900 mV both at 300 K. If the junction area is 0.1 mm^2 calculate the dopant densities of the p and n-regions. What is the depletion layer width of the junction at a reverse bias of 10 V and what proportion of the total layer width is in the n-region?
(Answers: 2.72×10^{25}, 8.94×10^{21} atoms/m^3, 1.27 μm, 99.97 per cent)

1.19 A voltage reference diode with a breakdown voltage of 3.3 V is to be fabricated in silicon using an abrupt p^+n junction with uniform dopant densities. If breakdown occurs when the electric field reaches 10^8 V/m, determine suitable dopant densities for the anode and cathode regions.
(Specimen answers: 10^{26}, 7.6×10^{23} atoms/m^3)

1.20 The reverse leakage current of a rectifying junction is 10 nA at 300 K with a reverse bias of 20 V. Based on the theoretical conduction properties of such a junction, what forward bias would be necessary to cause a current of 10 mA to flow? (Answer: 359 mV)

1.21 An abrupt *pn* junction is formed in germanium with uniform dopant densities in the *p* and *n*-regions of 10^{22} and 10^{23} atoms/m^3 respectively. If the junction area is 0.1 mm^2 and the minority carrier lifetime in the neutral regions is 1 μs, calculate the reverse leakage current at 300 K for a reverse bias of 1 V and hence obtain the theoretical diode equation for the junction at this temperature. Repeat the calculation for a junction having the same parameters formed in silicon assuming that the minority carrier lifetime in the depletion layer is 100 μs.
(Answers: 0.21 μA, 2.4 pA)

11.22 An *RC* oscillator circuit has a frequency of oscillation given by $(2\pi\ CR)^{-1}$ where $R = 100$ kΩ. The frequency of oscillation is to be controlled electrically by using a reverse-biased silicon *pn* junction as the capacitive element. If the junction is abrupt, having an area of 0.1 mm^2 with uniform dopant densities $N_a = 10^{24}$ atoms/m^3 and $N_d = 10^{21}$ atoms/m^3, find the frequency range available for a reverse bias range 0–10 V at 300 K.
(Answer: 150–566 kHz)

1.23 A silicon *pn* junction has a reverse leakage current of 10 nA at 25 °C. Calculate the leakage current for the junction at 45 °C, 100 °C and 200 °C. What is the sensitivity of the forward voltage of a silicon junction to change of temperature at constant current and a forward voltage of 0.6 V, at 25 °C and 100 °C?
(Answers: 42.4 nA, 1.03 μA, 54.9 μA, −1.94, −1.6 mV/°C)

1.24 Plot to scale the theoretical *I–V* characteristic of a rectifying junction over the range of applied bias −200 to +200 mV at 300 K if the reverse leakage current of the junction is 10 nA at this temperature. By representing the reverse characteristic by a single resistance and the forward characteristic by a single resistance in series with a voltage source, produce a first-order network model representing the static performance of the junction over this voltage range. Repeat for the range −80 to +80 mV noting the different resistance values obtained.
(Specimen answers: reverse segment 20 MΩ, forward segment 3 kΩ in series with 140 mV; reverse segment 7 MΩ, forward segment 240 kΩ in series with 35 mV)

1.25 An abrupt *pn* junction in silicon has uniform dopant densities of 10^{24} and 10^{21} atoms/m^3 in the *p* and *n*-regions respectively. The junction has an area of 0.1 mm^2, the reverse saturation current at 300 K is 10 nA and the minority carrier lifetime in the *n*-region is 1 μs. Assuming the validity of the theoretical diode equation, produce network models representing the small-signal performance of the junction at a static forward bias of 300 mV and at a static reverse bias of 10 V, both at 300 K.
(Answers: 25.2 Ω in parallel with 39.6 nF; 2.8 pF)

2 Discrete Semiconductor Devices

This chapter is concerned with the wide range of discrete semiconductor devices used in modern electronic systems. For each device, the physical construction, electronic operation and terminal characteristics are discussed leading to a network model which can be used in the analysis of circuits using the device.

2.1 SEMICONDUCTOR DIODES

Semiconductor diodes (diode: *two*-elect*rode* device) are used in a wide variety of applications such as rectification, voltage reference, switching, signal generation and as electro-optical transducers utilising various properties as indicated in table 2.1.

2.1.1 Low-power pn Junction Diodes

A junction diode consists of a *pn* junction with ohmic contacts to the *p* and *n*-regions providing external connections. Consideration of the conduction properties of a *pn* junction (section 1.6.2) shows that it has a non-linear I–V characteristic, the flow of current being controlled by the polarity and magnitude of the applied voltage (bias) across the junction, as given theoretically by the diode equation (1.132)

$$I = I_0 \left[\exp\left(\frac{eV}{kT}\right) - 1 \right] \qquad (1.132)$$

This property enables the diode to be used as a voltage-controlled switch. [1,2,3,4]

Table 2.1 Semiconductor diode applications

	Type	Property used	Applications
Rectification	General-purpose *pn* junction diode/rectifier	Unilateral IV characteristic of *pn* junction	a.c.–d.c. conversion, low-speed switching
Voltage reference	Breakdown (Zener and avalanche) diode	Sharp reverse IV characteristic	Voltage reference, typically 2.7–200 V
	Stabistor	Sharp forward IV characteristic	Low voltage reference, typically 0.6–1.3 V
Switching	Fast switching *pn* diode	Low minority carrier storage due to small junction area and/or short carrier lifetime	High-speed switching (ns)
	Schottky-barrier diode		High-speed switching (ps), detection and mixing at UHF
Signal generation	Tunnel diode	Instability (oscillation) due to negative slope of IV characteristic	High-frequency (GHz) generation
	Gunn diode	Oscillations in bulk material in presence of electrical field	High-frequency (GHz) generation
	Varactor diode	Change of junction capacitance with bias	Voltage dependent capacitance, high-frequency (GHz) generation as a frequency multiplier (up-converter)
	Step-recovery (snap, pin) diode	Sharp switch-off transient	High-frequency (GHz) harmonic generator
	Avalanche transit-time (ATT) diode	Negative conductance due to time delay causes oscillation	High-frequency (GHz) generation
Optical	Light-emitting diode (LED)	Energy released during recombination is emitted in optical spectrum	Optical signal source, indicator and display
	Laser diode	As LED but higher power output due to multiplication in optical cavity	Optical signal source
	Photodiode (pin and avalanche)	Incident optical energy generates a photocurrent	Optical signal detector
	Solar cell	Incident optical energy converted to electrical output power	Electrical power source

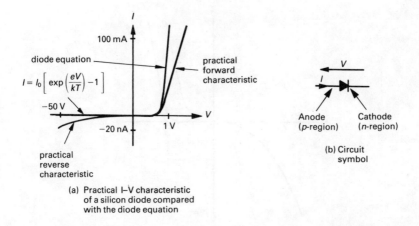

Figure 2.1   Semiconductor diode

The voltage that must be applied across a *practical* diode to cause a certain value of forward current is greater than that predicted by the theoretical diode equation (figure 2.1a). This is due to the ohmic voltage drop across the bulk p and n-regions which causes the forward characteristic to become approximately linear at moderate forward current instead of exponential as indicated by the junction equation. At very low forward currents ($< 100\ \mu$A) the ohmic voltage drop is negligible and the rise of current with voltage is approximately as given by the diode equation. 'Practical' forward conduction ($\geqslant 1$ mA) for a *silicon* diode occurs at a forward bias exceeding about 0.6 V at normal operating temperature and a silicon diode is therefore said to 'switch on' (that is, to conduct appreciably) at a forward bias of 0.6–0.7 V. As forward conduction increases, the ohmic drop becomes substantial and typically the applied bias for a forward current of 100 mA is of the order of 1.5 V for a low-power device.

Under reverse bias conditions, current flow (reverse leakage) is due to minority carrier drift across the depletion layer, thermal generation within the depletion layer and surface leakage. For a *silicon* diode at normal operating temperature, thermal generation in the depletion layer dominates and the leakage current therefore increases with reverse bias (equation 1.136), typically up to about 20 nA at 50 V for a low-power diode at 25 °C (figure 2.1a).

Germanium has a higher intrinsic carrier density n_i than silicon which results in a lower contact potential ψ (equation 1.102). A germanium diode, therefore, exhibits a lower forward voltage drop than its silicon equivalent, 'switching on' at a forward bias of 0.2–0.3 V. Thus a germanium diode is a closer approximation to an ideal conducting switch (that is, a short-circuit,

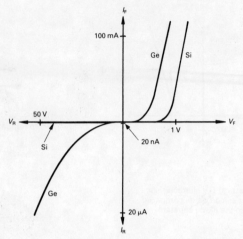

Figure 2.2 Comparison of the static I–V characteristics of low-power silicon and germanium diodes at 25 °C

$V = 0$) than a silicon diode (figure 2.2). The higher intrinsic carrier density, however, causes germanium diodes to have a reverse leakage current of the order of 1000 times that of a similar silicon diode at the same reverse bias and temperature. Reverse leakage of a *germanium* diode is dominated by minority carrier drift across the depletion layer as given by equation 1.133 and is typically 20 μA at 25 °C for a reverse bias of 50 V compared with only 20 nA for a corresponding silicon device (figure 2.2). Germanium diodes are therefore inferior to silicon types in the 'switched-off' state and they are now of only minor importance. The subsequent discussion of semiconductor diodes refers mainly to modern silicon types.

Physical Structure
The majority of *pn* junction diodes are produced by the epitaxial planar process (section 3.1) although some alloy junction types are still produced particularly for voltage-reference (section 2.1.3) and applications requiring high surge ratings and low noise. Figure 2.3 shows a range of modern diode package styles ranging from the low-power (75 mA, 50 V) BAX13 to the medium-power (60 A, 45 V) SD51 Schottky diode. The current and voltage ratings quoted are the maximum average value of *forward* current and the maximum repetitive peak *reverse* bias, mostly at 25 °C. Each package style or *outline* is described either by a JEDEC (Joint Electronic Device Engineering Council (U.S.)) diode outline (DO) number or a British Standard (BS 3934) semiconductor outline (SO) number. Where applicable, the outline numbers for the diode packages shown in figure 2.3 are included with the device type numbers. The semiconductor device type numbering systems are described in appendix E.

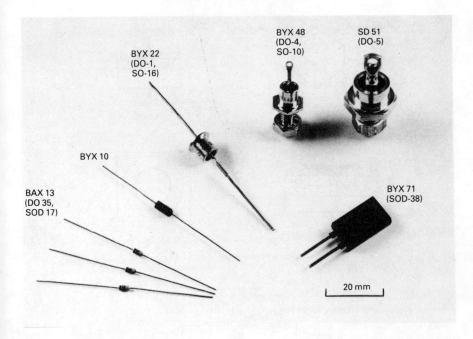

Figure 2.3 Various diode package styles: BAX 13, 75 mA, 50 V; BYX 10, 360 mA, 1600 V; BYX 22, 1.4 A, 800 V; BYX 71, 7 A, 600 V; BYX 48, 9 A, 1200 V; SD 51, 60 A, 45 V

The structure of a modern low-power *pn* junction diode such as a BAX13 is shown in figure 2.4. Basically the chip consists of a shallow *p*-region formed within the surface of an *n*-type substrate with metallisation of the surface of the *p*-region and the underside of the chip forming the anode and cathode contacts (figure 2.4a). A common packaging method is to sandwich the chip between two metal studs, the arrangement being held in compression by the shrinkage of glass encapsulation which also protects the junction from chemical contamination and provides mechanical strength. External connecting leads are welded to the metal studs. In some older types, connection from the chip surface to the metal stud is via an S-shaped spring often called a 'whisker'. The direct method of connection between chip and stud described above is therefore termed 'whiskerless'.

Details of a BAX13 diode are given in appendix H.1. It should be noted that diode I–V characteristics are often plotted as log I–V and the forward characteristic (I_F–V_F) presented separately from the reverse characteristic (I_R–V_R) to improve legibility.

Under static operating conditions the important properties of the diode are

(1) Forward voltage drop, which comprises the contact potentials of the junctions (metal–semiconductor contacts as well as the *pn* junction) and

98 Semiconductor Device Technology

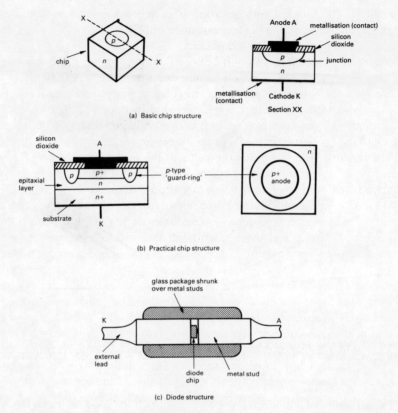

Figure 2.4 Low-power *pn* junction diode structure

the ohmic drop across the bulk semiconductor regions;
(2) forward current capability as determined by power dissipation and the thermal properties of the package;
(3) reverse leakage current;
(4) reverse breakdown voltage.

These properties are dependent on the dopant densities of the *p* and *n*-regions and the geometry of the structure. The lower the dopant densities, the lower the contact potential (equation 1.102) and hence the lower the forward voltage drop making the diode a better approximation of a short-circuit under forward bias conditions. Lower dopant densities also cause a wider depletion region under reverse bias conditions (equation 1.137) thus increasing the reverse breakdown voltage of the junction. However, lower dopant densities cause the bulk *p* and *n*-regions to have higher resistance and so the forward characteristic is less steep, thus, although the contact potential is reduced, the forward voltage drop varies over a wider range as the current changes. In addition, the higher resistance causes increased power dissipation and reduced forward current capability.

In practice there is a trade-off between diode properties and it is common to create a p^+n or pn^+ junction having one region heavily doped (10^{25}–10^{27} atoms/m^3 for silicon) increasing its conductivity and reducing power dissipation and the other region relatively lightly doped (10^{22}–10^{23} atoms/m^3) to provide the required reverse breakdown voltage. It should be noted that many of the properties of a p^+n or pn^+ junction are determined by the doping of the *lightly* doped side since the value of the term $(N_a + N_d)/N_a N_d$, which appears in several junction expressions such as depletion layer width, junction field and depletion layer capacitance (equations 1.137, 1.138 and 1.151), depends mainly on the lower dopant density. Power dissipation in the lightly doped region is reduced by making the region thin. The minimum thickness is determined by the depletion layer width corresponding to the breakdown voltage required, so as to avoid *punch-through* where the depletion layer spreads right across the region. A typical chip structure is shown in figure 2.4b in which a thin lightly doped n-type epitaxial region is grown on a more heavily doped n^+ substrate. The junction is formed by a shallow p^+ diffusion, the depth of which is controlled to provide the necessary n-region thickness. A heavily doped substrate is also important for another reason. The final structure contains three junctions, the pn junction and two metal–semiconductor junctions forming the anode and cathode contacts. Clearly the metal–semiconductor contacts are required to be ohmic, not rectifying (section 1.6.1), and if aluminium metallisation is to be used, contact with n^+ is imperative as an aluminium/lightly doped n-type silicon junction has rectifying properties.

Figure 2.4b also shows an additional lightly doped p-type 'guard-ring' diffusion which is necessary to avoid premature reverse breakdown where the junction meets the chip surface. In the absence of a guard-ring, surface conditions (section 1.5) and tapering of the p^+ region near the surface cause the depletion layer to be narrower, and hence the junction field higher, at the surface than in the bulk, resulting in premature breakdown at the surface. By surrounding the p^+ region with a more lightly doped p-region the depletion layer at the surface is widened thus reducing the field and preventing premature breakdown.

Practical Identification and Functional Verification
A diode has its type number either printed directly on it, known as *type branding*, or coded using coloured bands around the diode body conforming to the IEC/EIA colour code (appendix F). When the diode is type branded, polarity is indicated in one of the following ways

(1) by the shape of the diode body (refer to manufacturer's data sheet for details);

(2) by a diode symbol printed on the diode body;
(3) the cathode is indicated by a coloured band at one end of the diode body
(4) in the case of certain older low-power types, the cathode is indicated by a red-coloured end to the diode body, the significance of red being that the cathode is the positive output terminal when the diode is used in a conventional rectifier circuit.

In the case of colour coding, the serial number is normally read from the cathode, so that for a BAX13, the brown (1) band of the 13 is nearest to the cathode terminal (BAX13 data, appendix H.1).

Terminal polarity can be checked practically using an ohmmeter (that is, a multi-purpose meter on the resistance range), the indicated forward resistance being typically in the range 400 Ω to 2 kΩ and the reverse resistance infinite. The same test is also useful to check that a diode is functional, that is, has not been damaged. Some multimeters have a 'diode test' mode displaying

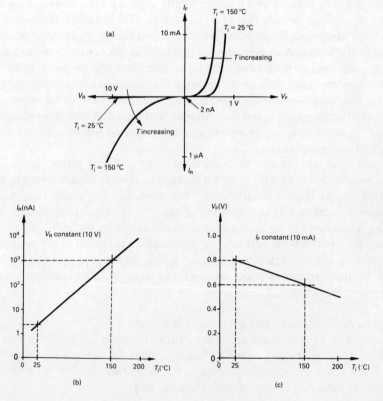

Figure 2.5 Change of characteristics with junction temperature for a typical low-power silicon diode

the forward voltage drop across the diode at a fixed current (typically in the range of 100 μA to 1 mA). Such a test shows a voltage drop of 500–800 mV for a silicon diode (100–400 mV for a germanium type) and zero for reverse bias.

Temperature Effects
Diode characteristics vary with temperature (figure 2.5a) due mainly to the variation of the intrinsic carrier density n_i with temperature. The relationship between n_i and T leads to theoretical expressions for the changes of reverse leakage current I_R and forward voltage V_F with temperature (section 1.6.2). The symbol I_R is widely accepted for practical reverse current in contrast to I_0 which is used to describe the theoretical reverse saturation current of a junction.

The fractional change of I_R for a silicon diode over a junction temperature range T_1 to T_2 (degrees kelvin) at constant reverse bias is given by equation 1.143 as

$$\frac{I_R(T_2)}{I_R(T_1)} = \left(\frac{T_2}{T_1}\right)^{3/2} \exp\left[6.3 \times 10^3 \left(\frac{T_2 - T_1}{T_1 T_2}\right)\right] \quad (2.1)$$

This relation enables calculation of the reverse leakage current at any temperature within the working range of the device based on the value at a single temperature obtainable from manufacturer's test data. Typical ranges of I_R for a low-power silicon diode at junction temperatures T_j of 25 °C and 150 °C are given in table 2.2. The approximate exponential relationship between reverse current and junction temperature at constant reverse bias is shown in figure 2.5b. For *small* changes of temperature in the region of 25 °C (298 K), equation 2.1 reduces to

$$\frac{I_R(T_2)}{I_R(T_1)} \approx \exp\left[0.06 \, (T_2 - T_1)\right] \quad (2.2)$$

Table 2.2 **Typical ranges of reverse leakage current and forward voltage for a low-power silicon diode at junction temperatures of 25° and 150°C**

Junction temperature T_j			25 °C	150 °C
Reverse leakage current (I_R)	@	$V_R = 10$ V	2–20 nA	1–10 μA
	@	$V_R = 50$ V	20–40 nA	10–40 μA
Forward voltage (V_F)	@	$I_F = 1$ mA	0.6–0.7 V	0.4–0.5 V
	@	$I_F = 10$ mA	0.7–0.9 V	0.5–0.7 V

which provides a simple method of calculating the increase in leakage current for a small change in ambient temperature, provided power dissipation is low so that the semiconductor temperature is approximately equal to the ambient temperature. The general relation (equation 2.1) shows that for a silicon diode at constant reverse bias, I_R *approximately doubles per 8 °C rise in temperature* for operating temperatures near 25 °C. At higher temperatures the fractional increase of I_R with temperature is reduced; in the region of 150 °C, equation 2.1 shows that I_R approximately doubles per 20 °C rise in temperature.

The change of forward voltage across the diode with temperature is given by equation 1.147 where the relevant temperature is that of the junction, thus

$$\left.\frac{\partial V_F}{\partial T_j}\right|_{I_F \text{ constant}} = \frac{V_F - W_g/e}{T_j} - \frac{3k}{e} \tag{2.3}$$

The change of V_F with T_j is approximately linear having a typical value in the range -1.5 to -2.5 mV/°C at low current levels as shown in figure 2.5c. At higher current levels where $V_F > W_g/e$, V_F increases with T_j. For example, the forward voltage temperature coefficient is of the order of $+0.15$ mV/°C for a BAX13 diode at a forward current of 75 mA (BAX13 data, appendix H.1).

As far as thermal limitations of a diode are concerned, it is the temperature of the junction T_j that is the limiting factor. Under static conduction, a power $V_F I_F$ is dissipated in the diode causing the junction temperature to be higher than the ambient T_{amb}. The basic thermal model of figure 2.6 is useful in assessing the safety of certain operating conditions. Maximum junction temperature $T_j\text{max}$ and the thermal resistance between junction and ambient $R_{\text{th}(j-\text{amb})}$ are quoted in the manufacturer's data. From the thermal equivalent of Ohm's law

$$T_j - T_{\text{amb}} = P \times R_{\text{th}(j-\text{amb})} \tag{2.4}$$

hence the junction temperature at a particular ambient temperature and power dissipation is given by

$$T_j = P \times R_{\text{th}(j-\text{amb})} + T_{\text{amb}} \tag{2.5}$$

and the maximum power dissipation corresponding to the maximum permissible junction temperature $T_j\text{max}$ is

$$P_{\text{max}} = \frac{T_j\text{max} - T_{\text{amb}}}{R_{\text{th}(j-\text{amb})}} \tag{2.6}$$

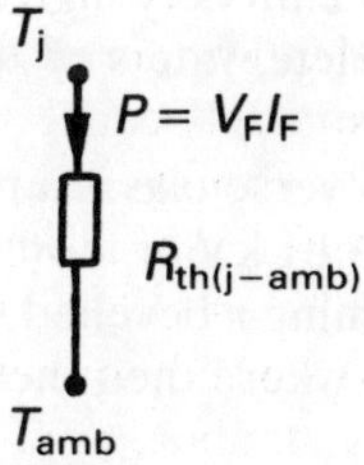

T_j

$P = V_F I_F$

$R_{th(j-amb)}$

T_{amb}

$$P_{max} = \frac{T_j max - T_{amb}}{R_{th(j-amb)}}$$

Figure 2.6 Basic thermal model of a diode for static conduction

V_F and I_F are related by the diode forward characteristic and it is therefore possible to define $I_F$max. As far as the user is concerned, it is more useful to know $I_F$max than P_{max} and manufacturers usually provide a plot of $I_F$max – T_{amb} (BAX13 data, appendix H.1) indicating the safe operating area (SOAR) for the diode in terms of current and temperature.

In rectification and switching, significant power dissipation occurs only while the diode is conducting and the heat generated has time to disperse during the OFF time. In such cases it is the average forward current $I_{F(AV)}$ which must be considered and plots of $I_{F(AV)}$max – T_{amb} are often provided for both sinusoidal and switching modes. In the case of switching, the ratio of ON time to the period of the switching waveform, termed the *duty cycle*, is the significant parameter (BAX13 data, appendix H.1).

2.1.2 High-power Diodes

The term 'high-power' is used here to describe diodes with a high forward current capability (for example, $I_{F(AV)}$ max $\geq$ 1 A) and/or a high reverse breakdown voltage (for example, V_R max $\geq$ 500 V). The forward current capability is specified either as the maximum static forward current (that is the d.c. value) $I_F$max or more commonly as the maximum average value $I_{F(AV)}$max under dynamic conditions. Similarly the maximum safe reverse voltage is given as the maximum continuous reverse voltage V_R max and/or the maximum repetitive peak reverse voltage V_{RRM} max. At the extreme high-power end of the range, diodes are available with $I_{F(AV)} > 3000$ A at a case temperature of up to 55 °C and $V_{RRM} > 4$ kV. Such diodes are normally encapsulated in a double-sided 'hockey puck' capsule (for example, JEDEC DO-200) similar to that shown in figure 2.89a for a high-power SCR.

As the forward current rating is increased, the most significant design factor is the high current density in the chip and the resultant power dissipation and temperature rise. To reduce current density the chip size is increased and heat flow from the package is improved by providing a metal mounting base or stud fixing as shown in figure 2.3. Very high current diodes (for example, $I_F > 200$ A) commonly use complete wafers of 20 mm or more diameter.

Premature breakdown at the surface under reverse bias is a major problem in diodes required to have breakdown voltages of 1 kV or more and it is usual for such diode chips to be physically etched forming a bevelled *mesa* structure (section 1.5) to alter the depletion layer width where the junction meets the chip surface.

The reverse leakage of a silicon *pn* junction (equation 1.136) is a function of junction area and depletion layer width. Since a high forward current device requires a large junction area for thermal reasons and operation at a reverse bias of several hundred volts causes a relatively wide depletion layer, leakage currents for high-power diodes are orders of magnitude greater than for low-power types. For example a 2 A, 800 V diode typically has a reverse leakage of up to 2 μA 25 °C increasing to 100 μA at 125 °C, while a 1000 A, 4.4 kV device has a typical reverse leakage of 30 mA at 55 °C.

2.1.3. Voltage-reference Diodes

Reverse breakdown of the *pn* junction (section 1.7) due to either the Zener or the avalanche effect causes a high reverse current to flow and hence increased power dissipation but the voltage across the junction varies only slightly for a wide range of reverse current (figure 2.7a). This approximately constant voltage drop across the junction at breakdown is utilised as a voltage reference particularly for voltage stabilisation. Note that in the case of voltage-reference diodes it is common to use the symbols V_Z and I_Z for reverse voltage and reverse current in place of the general symbols V_R and I_R. The suffix Z derives from Zener breakdown and is used regardless of whether the diode is actually an avalanche or Zener device. Equation 1.138 shows that the junction field depends on the dopant densities particularly the density on the more lightly doped side of the junction. As breakdown occurs when the junction field reaches a certain value (section 1.7), junctions can be produced with a particular breakdown voltage by control of this dopant density.

Ultimately it is junction temperature and eventual melting of the semiconductor due to overcurrent that destroys a junction semiconductor device and so diodes intended for operation at breakdown, usually termed *voltage-reference*, *voltage-regulator* or simply *breakdown diodes*, are designed to withstand a certain power dissipation. As long as the maximum total power dissipation P_{tot} max and hence the maximum junction temperature is not exceeded, the diode will not be damaged and will 'come out of breakdown' when the reverse bias is reduced below V_B. Manufacturers produce ranges of voltage-reference diodes such as the Mullard BZY 88 series which has a

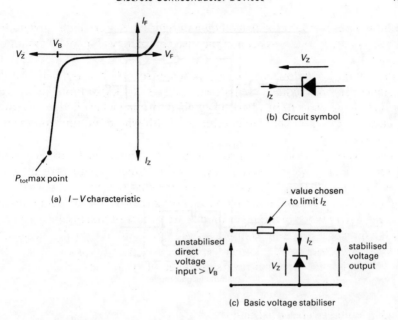

Figure 2.7 Voltage–reference (breakdown) diode

power dissipation limit of 400 mW and nominal breakdown voltages in the range 2.7 to 33 V with intermediate values according to the E24 series (appendix G) designated C2V7 to C33, where V represents a decimal point and C indicates a 5 per cent tolerance in the nominal breakdown voltage (appendix E). The static reverse characteristics of this range are given in appendix H.2 together with the forward characteristic which is similar to that of a general-purpose *pn* junction diode.

Breakdown diodes can be used in series to provide reference voltages not obtainable using a single diode. In addition it is interesting to note that the temperature coefficient (variation of breakdown voltage with temperature S_Z) is negative for diodes with V_B below 5.6 V and positive for those with V_B above 5.6 V. This is due to the fact that different mechanisms, namely Zener and avalanche, are responsible for breakdown on either side of this breakpoint. The change of sign of the temperature coefficient provides a method for compensating for voltage variations due to temperature change, for example, a series combination of 3.3 V and 6.8 V devices provides a more stable reference than a single 10 V device for changes in temperature. The working voltage obtained across a voltage-reference diode, that is, the position of the operating point on the $I_Z - V_Z$ characteristic, is set by controlling the reverse current flow using an external resistor (figure 2.7c).

Another range of voltage-reference diodes known as *stabistors* or *controlled-conductance diodes*, uses the sharp *forward* characteristic of a *pn*

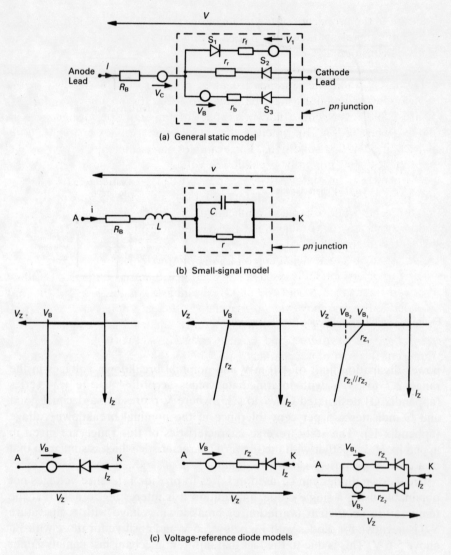

Figure 2.8 Network-modelling of diode performance

junction and provides low voltage references below 1.5 V (for example, BZY 88 C0V7 and C1V3 diodes).

2.1.4 Diode Modelling

Representation of the conduction properties of a *pn* junction under static and dynamic conditions is considered in section 1.8. For a practical diode, in addition to the properties of the *pn* junction, the bulk semiconductor regions, the ohmic metal–semiconductor contacts and the external connecting leads must be considered.

For static operation, the combined bulk resistance of the 'neutral' p and n-regions R_B and the combined contact potential V_C of the two metal–semiconductor junctions must be added to the static *pn* junction model of figure 1.29b to provide a network model of the static performance of the diode (figure 2.8a). In practice this model can often be simplified for particular operating conditions. For example, unless the applied reverse bias approaches the junction breakdown voltage, the branch representing performance at breakdown (S_3, r_b and V_B) can be removed. In addition, except for very low values of forward bias, the contact potential V_C is insignificant, while for many applications the reverse leakage of a silicon diode (branch S_2, r_r) is negligible. Piecewise-linear modelling techniques are presented in appendix A.1.

Under small-signal dynamic conditions the model of figure 2.8b applies. The *pn* junction model is discussed in section 1.8 and to this must be added the lead inductance L and the bulk resistance R_B of the p and n 'neutral' regions. In practice, however, R_B is usually sufficiently small to be neglected.

In most applications of voltage-reference diodes, only the breakdown properties are of interest and in such cases the network model need only represent this part of the characteristic. Figure 2.8c shows network models of the static reverse characteristic of a voltage-reference diode, derived from the general junction model of figure 1.29b, giving different degrees of accuracy.

The unblanked diode symbols used in the network models of figure 2.8 represent hypothetical voltage-controlled switches which are short-circuit for V positive and open-circuit for V negative, where V is defined in the same way as for a real diode (figure 1.24a).

Network-modelling has very limited use in switching (that is, large-signal dynamic conditions), a more convenient representation being the *charge-control model* introduced in section 1.8. This representation is considered in detail in section 2.4.1 concerned specifically with diodes intended for fast switching applications.

2.2 BIPOLAR JUNCTION TRANSISTORS (BJTs)

The transistor is a three-terminal solid-state device in which current flow through the device between two of the terminals can be controlled by the signal applied to the third terminal. Such properties enable the device to be used both as an amplifier and a switch. Possibly the earliest suggestion of such a device was made in 1935 by Heil, in Germany, in which he suggested that the conductivity of a semiconductor layer could be modulated by an electric field provided by the potential applied to a metallic layer adjacent to, but

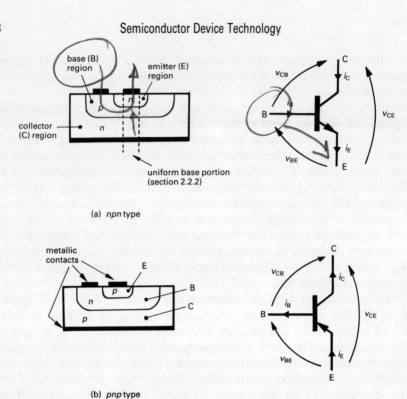

Figure 2.9 Basic structure and circuit symbols of *npn* and *pnp* BJTs

insulated from, the semiconductor layer. Several researchers were involved in the theory and application of semiconducting materials during the late 1930s and 1940s, most significantly in the use of metal–semiconductor contacts (for example, copper–copper oxide) and *pn* junctions in silicon and germanium as rectifiers.

The first working transistor is attributed to Bardeen, Brattain and Shockley, researchers at the Bell Telephone Laboratories in the United States. The development was reported [5] in June 1948 and eight years later they received the Nobel Prize for Physics. Their original attempts to construct a 'semiconductor triode' were based on the field-effect principle previously suggested by Heil but results were not encouraging. Subsequently Brattain and Bardeen used a rectifying metal–semiconductor contact as a means of applying the controlling voltage to the device; results were more successful and led to the *point–contact* arrangement. Subsequent investigation to explain the mechanism of operation of the point–contact transistor showed that the new device operated on a principle entirely different from the field-effect originally proposed. Current flow through the device was controlled by the current flow from the controlling electrode rather than by the electric field created by an applied control voltage. During the attempt to explain the operation, Shockley [6] suggested a different structure formed from two *pn* junctions. Although this was originally only intended to aid the understanding of the operation of the point–contact transistor, later implementation showed that the new *junction* structure overcame the limitations of the point–contact arrangement, namely, noise and low current capability. The resulting *bipolar junction transistor* (BJT) became the major factor in the introduction of solid-state technology and with considerable development in structure and fabrication has continued to be of major importance. The field-effect principle was initially discarded owing to the success of junction devices and it was not until 1958 that the first commercial field-effect transistor was produced by Teszner in France. During the 1960s field-effect transistors (FETs) developed rapidly and since then there has been continual competition between the bipolar and field-effect technologies, in terms of electrical performance and cost, which later escalated by their competition in integrated form.

Junction transistors are described as *bipolar* since both sign of charge carrier, electrons and holes, contribute to the current flow. They consist of two back-to-back *pn* junctions formed typically one micron (10^{-6} m) apart in a single *chip* of semiconductor, usually silicon. The junctions are formed using the epitaxial planar process (section 3.1) by alternately diffusing acceptor and donor-dopants into a substrate of suitable type so forming a three-layer *npn* or *pnp* structure. The three regions are termed the *emitter, base* and *collector*; the basic cross-sectional structure together with circuit symbols for the two configurations is shown in figure 2.9. The directions defined for the emitter, base and collector currents correspond to the actual directions of conventional current flow (that is, from high to low potential) for *normal* operation, that is, when operating as an amplifier. The emitter arrow on the BJT symbol indicates the direction of emitter current. Figure 2.10 shows a range of modern BJT package styles ranging from the low-power (300 mW, 100 mA, 45 V) BC107 to the high-power (115 W, 15 A, 60 V) BCY20. The quoted ratings are the maximum total power dissipation P_{tot} max, the maximum static collector current I_C max and the maximum static collector-emitter voltage with zero base current V_{CEO} max at 25 °C. Voltage and current ratings and maximum power dissipation for a BJT are considered in sections 2.2.19 and 2.2.21.

Practical Identification and Functional Verification
BJTs are type branded, the type number (appendix|E) being printed directly on the transistor body. Various styles of encapsulation package, known as the transistor *outline*, are used in both plastic and metal-can form and details such as lead configuration can be obtained from the manufacturer's data sheet. Most package styles have a JEDEC (U.S.) transistor outline (TO) number

Figure 2.10 Various BJT package styles: BC 107, 300 mW, 100 mA, 45 V; BC 149, 350 mW, 100 mA, 20 V; BF 594, 250 mW, 30 mA, 20 V; BFY 52, 800 mW, 1 A, 20 V; BD 131, 15 W, 3 A, 45 V; TIP 31 A, 40 W, 3 A, 60 V; BDY 20, 115 W, 15 A, 60 V

and/or a British Standard (BS 3934) semiconductor outline (SO) number. The outline numbers of some popular packages are given in figure 2.10.

Under experimental conditions it is often necessary to check that a BJT is functional, that is, has not been damaged. This can be easily done using an ohmmeter (that is, a multi-purpose meter on the resistance range) by measuring the forward and reverse resistances between each pair of terminals. The B-E and B-C junctions typically have an indicated forward resistance between 400 Ω and 2 kΩ and infinite resistance with the meter probes interchanged. Between the C-E terminals are two back-to-back diodes and so the indicated resistance should be infinite in both directions. Alternatively the 'diode test' mode available on some multimeters can be used, the indicated junction voltage being typically in the range 600–700 mV for forward bias and zero for reverse bias. These simple tests enable a major failure such as an open-circuit junction to be readily detected. Device characteristics can be conveniently displayed using a commercial 'curve-tracing' oscilloscope.

Discrete Semiconductor Devices

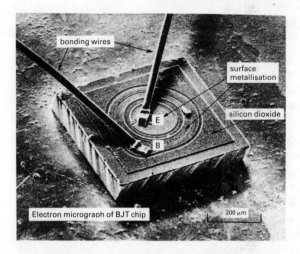

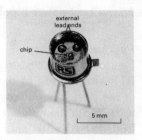

(a) Structural details of a BC107 npn silicon planar epitaxial BJT

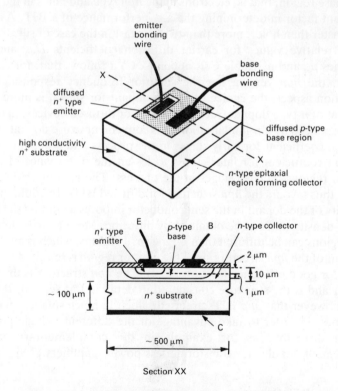

(b) Typical *npn* chip structure

Figure 2.11 Low-power BJT structure

2.2.1 Low-power Bipolar Junction Transistors (BJTs)

Physical Structure

Figure 2.11a shows structural details of a typical low-power planar epitaxial BJT with the top of the can removed together with an electron micrograph of the chip. The micrograph shows the metallised emitter and base contacts with 25 μm diameter gold bonding wires which make contact to the package leads. The BC107 chip shown is approximately $600 \times 500 \times 100\mu$m thick. Figure 2.11b gives a schematic representation of the chip. For an *npn* BJT the *n*-type collector is formed by an epitaxial layer grown on the surface of a heavily doped (n^+) substrate and the base and emitter regions are formed by selective doping of a thin surface region of the epitaxial layer. The active part of the device thus lies within 5 μm of the surface, the remainder of the chip merely acts as a carrier and provides mechanical strength.

For both performance and fabrication reasons, the *npn* configuration has become more popular than the *pnp* type. The transit time of minority carriers in the base region, that is, electrons in the *npn* type and holes in the *pnp*, is an important factor in determining the a.c. performance of a BJT. As electrons move faster than holes, more than twice as fast in the case of silicon as shown by the relative values for carrier diffusion coefficients D_n and D_p and mobilities μ_n and μ_p (table C.1, appendix C) it follows that, for comparable devices, the *npn* version has the superior frequency response. From the fabrication aspect, the choice of *p*-type dopant for silicon is more restricted than that of *n*-type dopant (section 1.2.1). For diffused devices, in contrast to those produced by ion-implantation, it is convenient to use dopants with a low diffusion coefficient for early stages in the process and a dopant with a high diffusion coefficient for later stages to lessen the disturbance of previously diffused regions by later stages in the process. The greater choice of *n*-type dopant thus favours the *npn* structure for diffused BJTs. In addition, the solid solubility of the dopant in the semiconductor imposes a limit on the maximum dopant density that can be obtained by diffusion. More heavily doped *n*-type (n^+) regions can be diffused than *p*-type (p^+) regions, which is another factor in favour of the *npn* structure since the emitter region is required to be heavily doped for good efficiency (section 2.2.2). The *npn* structure is thus the most popular and is the structure considered in detail in this section. It should be noted however that *pnp* BJTs are used, albeit in much smaller quantity than *npn* types, in order to take advantage of the different voltage polarity and current direction, as for example in the complementary 'push–pull' arrangement popular for transformerless power amplifiers.[7,8]

2.2.2 npn BJT Operation

Depending on the polarity and magnitude of the applied voltages there are three distinct *modes* or *regions* of operation of a BJT known as the active

Discrete Semiconductor Devices

Table 2.3 Modes of operation of BJTs and their practical application

Mode	Operation	Application	
Active	Intermediate conduction	Amplification	
Saturation	Fully conducting	ON state	⎫ switch
Cut-off	Fully non-conducting	OFF state	⎭

(normal), saturation and cut-off modes (table 2.3). It is convenient to discuss the active mode of operation initially as the saturation and cut-off modes are the conditions that exist at either end of the active range. For simplicity a uniform base structure (figure 2.12) is considered which is a portion of the planar structure (figure 2.9a). Additional factors arising from the practical non-uniform geometry are discussed in sections 2.2.23 and 2.2.24 in connection with high power operation and response at high frequency. The voltage and current symbols used in figure 2.9 (for example, v_{BE} and i_E) indicate the general case of a total value composed of a constant (d.c.) component (for example, V_{BE} and I_E) and a signal (a.c.) component (for example, v_{be} and i_e) in accordance with the system of symbols described in appendix B.2. Initial discussion of BJT operation considers the static case of constantly applied voltages causing constant current flow and so the appropriate upper case symbols are used in figures 2.12 and 2.13.

Active operation can be described under the headings injection, diffusion and collection.

Injection
In order to initiate current flow the emitter–base (E-B) junction is forward-biased by applying a positive value of V_{BE}, thus making the *p*-type base region positive with respect to the *n*-type emitter. Electrons are thus *emitted* (hence the name emitter) or *injected* into the base region where they become minority carriers (figure 2.12b). These injected electrons are termed *excess* electrons as they are in addition to the thermally generated electrons already in the base as determined by the dopant density. The density of electrons just inside the base thus increases.

Diffusion
The non-uniform density of electrons in the base due to injection initiates diffusion whereby the excess electrons move away from the E-B junction tending to produce a uniform density. As the electrons move across the base some recombine with holes and are annihilated and, if the base region were wide, the diffusion–recombination process would result in an exponential decay of the excess electron density with distance (figure 2.12c) having a diffusion length L_n, as given by the solution of the electron continuity equation

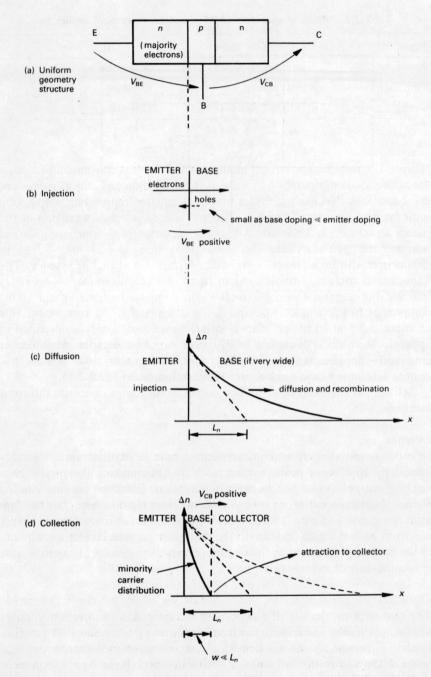

Figure 2.12 Basic BJT operation

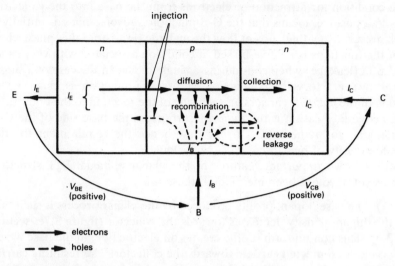

Figure 2.13 Charge movement in an *npn* BJT for active operation

with zero electric field (section 1.3.6, figure 1.12). At a distance of five times the diffusion length ($5L_n$) from the E-B junction less than 1 per cent of the injected electrons would exist as free carriers (equation 1.70).

Collection
If the collector region is biased positively with respect to the base, any free electrons reaching the collector–base (C-B) junction, having avoided recombination in the base, are attracted across the junction into the collector region by the influence of the electric field, thus forming a collector current. For an efficient transfer of charge from emitter to collector, the loss due to recombination in the base must be very small. This is achieved in practice by making the distance between the two junctions, the base width w, small compared with the diffusion length (figure 2.12d) so that charge movement across the base occurs in a very short time, drastically reducing the probability of recombination before reaching the C-B junction. For silicon with a recombination lifetime of 1 μs, the diffusion length for electrons is of the order of 60 μm (section 1.3.6). Practical base widths for modern BJTs are of the order of 1 μm and so the degree of recombination is very small. *The narrow base is thus fundamental to transistor action and explains why a BJT cannot be formed by the back-to-back connection of two discrete diodes.* Figure 2.12d shows the influence of the collector voltage on the minority carrier distribution in the base. Since all electrons reaching the C-B junction are accelerated across the junction into the collector region, the minority carrier density at the C-B junction falls to zero.

The condition for attraction of electrons from the base into the collector, that is V_{CB} positive, means that the C-B junction is reverse-biased. Initially it is perhaps surprising that current flow through the transistor takes place when one of the junctions is reverse-biased, a condition associated with very small conduction (leakage) when considering a single diode. In the case of a single diode, however, forward conduction across the junction is due to the diffusion of the *majority* carriers on each side of the junction to the other side. In the active BJT case it is the *minority* carriers on the base side of the C-B junction that form the main current flow hence the requirement for the opposite polarity of bias to create current flow.

There are two important features of the planar epitaxial BJT structure which as yet have not been mentioned, these are

(1) The base dopant profile produced by the planar process is such that the dopant density decreases towards the collector (figure 3.7c, section 3.3). This non-uniform profile creates an electric field in the base which accelerates injected carriers towards the collector. The resulting carrier drift is in addition to the diffusion mechanism described above and has the effect of reducing the transit time of carriers crossing the base thus improving the frequency response/switching speed of the BJT (transition frequency f_T, section 2.2.26, is increased) relative to a structure with a uniformly doped base.

(2) The collector region (epitaxial layer) has a lower dopant density than the base so that it absorbs most of the C-B depletion layer under reverse bias (equation 1.112) which is essential as the base width is small (typically < 1 μm). However, low dopant density would cause the collector region to have low conductance resulting in a high voltage drop across the collector, high $V_{CE(sat)}$ (section 2.2.13), high power dissipation and poor charge/discharge time for the C-B depletion layer under dynamic conditions. Therefore the lightly doped collector region (the epitaxial layer) is formed thick enough to absorb the C-B depletion layer corresponding to V_{CB} max required and the remainder of the collector region (the substrate) is heavily doped to provide high conductance.

The above discussion identifies the main features of *npn* BJT conduction in the active mode (V_{BE} and V_{CB} positive) but there are several secondary effects (figure 2.13) which must be considered.

(1) Injection of Holes from Base to Emitter
Forward-biasing of the E-B junction causes injection of electrons from the emitter into the base as required to initiate conduction but also causes hole injection from base to emitter. The emitter current is made up of these two components. As the holes injected into the emitter do not come from the collector they cannot form part of the main current flow through the device

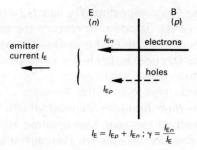

(a) Charge movement at the forward-biased E – B junction

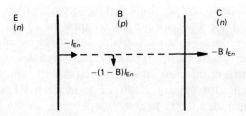

(b) Loss of electrons due to recombination in the base

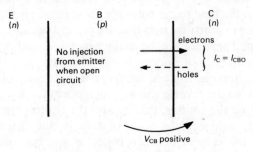

(c) Collector–base reverse leakage

Figure 2.14 Secondary components of charge movement in *npn* BJT active operation

and thus form a parasitic component. If I_{En} is the component of emitter current due to electron injection E → B (figure 2.14a), the ratio of I_{En} to the total emitter current is termed the *emitter injection efficiency*, γ , where

$$\gamma = \frac{I_{En}}{I_E} = \frac{I_{En}}{I_{Ep} + I_{En}} < 1 \tag{2.7}$$

The relative magnitudes of the two components I_{En} and I_{Ep} depend on the relative doping on either side of the junction. In practice the hole injection component I_{Ep} is reduced to the order of 0.3 per cent of I_{En} by heavily doping the emitter relative to the base (section 3.3) giving a value of γ of typically 0.997.

(2) Removal of Holes from the Base due to Recombination
As injected electrons diffuse across the base region, some are lost due to recombination. This not only reduces the collector current but also removes holes from the base region which are replenished via the base contact. Holes effectively flow into the base region from the external circuit although due to the high free-electron population, holes do not exist in metals, and the flow of *holes into* the base to replace those lost by recombination appears as a flow of *electrons out* of the base forming a component of the base current. This component is reduced to an acceptably low level by making the base very narrow. The term *base transport factor B* is used to describe the proportion of the electron current injected into the base from the emitter I_{En} that actually reaches the collector (figure 2.14b). For modern BJTs the value of B is typically in the range 0.993 to 0.999 depending on the base width and the quality of the epitaxial layer.

(3) Collector–base Leakage
The reverse bias applied to the C-B junction to attract the *excess* minority charge from the base to the collector also causes a leakage current due to the drift of those minority carriers that are present because of the doping levels in each region. These are the thermally generated minority carriers that exist in the base and collector regions in the unbiased state (figure 2.18a). This is termed the *reverse leakage* component between collector and base and corresponds to the reverse leakage component for a single junction. Without injection from the emitter, that is, with the emitter open-circuit, this leakage component would form the whole of the collector current and hence it may be defined as the current between collector and base with the emitter open-circuit I_{CBO} (figure 2.14c). For silicon junctions, reverse leakage is due mainly to thermal generation in the depletion region (section 1.6.2, equation 1.136) and I_{CBO} therefore increases with V_{CB}. Typically I_{CBO} is of the order of 2–20 nA at 25 °C for low-power silicon BJT.

(4) Base Current
The three secondary effects described above combine to create a flow of current into the base contact (figure 2.13). The relative significance of the three components, namely, hole injection into the emitter, recombination and C-B reverse leakage, depends on the particular operating conditions but for typical operation of a low-power silicon BJT as a small-signal amplifier with $I_C = 2$ mA, the base current would probably be of the order of 5 to 15

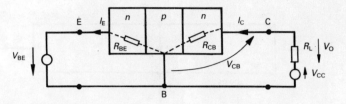

Figure 2.15 Static amplification using a BJT

μA indicating that the reverse leakage component of up to 20 nA is negligible. Under such conditions the major component of base current is due to the injection of holes from the base into the emitter.

2.2.3 BJT Action in Voltage Amplification

If an external resistor is connected in series with the collector (figure 2.15) an output voltage is developed across the resistor and the ratio of output voltage to input (E-B) bias is a measure of the static voltage gain of the arrangement. The C-B reverse bias is provided by the V_{CC} supply in the collector circuit. As the C-B junction is reverse-biased, its effective resistance R_{CB} is high and so a moderate value of R_L can be connected in series without a significant drop in I_C taking place. It is this situation that allows the transistor to force the collector current through the load and so establish the output voltage. Indeed it is the change of resistance between the input and output circuits that enables the device to act as an amplifier. R_{BE} is low due to the forward bias which allows I_E to be established for only a small value of V_{BE} (0.6–0.7 V). R_{CB} is high due to the reverse bias, allowing I_C to be forced through a load resistor of moderate value enabling a substantial output voltage V_O (several volts) to be developed.

$$\text{Static voltage gain} = \frac{V_O}{V_{BE}} = \frac{I_C R_L}{I_E R_{BE}} \qquad (2.8)$$

BJT operation ensures $I_C \approx I_E$ and, as R_L can be much greater than R_{BE} by virtue of the high value of R_{CB}, it follows that the static voltage gain can be much greater than unity.

The change of resistance between the input and output of the BJT for this mode of operation, that is, the *trans*fer re*sist*ance between input and output, which enables voltage amplification to take place is the origin of the name *transistor*.

In practice it is the a.c. voltage gain of the arrangement that is of most interest. Here the time-varying signal to be amplified is superimposed on V_{BE} causing the emitter and collector currents to vary, and the varying collector current establishes a corresponding time-varying voltage across R_L. The same

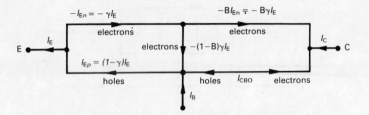

Figure 2.16 Current components in *npn* BJT active operation

basic explanation applies to the a.c. case, although as the effective resistance of the C-B junction is higher, the voltage gain can be larger.

2.2.4 Terminal Current Relations

Having defined the terms emitter injection efficiency γ and base transport factor B (figure 2.14), the components of charge movement shown in figure 2.13 which form the terminal currents I_E, I_C and I_B can be described quantitatively (figure 2.16). Thus

$$I_C = B\gamma I_E + I_{CBO}$$
$$= \alpha_{dc} I_E + I_{CBO} \qquad (2.9)$$

and
$$I_B = (1 - \gamma)I_E + (1 - B)\gamma I_E - I_{CBO}$$
$$= (1 - \gamma B)I_E - I_{CBO}$$
$$= (1 - \alpha_{dc})I_E - I_{CBO} \qquad (2.10)$$

where
$$\alpha_{dc} = \gamma B \qquad (2.11)$$

In addition
$$I_E = I_C + I_B \qquad (2.12)$$

The parameter α_{dc} describes the relationship between output current (I_C) and input current (I_E) and is thus a *current gain* term for this form of connection of the BJT which is referred to as *common-base* as the base terminal is common to both the input and output circuits. As the leakage term I_{CBO} is negligible in comparison with practical values of I_E and I_C for silicon BJTs at normal temperatures

$$\alpha_{dc} = \frac{I_C}{I_E} \qquad (2.13)$$

α_{dc} is thus termed the *static (d.c.) common-base current gain* of the transistor. In representation of the performance of a BJT by the h-parameter terminal network model (section 2.2.26), all currents are defined as flowing into the model. Due to the different convention adopted for I_E in the case of the model compared with actual current flow (figure 2.17), the corresponding h-parameter describing the static current gain in common base h_{FB} is related to α_{dc} by

$$\alpha_{dc} = -h_{FB} \qquad (2.14)$$

whereby equation 2.9 becomes

$$I_C = -h_{FB}I_E + I_{CBO} \qquad (2.15)$$

Typical values of γ and B for a low-power silicon BJT are 0.997 and 0.999 giving a value of 0.996 for α_{dc} ($h_{FB} = -0.996$). This value indicates that I_C is 99.6 per cent of I_E and the base current I_B, due to the secondary effects described above, is only 0.4 per cent of I_E. Thus for $I_E = 2$ mA, I_C would be 1.992 mA and the difference gives I_B as 8 μA.

2.2.5 Collector Multiplication

The reverse bias applied to the C-B junction causes all minority carriers arriving at the junction to be swept into the collector forming the collector current which, due to factors discussed above, is slightly less than the emitter current. If, however, the reverse bias V_{CB} is increased sufficiently to cause breakdown of the C-B junction, avalanche multiplication (section 1.7) occurs and the number of carriers flowing into the collector becomes greater than the number arriving at the C-B junction from the emitter, so that $I_C > I_E$. The

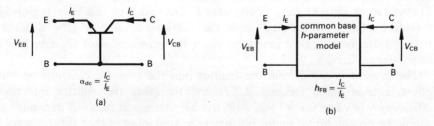

Figure 2.17 Common-base (CB) connection and h-parameter convention

multiplication of carriers at the C-B junction due to this effect is described quantitatively by the *collector multiplication factor M* given approximately by

$$M \approx \frac{1}{1 - \left(\frac{V_{CB}}{V_B}\right)^3} \quad (2.16)$$

where V_B is the breakdown voltage of the C-B junction. If V_{CB} is sufficiently high for collector multiplication to be significant then

$$h_{FB} = -\alpha_{dc} = -\gamma BM \quad (2.17)$$

A low-power silicon BJT typically has a limiting value of V_{CB} of 40–50 V, so that a C-B reverse bias of at least 15–20 V can be used without collector multiplication having a significant effect.

2.2.6 Depletion Regions at the Junctions

Section 2.2.2 refers to the injection of electrons at the E-B junction and their collection at the C-B junction with no mention of the presence or effect of the depletion regions at each junction. The depth of penetration of these regions on either side of each junction depends on the dopant densities and the applied bias (equations 1.111, 1.112 and 1.137). For a modern epitaxial planar BJT the dopant densities of the emitter, base and collector regions are typically 10^{25}–10^{27}, 10^{22}–10^{23} and 10^{21}–10^{22} atoms/m^3 respectively (section 3.3) and the corresponding depletion regions and free carrier densities with no applied biases are as shown in figure 2.18a.

For active region operation the E-B junction is forward-biased causing the depletion region at that junction virtually to disappear and the C-B junction is reverse-biased causing its depletion region to widen considerably. The effect of the C-B bias in attracting electrons from the base to the collector causes the electron density in the base to fall from the injected value on the emitter side (point O, figure 2.18b) to zero at the edge of the C-B depletion region (point S). As far as current flow is concerned, the width of the base region is therefore effectively reduced from the fabricated value w (the distance between the metallurgical junctions where the semiconductor changes type) to w_B (figure 2.18b).

The current injected from the emitter into the base I_{En} depends on the effective base width (section 2.2.7) and therefore the emitter injection efficiency γ is a function of w_B. Also the base transport factor B depends on w_B as the probability of minority carriers recombining as they diffuse across the base depends on the distance they have to travel. It follows therefore that

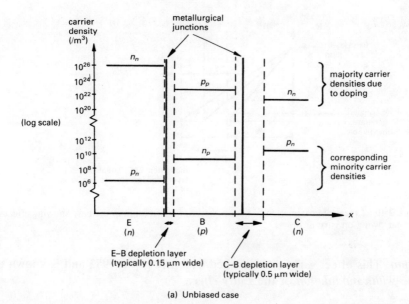

Figure 2.18 Effect of bias on junction depletion regions and minority carrier distribution in the base

the static gain α_{dc} ($= \gamma B$, equation 2.11) depends on w_B and is thus a function of V_{CB}.

The dependence of effective base width on C-B reverse bias is a *feedback effect* within the BJT as the *input* current is affected by the voltage at the

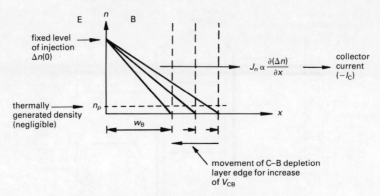

Figure 2.19 Linear approximation of minority carrier distribution in the base showing effect of base narrowing on current flow

output. This effect was first described by Early [9] in 1952 and is known as *base–width modulation* or the *Early effect*.

2.2.7 Minority Charge Distribution in the Base

The movement of minority charge across the base, electrons in an *npn* BJT, is largely by diffusion. At normal levels of conduction, the number of excess electrons in the base, that is, those injected from the emitter, far exceeds the number due to thermal generation and the current density is therefore given (equation 1.49) by

$$J_n \text{ (diffusion)} = eD_n \frac{\partial(\Delta n)}{\partial x} \tag{2.18}$$

where $\partial(\Delta n)/\partial x$ is the gradient of the electron distribution in the base.

Not all the injected electrons reach the collector with some being lost by recombination and so the current flow decreases towards the collector. The corresponding electron distribution is non-linear as shown in figure 2.18b, slope RS being less than slope OP. In practice, however, the proportion of injected electrons lost by recombination is only about 0.2 per cent for modern BJTs and therefore the electron distribution in the base is approximately linear (OS).

As the current flow across the base is proportional to the electron gradient (equation 2.18), it can be seen from figure 2.19 that an increase in V_{CB}, which causes a reduction in w_B (Early effect, section 2.2.6), results in an increased current flow.

Assuming a linear distribution, the electron gradient in the base $\partial(\Delta n)/\partial x$ is $-\Delta n(0)/w_B$ and the corresponding diffusion current density from equation 2.18 is

$$J_n \text{ (diffusion)} = -\frac{eD_n \Delta n(0)}{w_B} \qquad (2.19)$$

This electron flow is the current component I_{En} (figure 2.14a, b) which for the practical case of low recombination ($B \approx 1$) provides the collector current, thus

$$I_C \approx I_{En} = -J_n \text{ (diffusion)} \times A = \frac{eD_n A \Delta n(0)}{w_B} \qquad (2.20)$$

where A is the effective cross-sectional area of the base. The negative sign in equation 2.20 is due to the definition of electron gradient in the x direction whereas conventional current flow through the device is in the opposite direction. Equation 2.20 ignores C-B leakage (I_{CBO}) which has been shown (section 2.2.2) to be negligible for silicon BJTs under normal operating conditions.

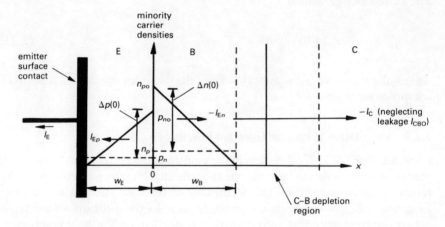

Figure 2.20 Emitter current components for active region operation

The injected electron density $\Delta n(0)$, figure 2.20, is a function of the B-E bias V_{BE}. From consideration of the pn junction under forward bias (equation 1.123)

$$\begin{aligned}
\Delta n(0) &= n_{po} - n_p \\
&\approx n_{po}, \text{ as } n_{po} \gg n_p \text{ in practice} \\
&= n_p \exp\left(\frac{eV_{BE}}{kT}\right)
\end{aligned} \qquad (2.21)$$

Thus from equation 2.20, the static collector current is

$$I_C \approx I_{En} = \frac{eD_n A n_p}{w_B} \exp\left(\frac{eV_{BE}}{kT}\right) \qquad (2.22)$$

2.2.8 Stored Charge in the Base

Although charge is continually being injected into and removed from the base during active region operation, the linear distribution of figure 2.19 exists at steady state so that the base is effectively a *store* of free charge. This is important with regard to the dynamic performance of a BJT as the base acts as a capacitor which must be charged and discharged as conditions vary.

If A is the area of the base perpendicular to the flow of current between emitter and collector, the effective base volume is $w_B A$ and the average charge density in this volume is $-\frac{1}{2} e \Delta n(0)$ assuming a linear distribution. The stored charge is then

$$q_B = -\tfrac{1}{2} e \, \Delta n(0) \, w_B \, A \qquad (2.23)$$

where the negative sign indicates that the base charge in an *npn* BJT comprises electrons.

2.2.9 Calculation of Emitter Injection Efficiency

Consideration of carrier diffusion in the emitter and base regions (figure 2.20) enables an expression to be developed for the emitter injection efficiency γ in terms of the conductivities and dimensions of the base and emitter regions. Equation 2.22 gives the static current I_{En} due to the diffusion of electrons across the base assuming recombination is negligible. The hole component I_{Ep} due to the injection of holes from the base into the emitter could be obtained from equation 1.128 if the distance (w_E) from the E-B junction to the emitter contact was sufficiently large to allow the normal diffusion–recombination exponential distribution, as described by the hole diffusion length L_p, to be established. However, the emitter region is heavily doped giving a lifetime of the order of 100 ns for which the corresponding diffusion length is (equation 1.72) 11 μm. As the emitter region is typically less than 3 μm thick, the exponential distribution does not form and the situation is therefore similar to the diffusion of minority carriers in the base, resulting in an approximately linear distribution of holes between the junction and emitter contact (figure 2.20). The hole density is zero at the contact as holes cannot exist in a metal.

If $\Delta p(0)$ is the excess hole density at the edge of the emitter region

$$\Delta p(0) \approx p_n \exp\left(\frac{eV_{BE}}{kT}\right) \tag{2.24}$$

from which

$$I_{Ep} \approx \frac{eD_p A p_n}{w_E} \exp\left(\frac{eV_{BE}}{kT}\right) \tag{2.25}$$

using a similar procedure to that used to obtain the diffusion current in the base (equation 2.22), making use of equation 1.124.

From equation 2.7 the emitter injection efficiency is

$$\gamma = \frac{I_{En}}{I_E} = \frac{I_{En}}{I_{Ep} + I_{En}} = \frac{1}{1 + \dfrac{I_{Ep}}{I_{En}}} \approx 1 - \frac{I_{Ep}}{I_{En}} \tag{2.26}$$

as, in practice, $I_{Ep}/I_{En} \ll 1$ and $(1 + x)^{-1} \approx 1 - x$ for $x \ll 1$. Substituting for I_{En} and I_{Ep} from equations 2.22 and 2.25

$$\gamma \approx 1 - \frac{w_B D_p p_n}{w_E D_n n_p} \tag{2.27}$$

It is useful to write this relationship in terms of the conductivities of the emitter and base regions σ_E and σ_B. From the Einstein relations, the ratio of carrier diffusion coefficients is equal to the ratio of carrier mobilities for the same semiconductor, $D_p/D_n = \mu_p/\mu_n$ (equation 1.57). Also the product of majority and minority carrier densities is constant at a fixed temperature (equations 1.29 and 1.31) and so $p_n n_n = p_p n_p$ from which $p_n/n_p = p_p/n_n$. Substituting these relationships into equation 2.27 gives

$$\gamma \approx 1 - \frac{w_B \mu_p p_p}{w_E \mu_n n_n} \tag{2.28}$$

In moderately doped p and n-type semiconductors, majority carrier conduction dominates and the corresponding conductivities are given by $\sigma_p = p_p e \mu_p$ and $\sigma_n = n_n e \mu_n$ (from equation 1.46), whereby

$$\gamma \approx 1 - \frac{w_B \sigma_B}{w_E \sigma_E} \tag{2.29}$$

The emitter and base widths w_E and w_B for a modern epitaxial structure are of similar order and therefore the emitter injection efficiency is determined largely by the ratio of the conductivities (and hence by the ratio of the dopant levels) of the two regions.

For a typical low-power silicon BJT with emitter and base dopant levels of 10^{25} and 10^{23} atoms/m^3 respectively, the conductivities of the two regions at room temperature are $\sigma_E \approx 2 \times 10^5$ S/m and $\sigma_B \approx 800$ S/m (resistivities of 5 $\mu\Omega$ m and 1.25 mΩ m). The widths of the base and emitter regions w_B and w_E are typically 1 μm and 2 μm giving an emitter injection efficiency of 0.998.

2.2.10 Calculation of Base Transport Factor

During diffusion of minority carriers across the base region some are lost by recombination. The proportion of the electron current injected into the base from the emitter that actually reaches the collector is described by the base transport factor B (figure 2.14b). At steady state, the free charge in the base is given (equation 2.23) by

$$q_B = -\tfrac{1}{2} e \, \Delta n(0) \, w_B A$$

If the lifetime of these injected electrons in the base is τ_B, charge is lost at a rate of q_B/τ_B due to recombination. This loss, which may be considered as an 'absorbed' current, is replenished by a component of the base current so maintaining the base charge.

The recombination of electrons in the base is thus equivalent to a component $e\Delta n(0)w_B A/2\tau_B$ of the injected current I_{En} that does not reach the collector, that is, the component $(1-B)I_{En}$ in figure 2.14b thus

$$(1-B)I_{En} = \frac{e\Delta n(0) w_B A}{2\tau_B} \tag{2.30}$$

Substituting for I_{En} from equation 2.20

$$1 - B = \frac{e\Delta n(0) w_B A}{2\tau_B} \times \frac{w_B}{eD_n A \Delta n(0)} = \frac{w_B^2}{2D_n \tau_B}$$

hence

$$B = 1 - \frac{w_B^2}{2D_n \tau_B} = 1 - \frac{w_B^2}{2L_n^2} \tag{2.31}$$

where $L_n = \sqrt{(D_n \tau_B)}$ is the electron diffusion length in the base. Equation 2.31 shows the significance of a narrow base in relation to the minority carrier diffusion length in providing efficient base transport.

For a low-power silicon BJT, the lifetime of minority carriers in the base τ_B is typically 1 μs giving a diffusion length L_n of 60 μm. With an effective base width w_B of 1 μm, equation 2.31 gives the base transport factor as 0.999.

2.2.11 Theoretical Common-base Static Current Gain of an npn BJT

For values of V_{CB} well below the C-B junction breakdown voltage, collector multiplication is insignificant and the common-base static current gain h_{FB} ($= -\alpha_{dc}$) is equal to $-\gamma B$ (section 2.2.4). Equations 2.29 and 2.31 give theoretical expressions for γ and B in terms of the dimensions and electrical properties of the emitter and base regions, from which

$$h_{FB} = -\alpha_{dc} \approx -\left(1 - \frac{w_B \sigma_B}{w_E \sigma_E}\right)\left(1 - \frac{w_B^2}{2L_n^2}\right)$$

$$\approx \frac{w_B \sigma_B}{w_E \sigma_E} + \frac{w_B^2}{2L_n^2} - 1 \qquad (2.32)$$

since in practice

$$\frac{w_B \sigma_B}{w_E \sigma_E} \ll 1 \text{ and } \frac{w_B^2}{2L_n^2} \ll 1$$

2.2.12 Energy Model of a BJT

The energy-band model of electronic energies within a semiconductor (sections 1.1 and 1.2) can be applied to a BJT as shown in figure 2.21a. With no externally applied biases the distribution of energy within the structure is such that the Fermi level of each region is at the same energy. If the energy distribution was such that the Fermi levels were not at the same energy, charge movement would take place causing a redistribution of energy until the equilibrium situation of uniform Fermi energy existed. The charge distribution necessary to produce this equilibrium situation results in contact potentials ψ_{EB} and ψ_{CB} across the two junctions. For a silicon *npn* BJT with emitter, base and collector dopant densities of 10^{25}, 5×10^{22} and 5×10^{21} atoms/m^3 respectively, the E-B and C-B contact potentials are 880 and 690 mV at room temperature assuming abrupt junctions in each case (equation 1.102). Taking account of the contact potentials at the metal–semiconductor contacts, the potential distribution throughout the structure is as shown in figure 2.21b. The corresponding electric field distribution together with typical values of the fields at the junctions (equation 1.118) is given in figure 2.21c. The fields at the two junctions are different due to the asymmetrical dopant profile of the structure (emitter doping > collector doping).

Figure 2.22 shows the effect of active region applied biases on an *npn* BJT. The B-E junction is forward-biased to cause injection by applying a positive V_{BE} and the collector is biased positively relative to the base (that is, V_{BC} is negative) to enable collection. The potential difference across a *pn* junction under the influence of a bias V is $\psi - V$ where V is the potential applied to the *p* side relative to the *n* (section 1.6.2), and so for active region operation the

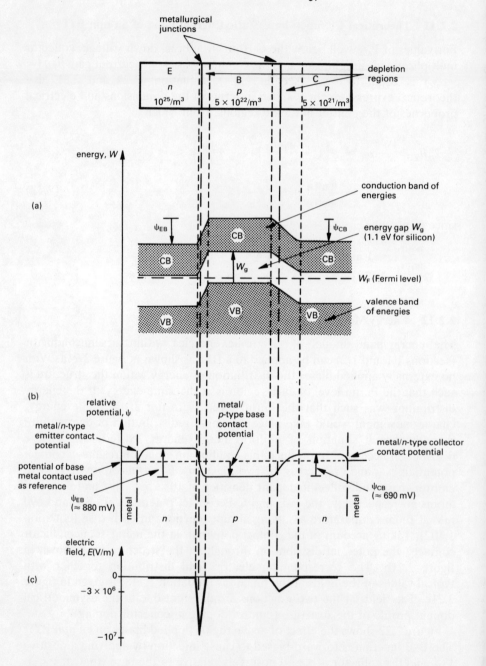

Figure 2.21 Energy model for an unbiased *npn* BJT with typical potential and electric field distributions

Discrete Semiconductor Devices

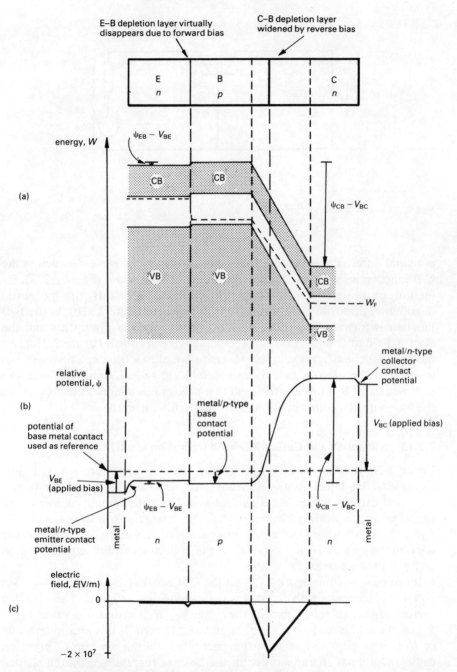

Figure 2.22 Energy model for an *npn* BJT biased for active region operation with typical potential and electric field distributions

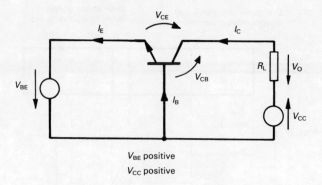

Figure 2.23 BJT biased for active region operation

potential difference across the E-B junction is reduced while that across the C-B junction is increased compared with the unbiased case (figure 2.22a). By including the metal–semiconductor contact potentials the potential distribution throughout the structure is as shown in figure 2.22b. As the E-B junction is forward-biased the depletion layer virtually disappears and the electric field at this junction is correspondingly very small (figure 2.22c). At the C-B junction, however, the reverse bias causes the depletion layer to widen and the electric field is considerably increased compared with the unbiased case. It is this electric field that accelerates minority electrons from the base into the collector forming the collector current.

2.2.13 Saturation and Cut-off Modes of Operation of a BJT

Figure 2.23 shows an *npn* BJT biased for active region operation. If V_{BE} is increased, the injection of electrons into the base from the emitter is increased causing the collector current and hence the output voltage V_O to rise. The corresponding change in the base charge distribution is given in figure 2.24a showing that the increased injection causes the slope of the distribution (which is proportional to the collector current, equations 2.18 and 2.20) to increase.

It can be seen from figure 2.23 that the C-B bias V_{CB} is equal to $V_{CC} - V_O$ and as V_O increases with I_C, V_{CB} is correspondingly reduced. The collector current cannot therefore exceed the value V_{CC}/R_L since at this value $V_O = V_{CC}$ and the C-B reverse bias is then reduced to zero. If this maximum value of I_C is reached and V_{BE} is further increased, the higher rate of injection causes additional charge storage in the base as represented by the shaded region in figure 2.24b. Slope OP corresponds to the maximum collector current as determined by V_{CC} and R_L. As I_C has then reached a limiting or saturation value, the BJT is said to be operating in the *saturation mode*,

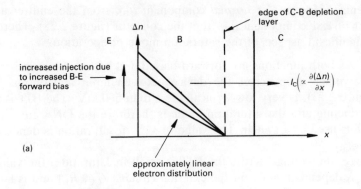

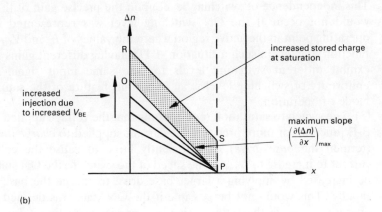

Figure 2.24 Overdriving a BJT into saturation

distribution OP being at the onset of saturation. Further increase of V_{BE}, termed *overdriving*, drives the BJT further into saturation whereby the minority charge distribution in the base changes typically to RS. A consequence of increased charge storage in the base after I_C has reached its saturation value $I_{C(sat)}$ (that is, the slope of the base charge distribution cannot be further increased) is that the minority carrier density at the edge of the C-B depletion region is no longer zero as it is in active operation, since electrons cannot then be attracted into the collector at a rate corresponding to their injection into the base from the emitter. The minority carrier density therefore rises (from P to S) causing the C-B junction to effectively become *forward*-biased. Thus in saturation both the B-E and C-B junctions are forward-biased so that both junctions have a low resistance. The saturation distribution RS (figure 2.24b) can be considered as the sum of two injected components, a *forward* or *normal* component q_{BN} from the emitter and a *reverse* or *inverse* component q_{BI} from the collector (figure 2.25). There are several significant aspects of the saturation mode of operation

(1) As both junctions are forward-biased and hence of low resistance, the voltage drop across the device between input and output, V_{CE} (figure 2.23). is very low, typically less than 500 mV. The BJT is fully conducting and the saturation mode is therefore the ON state when a BJT is used as a switch. The value of V_{CE} in saturation is denoted by $V_{CE(sat)}$.

(2) As the voltage across the BJT is small in saturation, the value of $I_{C(sat)}$ depends mainly on the external circuit (V_{CC}, R_L) and is largely independent of the precise properties of the BJT although the gain must be sufficiently high to ensure saturation for the base drive provided. This independence of switching level upon the precise gain of the BJT would not occur if the ON switching level was represented by an operating point in the active region where the values of I_C and V_{CE} are a function of gain. In such a situation, BJTs having different gains would exhibit different switching levels for the same input signals. The uniformity of switching level is an important feature of the saturation mode of operation.

(3) The switch to saturation requires only that the voltage applied to the B-E junction, or more precisely the current supplied to *charge* the base (termed the *base drive*), is sufficiently large to cause the collector current to increase to $I_{C(sat)}$. The speed of the switch to the ON state can be increased by applying a larger base drive to charge the base more rapidly. This would not be possible if the ON state was defined as an operating point in the active region as a precise operating point would require a precise base drive.

(4) As the voltage across the BJT (V_{CE}) is low in saturation, the power dissipation in the device (equation 2.65) is low in this mode.

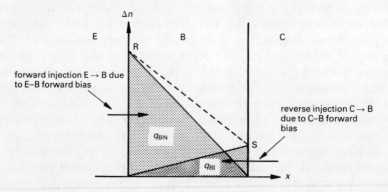

Figure 2.25 Components of BJT base charge at saturation

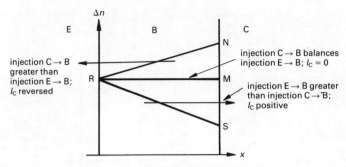

Figure 2.26 Dependence of collector current on degree of C-B forward bias in the saturation mode

These properties of the saturation mode, namely, low resistance, independence of precise BJT properties, low power dissipation and rapid switching by overdriving make it suitable as the ON switching state.

If the C-B forward bias is increased sufficiently by external means, injection into the base from the collector can equal that from the emitter whereby the balance causes the collector current to fall to zero (distribution RM, figure 2.26). Further increase of the C-B forward bias causes the injection from the collector to exceed that from the emitter causing the collector current to reverse, the minority carrier distribution in the base changing to RN. Figure 2.31a shows the base charge distributions corresponding to the various regions of the common-base output characteristic of a BJT.

Returning to figure 2.23 with the BJT biased for active region operation, if V_{BE} is reduced, the injection of electrons from the emitter into the base falls with corresponding reduction in I_C and V_O. When V_{BE} is zero or made negative so as to reverse bias the B-E junction, injection ceases and I_C falls to the leakage value I_{CBO} (equation 2.9) which is practically zero for a low-power silicon BJT (typically 2–20 nA at 25 °C). The BJT therefore ceases to conduct (termed the *cut-off* mode), the output voltage V_O falls to zero and the C-B reverse bias V_{CB} rises to the supply value V_{CC}.

The change of the base charge distribution during the switch to cut-off is shown in figure 2.27. Reduction of V_{BE} during active region operation causes the distribution to fall typically from AP to DP. If V_{BE} is reduced to zero, injection from the emitter ceases and the minority carrier density at the edge of the B-E depletion layer is the density n_p due to thermal generation consistent with the base doping, the corresponding distribution being FP. If the B-E junction is reverse-biased by making V_{BE} negative, electrons are removed from the base into the emitter causing the electron density at the edge of the B-E depletion region to fall to zero (point H). Removal of base charge via both the collector and emitter causes the base charge to be rapidly

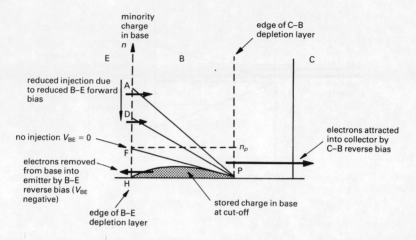

Figure 2.27 Switching a BJT to cut-off

reduced and the collector current correspondingly falls to zero. Some minority charge remains in the base at cut-off (figure 2.27) due to thermal generation. Significant aspects of the cut-off mode are

(1) As the collector current is reduced almost to zero under these conditions, this mode is the OFF state when the BJT is used as a switch. Reduction of the voltage applied to the B-E junction to zero or to a reverse bias value causes the current delivered to the load R_L (figure 2.23) and hence the output voltage V_O to fall to zero.

(2) The switch to the OFF state is achieved more rapidly if the B-E junction is reverse-biased (termed *reverse base drive*) as base charge is then removed via the emitter as well as the collector.

(3) Although the switch to the ON state (saturation) is faster if the BJT is overdriven, this results in appreciable charge storage in the base which worsens the switch OFF time as the collector current cannot fall until the excess base charge has been removed. In order to improve the over-all switching speed, some circuits are designed to limit the depth of saturation (for example, Schottky-clamping in TTL, section 4.7.5)

(4) Switch-back to the ON state from cut-off cannot be achieved until the base has been recharged to re-establish the minority carrier distribution in the base.

(5) Power dissipation in the BJT (equation 2.65) is low at cut-off as the collector current is small (leakage).

The saturation and cut-off modes thus provide the ON and OFF switching states required to enable either the supply of power to an external load to be

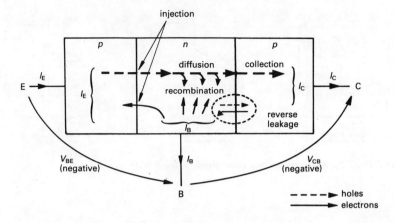

Figure 2.28 Charge movement in a *pnp* BJT for active operation

switched or to enable use of the two distinct output voltage levels from the BJT, $V_{CE(sat)}$ and V_{CC}, to be used as the two binary states for logic applications. The switching of BJTs is considered quantitatively in section 2.2.26 as an application of the charge-control model of the device.

2.2.14 pnp BJTs

The discusssion of *npn* BJT operation in sections 2.2.2 to 2.2.13 applies equally to *pnp* BJTs provided that

(1) the roles of electrons and holes are exchanged;
(2) voltage polarities and current directions are reversed.

pnp BJTs are produced by the epitaxial planar process using a *p*-type epitaxial layer grown on a high conductivity p^+ substrate. An *n*-type base is diffused into the epitaxial layer and a heavily doped (p^+) emitter is diffused into the base forming a structure complementary to that of the *npn* device shown in figure 2.11b.

Current flow is initiated by applying a negative V_{BE} to forward bias the E-B junction causing holes to be injected into the base. The injected holes diffuse across the base and are attracted across the C-B junction into the collector by application of a negative V_{CB}. Base current comprises components due to recombination of holes in the base, injection of electrons from base to emitter and reverse leakage at the C-B junction. Figure 2.28 shows the components of charge movement in a *pnp* BJT for active region operation and comparison with figure 2.13 shows that operation is complementary to that of an *npn* BJT.

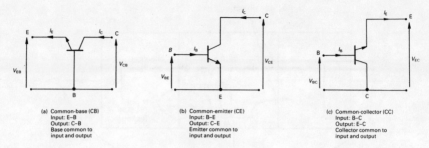

Figure 2.29 Alternative connection configurations for a BJT

2.2.15 Connection Configurations and Static Characteristics of BJTs

The static characteristics are the relationships between the static (that is, constant) terminal currents and voltages of a device and are usually presented in graphical form. In the case of a two-terminal device such as a diode, the static I–V characteristic (figure 2.1a) describes the current flow through the device for a range of forward and reverse-applied bias. For a three-terminal device such as a BJT, the situation is complicated by the fact that there are three currents (I_E, I_B and I_C) and three voltages (V_{BE}, V_{CB} and V_{CE}). Which of the relationships between these variables are the most significant depend on how the device is to be used.

Figure 2.29 shows the three alternative connection configurations for a BJT, termed the *common-base* (CB), *common-emitter* (CE) and *common-collector* (CC) arrangements according to which of the three terminals is common to both input and output. The three arrangements have different circuit properties regarding voltage gain, current gain, power gain, input resistance and output resistance as discussed in texts concerned with electronic circuits. [10,11,12,13]

Terminal properties of a BJT are described by three static characteristics

(1) input characteristic—the relationship between input current and input voltage;
(2) output characteristic—the relationship between output current and output voltage;
(3) transfer characteristic—the relationship between output current and input current.

Table 2.4 lists the static input and output variables and the three characteristic relationships for each configuration. In practice the common-collector characteristics are rarely used as it is more convenient to adapt the common-emitter relationships (section 2.2.26, figure 2.57b).

Discrete Semiconductor Devices

Table 2.4 Static input and output variables and characteristic relationships for the CB, CE and CC BJT connection configurations

Configuration	CB	CE	CC
Input voltage	V_{EB}	V_{BE}	V_{BC}
Input current	I_E	I_B	I_B
Output voltage	V_{CB}	V_{CE}	V_{EC}
Output current	I_C	I_C	I_E
Input characteristic	I_E–V_{EB}	I_B–V_{BE}	I_B–V_{BC}
Output characteristic	I_C–V_{CB}	I_C–V_{CE}	I_E–V_{EC}
Transfer characteristic	I_C–I_E	I_C–I_B	I_E–I_B

2.2.16 Common-base Characteristics

Typical CB characteristics for a low-power silicon *npn* BJT are given in figures 2.30, 2.31 and 2.32.

CB Input Characteristic $(I_E$–$V_{EB})$

For constant C–B bias V_{CB}, the input characteristic is the I–V characteristic of the E–B junction diode. For active region operation the E–B junction is forward-biased and the I_E–V_{EB} relationship is as shown in figure 2.30a. As far as a.c. signals are concerned, the effective resistance is given by the reciprocal of the slope of the characteristic at the operating point. If the E–B junction is biased to operate at a point Q (figure 2.30a), the effective resistance presented to an a.c. signal superimposed on the bias is the ratio of the small change in emitter–base voltage ΔV_{EB} to the resulting change in emitter current ΔI_E. This resistance is termed the *small-signal common-base input resistance* and is given the symbol h_{ib} (reference *h*-parameters, section 2.2.26), thus

$$h_{ib} = \frac{\Delta V_{EB}}{\Delta I_E}\bigg|_{V_{CB} \text{ constant}} \tag{2.33}$$

The value of h_{ib} for typical operating conditions in the active region is 5–150 Ω depending on the position of the operating point Q. At cut-off the E–B junction is reverse-biased and so the emitter current then comprises the reverse leakage component of the junction.

The value of V_{CB} affects the input characteristic (figure 2.30b) as V_{CB} alters the effective width of the base (section 2.2.7) resulting in a change in the slope of the base–charge distribution and hence a change in the emitter current. For the C–B junction open circuit, none of the injected electrons flow into the collector and the input characteristic is that of E–B junction diode. If the C–B

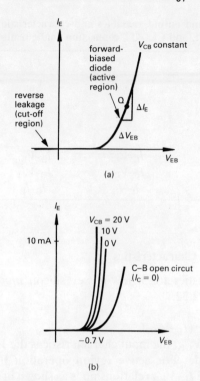

Figure 2.30 Static common-base input characteristics for a low-power silicon *npn* BJT

junction is short circuited ($V_{CB} = 0$), electrons are attracted into the collector by the junction field causing the electron density just inside the base to fall to zero. The slope of the base charge distribution therefore increases resulting in a higher I_E for the same V_{EB}. Increase of positive V_{CB} narrows the effective base further increasing I_E.

CB Output Characteristic (I_C–V_{CB})
Figure 2.31a shows the form of the CB output characteristic for constant emitter current I_E. Active region operation corresponds to the first quadrant of the characteristic (V_{CB} positive), the collector current being given by equation 2.15. For values of V_{CB} well below breakdown, h_{FB} is approximately -1 (typically -0.990 to -0.997) and the leakage component I_{CBO} is negligible for silicon BJTs in comparison with practical values of I_E, thus $I_C \approx I_E$. However $|h_{FB}|$ increases slightly with V_{CB} due to base width modulation (section 2.2.7) and so the output characteristic has a slight positive gradient in this region of operation. The slope of the characteristic is the *small-signal common-base output conductance* (h_{ob}) of the device

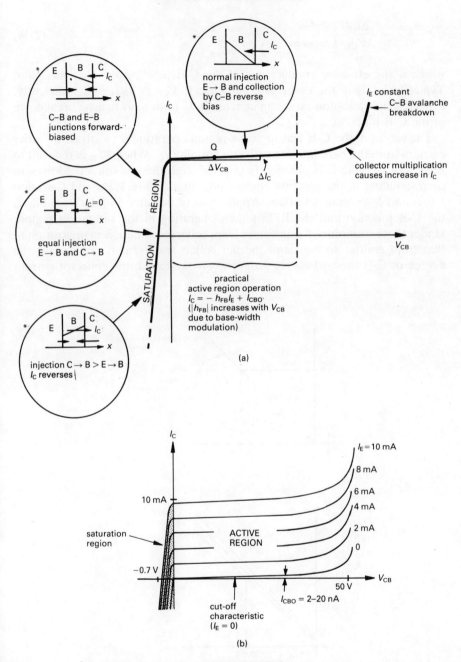

Figure 2.31 Static common-base output characteristics for a low-power silicon *npn* BJT

$$h_{ob} = \frac{\Delta I_C}{\Delta V_{CB}}\bigg|_{I_E \text{ constant}} \qquad (2.34)$$

which is the effective conductance of the C-B junction for a.c. operation. Typically h_{ob} is in the range of 0.05–5 μS. If V_{CB} is increased sufficiently, collector multiplication causes an increase in I_C prior to avalanche breakdown of the C-B junction.

The action of the C-B bias in active region operation is to attract minority carriers from the base into the collector forming I_C. When V_{CB} is reduced to zero, although the C-B junction field is reduced, there is still a field present corresponding to the junction contact potential (figure 1.21e and f, section 1.6.2) and I_C continues to flow. Application of negative V_{CB} forward biases the C-B junction and the BJT is then operating in the saturation region. Under these conditions the injection from collector to base is in opposition to that from emitter to base and the net collector current is reduced. If the degree of C-B forward bias is progressively increased, the collector current

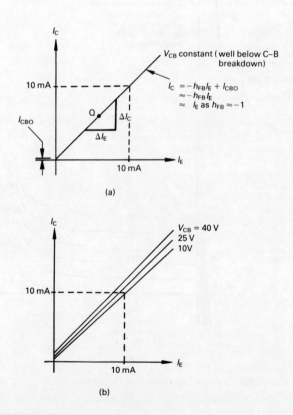

Figure 2.32 Static common-base transfer characteristics for a low-power silicon *npn* BJT

falls to zero and then reverses (figure 2.26, section 2.2.13). The inserts in figure 2.31a show the base charge distributions corresponding to the various modes of operation.

Variation of the E-B forward bias causes a change in the injection from emitter to base resulting in a *family* of output characteristics as shown in figure 2.31b for a range of emitter current from zero to 10 mA. Although six specific characteristics are shown, for I_E = 0,2,4,6,8,10 mA, it should be appreciated that the family of characteristics is continuous as I_E can take any intermediate value.

The cut-off characteristic is for zero emitter current, the E-B junction being reverse-biased in which case the collector current is the C-B reverse leakage value I_{CBO} (typically 2–20 nA at 25 °C for a low-power BJT). If V_{CB} is increased, collector mulitiplication eventually causes the leakage value to rise prior to C-B breakdown.

CB Transfer Characteristic (I_C–I_E)
The relationship between I_C and I_E for active region operation is given by equation 2.15. For constant C-B reverse bias, the static current gain h_{FB} is approximately constant, typically in the range −0.990 to −0.997 for V_{CB} well below breakdown, and the transfer characteristic is therefore linear (figure 2.32a). A small change of emitter current (ΔI_E) about an operating point Q due to a small change in emitter–base bias ΔV_{EB}, causes a small change in the collector current ΔI_C. The relationship between ΔI_C and ΔI_E, that is, the slope of the transfer characteristic at the operating point Q, is termed the *small-signal common-base current gain* of the BJT at Q and is denoted by α or $-h_{fb}$, thus

$$\alpha = -h_{fb} = \left.\frac{\Delta I_C}{\Delta I_E}\right|_{V_{CB}\text{ constant}} = \left.\frac{i_c}{i_e}\right|_{v_{cb}=0} \tag{2.35}$$

where the lower case symbols i_c, i_e and v_{cb} refer to signal components.

Although this definition is for V_{CB} constant, the uniform nature of the output characteristics shows that $\Delta I_C/\Delta I_E$ remains constant for a wide range of V_{CB} well below breakdown. The negative sign in equation 2.35 is due to the convention adopted in the *h*-parameter terminal model (figure 2.56a) whereby both the emitter and collector currents are defined as flowing *into* the BJT. A similar situation occurs in the relationship between the static values α_{dc} and h_{FB} (figure 2.17 and equation 2.14).

It is important to appreciate the difference between h_{fb} ($-\alpha$) and h_{FB} ($-\alpha_{dc}$). The small-signal gain h_{fb} describes the relationship between the *change* of emitter current and the corresponding *change* of collector current and is significant in representation of the a.c. performance of the BJT, for example in amplification. The static gain h_{FB} relates the static values I_E and I_C at the operating point and thus applies to the d.c. performance of the BJT.

Although h_{fb} and h_{FB} describe quite different properties they have similar numerical values due to the degree of proportionality between I_C and I_E (figure 2.32a).

As V_{CB} approaches the breakdown voltage for the C-B junction, $|h_{FB}|$ increases above unity due to collector multiplication and correspondingly the slope of the transfer characteristic increases (figure 2.32b).

The common-base configuration is capable of providing a voltage gain much greater than unity as $I_C \approx I_E$ and the external load R_L can be much greater than R_{BE} (figure 2.15). The current gain is approximately unity and therefore the power gain, as given by the product of voltage and current gains, is greater than unity.

2.2.17 Common-emitter Configuration and Characteristics

For active region operation, a considerably increased power gain can be obtained if the base is used as the input terminal (figure 2.29b) in place of the emitter. The voltage gain of this arrangement (termed the *common-emitter* (CE) configuration as the emitter is common to the input and output circuits) is similar to that of the common-base configuration. However, a substantial change in collector current corresponds to only a small change in base current giving a high current gain and hence a high power gain. For this reason the CE configuration is the most widely used of the three arrangements shown in figure 2.29. The input and output variables together with the relevant characteristic relationships are listed in table 2.4

As the input and output currents for the CE configuration are I_B and I_C, it is convenient to express I_C in terms of I_B. Using equations 2.9 and 2.12

$$I_C = \alpha_{dc}(I_C + I_B) + I_{CBO}$$

$$= \left(\frac{\alpha_{dc}}{1 - \alpha_{dc}}\right) I_B + \frac{I_{CBO}}{1 - \alpha_{dc}} \qquad (2.36)$$

$$= \beta_{dc} I_B + I_{CEO} \qquad (2.37)$$

where

$$\beta_{dc} = \frac{\alpha_{dc}}{1 - \alpha_{dc}} \qquad (2.38)$$

Thus

$$\alpha_{dc} = \frac{\beta_{dc}}{1 + \beta_{dc}} \qquad (2.39)$$

and

$$\frac{1}{1 - \alpha_{dc}} = \beta_{dc} + 1 \tag{2.40}$$

Also from equations 2.36, 2.37 and 2.40

$$I_{CEO} = \frac{I_{CBO}}{1 - \alpha_{dc}} \tag{2.41}$$

$$= (\beta_{dc} + 1) I_{CBO} \tag{2.42}$$

$$\approx \beta_{dc} I_{CBO}, \text{ as } \beta_{dc} \gg 1 \text{ in practice} \tag{2.43}$$

The parameter β_{dc} is the *static (d.c.) common-emitter current* gain of the BJT. I_{CEO} may be regarded as the C-E leakage current; physically it is the current flowing between collector and emitter when the base current is zero (from equation 2.37), that is, when the base is open-circuit, hence the suffix CEO.

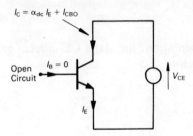

Figure 2.33 Common-emitter leakage

This situation is shown in figure 2.33. When the base current is zero

$$I_E = I_C$$
$$= \alpha_{dc} I_E + I_{CBO}$$

in which case

$$I_E = \frac{I_{CBO}}{1 - \alpha_{dc}} = I_{CEO} \tag{2.44}$$

For a silicon BJT under normal operating conditions (current levels and temperature)

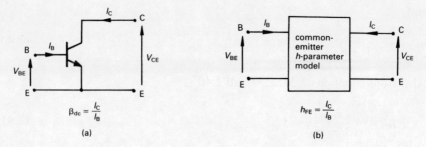

Figure 2.34 Common-emitter (CE) connection and h-parameter convention

$$I_{CEO} \ll \beta_{dc} I_B \tag{2.45}$$

and so from equation 2.37

$$\beta_{dc} = \frac{I_C}{I_B} \tag{2.46}$$

In h-parameter representation, the static CE current gain is denoted by h_{FE} (figure 2.34) and therefore

$$\beta_{dc} = h_{FE} \tag{2.47}$$

Equations 2.37, 2.38, 2.39, 2.41, 2.42 and 2.43 can therefore be written as

$$I_C = h_{FE} I_B + I_{CEO} \tag{2.48}$$

$$h_{FE} = \frac{-h_{FB}}{1 + h_{FB}} \text{ as } \alpha_{dc} = -h_{FB} \text{ (equation 2.14)} \tag{2.49}$$

from which

$$h_{FB} = \frac{-h_{FE}}{1 + h_{FE}} \tag{2.50}$$

and

$$I_{CEO} = \frac{I_{CBO}}{1 + h_{FB}} \tag{2.51}$$

$$= (h_{FE} + 1) I_{CBO} \tag{2.52}$$

$$\approx h_{FE} I_{CBO}, \text{ as } h_{FE} \gg 1 \text{ in practice} \tag{2.53}$$

Based on the theoretical value of h_{FB} for an *npn* BJT given by equation 2.32 in terms of the dimensions and conductivities of the base and emitter regions w_B, σ_B, w_E, σ_E and the diffusion length for electrons in the base L_n, the corresponding expression for h_{FE} is

$$h_{FE} \approx \frac{1}{\dfrac{w_B \sigma_B}{w_E \sigma_E} + \dfrac{w_B^2}{2L_n^2}} \qquad (2.54)$$

In device data supplied by manufacturers, the *h*-parameter gain symbols are used in preference to α and β. Typical ranges of values of h_{FB} ($-\alpha_{dc}$) and I_{CBO} for a low-power silicon BJT have previously been quoted as -0.990 to -0.997 and 2 to 20 nA respectively at 25 °C. From equations 2.49 and 2.51 the corresponding ranges for h_{FE} (β_{dc}) and I_{CEO} are approximately 100–330 and 0.2–7 μA. Note that I_{CEO} is dependent on I_{CBO} *and* h_{FB}.

As $|h_{FB}|$ is only slightly less than unity, the term $(1 + h_{FB})$ is very small and a slight change in h_{FB} thus produces significant changes in h_{FE} and I_{CEO}. For example, if h_{FB} is increased to -0.999, h_{FE} increases to approximately 1000 and I_{CEO} to 2–20 μA. In physical terms this change of sensitivity is due to the fact that a small change of h_{FB} produces only a small change in I_C relative to I_E but a large change in I_B.

Typical CE characteristics for a low-power silicon *npn* BJT are given in figures 2.35, 2.36 and 2.37.

CE Input Characteristic ($I_B - V_{BE}$)

For constant V_{CE} the input characteristic has the form of the I–V characteristic of a *pn* junction (figure 2.35a). As in the CB case, conditions at the collector affect the input characteristics (figure 2.35b). With the collector open-circuit, $I_B = I_E$ and the corresponding CE input characteristic is therefore the same as the CB input characteristic for open-circuit collector. For $V_{CE} = 0$, that is C-E short-circuit, $V_{BC} = V_{BE}$ and thus increase of V_{BE}, causing E-B forward bias, also forward biases the C-B junction creating two forward-biased junctions in parallel hence increasing the base current. For a certain value of V_{BE}, the base current for $V_{CE} = 0$ is therefore greater than for the collector open-circuit.

As V_{CE} is increased with V_{BE} constant, the bias across the C-B junction changes ($V_{CB} = V_{CE} - V_{BE}$). For low values of V_{CE} less than V_{BE}, V_{CB} is negative causing C-B forward bias. Increase of V_{CE} causes the C-B forward bias to be reduced with a corresponding reduction in I_B. When $V_{CE} > V_{BE}$ the C-B junction becomes reverse-biased and I_B drops to the normal active region component (figure 2.35b). Continued increase in V_{CE} further increases the C-B reverse bias, narrowing the effective base causing further marginal reduction in the base current.

Practical application of a BJT requires knowledge of suitable values of V_{BE}.

Figure 2.35c shows the typical useful active working range for a low-power silicon BJT. For $V_{BE} < 0.5$ V the current flow is very small and the gain is therefore low (figure 2.38). The voltage at which the slope of the characteristic increases sharply is termed the *threshold voltage* V_γ. If V_{BE} is increased to about 0.8 V, the slope of the characteristic becomes very high, the current increasing rapidly for only very small increase in V_{BE}. This limiting or *saturation* value of V_{BE} is termed $V_{BE(sat)}$ and is often represented by V_σ. The useful active range (the linear of amplifying region) is between V_γ and V_σ, and a static operating (quiescent or bias) point is usually chosen in the range $0.6 < V_{BE} < 0.7$ V.

It should be noted that as I_C is of more interest to the circuit designer than I_B, manufacturers often provide the I_C–V_{BE} characteristic in place of the true input characteristic (BC107 data, appendix H.3). Furthermore, due to the range of values involved, the collector current is often plotted on a logarithmic scale so that at first sight the characteristic does not appear to have the familiar 'exponential' form. If the true input characteristic (I_B–V_{BE}) is not supplied, it can be obtained if required, by combining the I_C–V_{BE} and I_C–I_B characteristics.

If the B-E junction is biased to operate at a point Q on the input characteristic (figure 2.35a), the effective resistance between base and emitter as far as an a.c. signal superimposed on the fixed bias is concerned, is given by the reciprocal of the tangent at Q. This effective resistance is the *small-signal common-emitter input resistance* and is given the symbol h_{ie} (reference *h*-parameters, section 2.2.26), thus

$$h_{ie} = \frac{\Delta V_{BE}}{\Delta I_B}\bigg|_{V_{CE} \text{ constant}} \quad (2.55)$$

For operation in the active region, the value of h_{ie} is typically in the range 1–10 kΩ depending on the position of the operating point Q.

CE Output Characteristics (I_C–V_{CE})

A typical CE output characteristic for constant base current I_B is shown in figure 2.36a. Normal active region operation requires the C-B junction to be reverse-biased and so V_{CE} must be greater than V_{BE}. The collector current for this condition is given by equation 2.48. As V_{CE} is increased, the effective base width reduces due to widening of the C-B depletion region, causing the collector current to increase. Mathematically this effect appears as an increase in the static CE current gain h_{FE} for increase in V_{CE} which corresponds to the increase in the CB current gain h_{FB} discussed in section 2.2.16. The output characteristic thus has a positive gradient for active region operation, the slope of which is the *small-signal common-emitter output conductance* (h_{oe}) of the BJT.

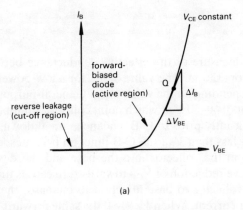

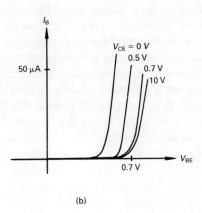

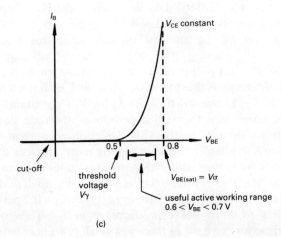

Figure 2.35 Static common-emitter input characteristics for a low-power silicon *npn* BJT

$$h_{oe} = \left.\frac{\Delta I_C}{\Delta V_{CE}}\right|_{I_B \text{ constant}} \tag{2.56}$$

This parameter is a measure of the effective conductance between collector and emitter for a.c. operation. The value of h_{oe} for a low-power BJT is in the range 10–1000 μS depending on the value of I_B. General-purpose BJTs of this type have a V_{CE} limit of 40–50 V at which point collector multiplication causes I_C to increase significantly prior to C-B avalanche breakdown.

For values of V_{CE} less than V_{BE} the C-B junction becomes forward-biased causing injection from the collector into the base and the device is then at saturation. Progressive reduction of V_{CE} towards zero causes the C-B forward bias and hence the collector to base injection to increase, thereby reducing the forward collector current. When $V_{CE} = 0$ the same forward bias is applied to both junctions since V_{BC} then equals V_{BE}. The collector current is then approximately zero as injection from the emitter into the base is approximately balanced by injection from the collector. However the BJT structure is not symmetrical, either in geometry or doping, and in practice when $V_{CE} = 0$ a small inverse (negative) collector current flows. The collector current is zero in fact when V_{CE} is of the order of 5 mV as indicated by the insert in figure 2.36a. If V_{CE} is increased negatively, the inverse injection from collector to base exceeds the forward injection from emitter to base causing the collector current to reverse.

Variation of the injection from emitter to base by alteration of the B-E forward bias results in a *family* of output characteristics as shown in figure 2.36b. Although six specific characteristics are shown for $I_B = 0, 10, 20, 30, 40, 50$ μA, the family of characteristics is continuous as I_B can also have any intermediate value. CE output characteristics are provided either for constant I_B or constant V_{BE} (BC107 data sheet, appendix H.3), typical corresponding values being shown in figure 2.36b.

Comparison of the quoted typical values of CE and CB output conductances h_{oe} and h_{ob} (10–1000 μS and 0.05–5 μS respectively) shows that the slope of the BJT output characteristic is very much greater for CE than for CB. This difference is due to the fact that the CB characteristic is for constant I_E while the CE characteristic is for I_B (or V_{BE}) constant. As V_{CE} and hence V_{CB} is increased in active region operation, the reduction in effective base width due to increase in the C-B depletion region causes the transfer of charge across the base to become slightly more efficient since recombination is reduced. Due to the vastly different magnitudes of I_E and I_B, the improved charge transfer from emitter to collector causes a small percentage increase in I_E but a large percentage decrease in I_B; typically a 0.2 per cent increase in I_E corresponds to a 70 per cent decrease in I_B. Thus on the CB output characteristic for constant I_E there is a small percentage increase in I_C as V_{CB} increases resulting in a low value of h_{ob}. In the CE case, however, due to the high percentage drop in I_B as V_{CE} increases, constant I_B corresponds to a considerable increase in I_C which results in the higher value of output conductance h_{oe}.

It can be seen from figure 2.36b that the slope of the CE active region output characteristic increases (that is, h_{oe} increases) with I_B. A certain change in collector current corresponds to a smaller change in base current at higher V_{CE} than at lower V_{CE} due to the improved charge transfer from emitter to collector resulting from a smaller effective base width. However, the CE output characteristics are for constant I_B and so in terms of a certain change in base current, the corresponding change in collector current is larger at higher V_{CE}. The spacing of the CE output characteristics thus increases with V_{CE}, resulting in an increase of slope (that is, h_{oe}) as I_B increases. A similar effect occurs in common-base but since the characteristics are plotted for constant I_E, which increases only slightly with V_{CB}, the spacing between characteristics in the active region increases only slightly with V_{CB} (figure 2.31b).

When the emitter to base injection is negligible, that is, for the base open-circuit or for B-E reverse bias, the collector current is the leakage value I_{CEO} and the transistor is said to be *cut-off*. At high values of V_{CE}, collector multiplication causes the leakage value to increase prior to C-B breakdown. Typically, I_{CEO} is in the range of 0.1–10 μA at 25 °C for a low-power silicon BJT well below breakdown.

The saturation region, corresponding to $V_{CE} < V_{BE}$ such that the C-B junction is forward-biased (V_{CB} negative), is shown shaded in figure 2.36b. The region is bounded by the locus of the 'knees' of the characteristics on one side and by the *saturation line* on the other. The knee is the term used to describe the onset of the linear active region and is the point where $V_{CE} = V_{BE}$ causing the C-B bias to be zero. Further reduction in V_{CE} causes reverse injection from the collector into the base and the collector current falls. If V_{CE} is further reduced it eventually reaches a limiting or saturation value $V_{CE(sat)}$ due to the resistance of the collector region R_{CS}. $V_{CE(sat)}$ is a function of collector current and is the minimum value of V_{CE} for a certain value of collector current. The variation of $V_{CE(sat)}$ with I_C is given by the *saturation line* on the output characteristics which is the line to which the characteristics tend as V_{CE} is reduced (figure 2.36b). Typically $V_{CE(sat)}$ is in the range 100–600 mV, the value being of particular interest in switching circuits as it is the voltage across the BJT when it is fully on or *bottomed*.

Inverse Operation

If V_{CE} is increased negatively and V_{BE} is made negative, the C-B junction becomes forward-biased and the E-B junction reverse-biased. This results in an *inverse active region* of operation having a set of output characteristics in the third quadrant (figure 2.36c). As the collector region is more lightly doped than the emitter, the efficiency of injection from collector to base is low. In addition, the area of the E-B junction is much less than that of the

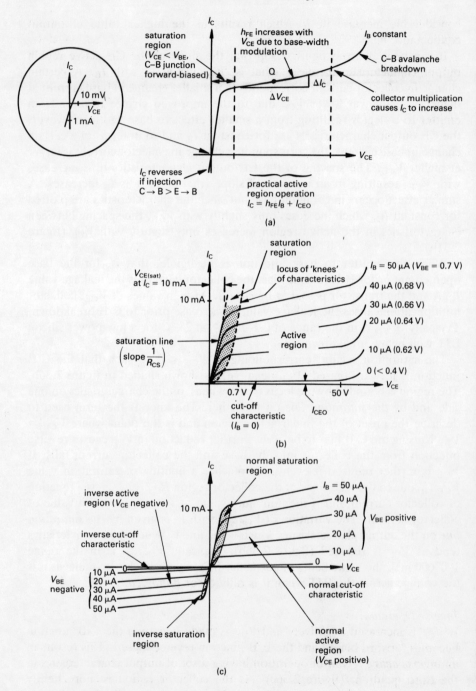

Figure 2.36 Static common-emitter output characteristics for a low-power silicon *npn* BJT

C-B junction making the emitter region an inefficient collector. The current gain for *inverse* operation is therefore considerably lower than for *normal* active region operation. In situations where inverse operation is to be considered, the symbols α_N and β_N are used to represent the static CB and CE current gains for *n*ormal operation (E-B forward-biased, C-B reverse-biased), previously represented by α_{dc} and β_{dc}, and the symbols α_I and β_I are used to represent the corresponding gains for *i*nverse operation. Typically α_I and β_I are 0.8 and 4 respectively compared with 0.995 and 200 for α_N and β_N. In each case α and β are related by equation 2.38. The inverse parameters are used in the Ebers–Moll large-signal BJT model (section 2.2.26).

Due to the low current gain it is inefficient to use a BJT in the inverted mode for active region operation (for example, amplification). The inverted mode, however, does have an important application in switching as $|V_{CE(sat)}|$ is lower than for normal operation, typically 50–200 mV compared with 100–600 mV respectively. This is because the collector is more lightly doped than the emitter so that the contact potential of the C-B junction ψ_{CB} is less than that of the E-B junction ψ_{EB} (equation 1.102 and figure 2.21) and saturation therefore occurs at a lower value of $|V_{CE}|$.

CE Transfer Characteristic $(I_C–I_B)$

Equation 2.48 shows that for active region operation, I_C and I_B are related by the CE static current gain h_{FE}. For a typical low-power BJT, h_{FE} varies with collector current as shown in figure 2.38a, resulting in the transfer characteristic of figure 2.37a. For small changes of collector current the characteristic may be considered to be linear, corresponding to an approximately constant value of h_{FE}.

A small change of base current (ΔI_B) about an operating point Q due to a small change in base–emitter bias ΔV_{BE}, causes a change in the collector current ΔI_C. The relationship between ΔI_C and ΔI_B (the slope of the transfer characteristic at the operating point Q) is the *small-signal common-emitter current gain* of the BJT at Q and is denoted by β or h_{fe}, thus

$$\beta = h_{fe} = \left.\frac{\Delta I_C}{\Delta I_B}\right|_{V_{CE}\text{ constant}} = \left.\frac{i_c}{i_b}\right|_{v_{ce}=0} \tag{2.57}$$

where the lower case symbols i_c, i_b and v_{ce} refer to signal components.

If V_{CE} is increased, base width narrowing causes h_{FE} to rise resulting in an increase in the slope of the transfer characteristic (figure 2.37b).

Under conditions of *low-level injection*, that is, low I_E and hence low I_C, transistor gain is reduced by the loss of a substantial proportion of I_E in the vicinity of the B-E junction by recombination. At high I_C, termed *high-level injection*, the increased electron density in the base causes an increased hole density tending to maintain charge neutrality. The increased hole density causes injection from base to emitter (I_{Ep}, figure 2.14a) to increase thereby

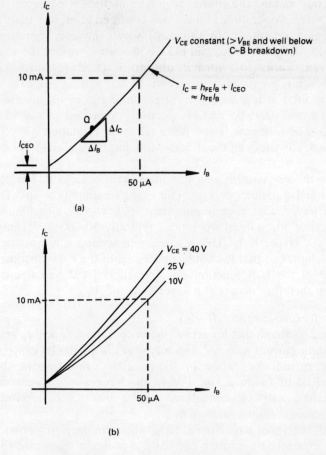

Figure 2.37 Static common-emitter transfer characteristics for a low-power silicon *npn* BJT

reducing the emitter injection efficiency γ and the transistor gain. Thus h_{FE} is reduced at both low and high values of static collector current.

The variation of h_{FE} and h_{fe} with I_C is of interest to the device user since operation at a higher value of h_{FE} means a lower base current bias I_B for linear applications or a lower base drive to cause saturation in switching applications. In terms of a.c. performance such as in signal amplification, a high value of h_{fe} provides a high amplifier gain and so the user is interested in selecting a static collector (bias) current corresponding to a high value of h_{fe}. Figure 2.38 shows typical variations of h_{FE} and h_{fe} with I_C for a low-power BJT at constant V_{CE} and constant junction temperature T_j. As the a.c. gain h_{fe} is also a function of the frequency of the applied signal, it is necessary to quote the test frequency in this case. Although h_{FE} and h_{fe} describe different

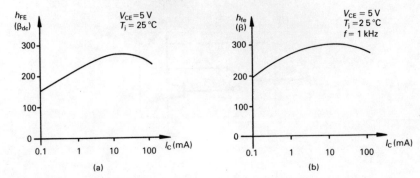

Figure 2.38 Variation of static and small-signal CE current gains with static collector current for a typical low-power BJT at constant V_{CE} and constant temperature

relationships, the former being a ratio of actual currents and the latter a ratio of changes in currents (figure 2.39), the degree of proportionality between the collector and emitter currents (figure 2.37a) causes the two parameters to have similar values.

Note Following consideration of the CB and CE BJT characteristics it should be realised that the mechanism of operation of the BJT is the same whatever the mode of connection. Thus the CE characteristics describe *exactly the same operation* as the CB characteristics; they are simply the relationships between different sets of variables for convenience.

The static characteristics of *pnp* BJTs have the same form as those for *npn* types. However, as the current directions and voltage polarities are reversed, CE *pnp* BJT characteristics show I_B, V_{BE}, I_C and V_{CE} as negative values.

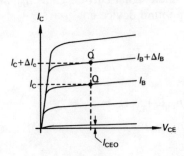

Figure 2.39 Difference between static and small-signal current gains:
Static case – operation at Q, assuming I_{CEO} leakage is insignificant

$$h_{FE} = I_C/I_B$$

Small-signal (a.c.) case – movement of operating point with V_{CE} constant, from Q to Q'

$$h_{fe} = \Delta I_C/\Delta I_B = i_c/i_b$$

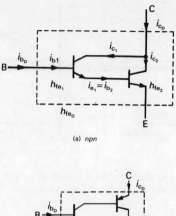

(a) npn

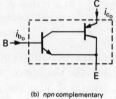

(b) npn complementary

Figure 2.40 Darlington arrangements

2.2.18 Darlington Transistor

If two BJTs are connected so that the first provides the base current for the second (figure 2.40), the combination provides high current gain and high input resistance. This configuration is produced in monolithic form as a single device and is known as a *Darlington pair* or *Darlington transistor*.

If the two BJTs have small-signal current gains h_{fe_1} and h_{fe_2}, the collector signal current of the compound device is

$$\begin{aligned}
i_{c_D} &= i_{c_1} + i_{c_2} \\
&\approx h_{fe_1} i_{b_1} + h_{fe_2} i_{b_2} \\
&\approx h_{fe_1} i_{b_1} + h_{fe_2}(h_{fe_1} i_{b_1}) \text{ as } i_{b_2} = i_{e_1} \approx i_{c_1} \\
&\approx h_{fe_1} h_{fe_2} i_{b_1} \text{ as } h_{fe_2} \gg 1 \\
&= h_{fe_1} h_{fe_2} i_{b_D}
\end{aligned}$$

Thus the small-signal current gain of the compound device is

$$h_{fe_D} = \frac{b1 i_{c_D}}{i_{b_D}} \approx h_{fe_1} h_{fe_2} \qquad (2.58)$$

Values of h_{fe_D} of up to 10^4 are typical. A consequence of the high gain is that the input current (i_{b_D}) is relatively low providing an input resistance considerably higher than that of a single BJT. By developing the h-parameter network model of the Darlington transistor from the h-parameter models of the individual BJTs (section 2.2.26), h_{ie_D} is found to be approximately $h_{fe_1} h_{ie_2}$. In practice it is usual for resistors to be included across $B_1 E_1$ and $B_2 E_2$ in the *npn* configuration (figure 2.40a) and across $C_1 E_2$ in the *npn* complementary version (figure 2.40b) to provide leakage current paths and so to improve thermal stability. In application the composite device can be considered as a single high-gain BJT.

2.2.19 Voltage and Current Ratings of BJTs

Punch-through
Spread of the C-B or E-B depletion layers right across the narrow base region is a limiting factor to the maximum reverse biases that can be safely applied. In active region operation, electrons injected into the base from the emitter are attracted across the C-B junction by the electric field created by the C-B reverse bias. If V_{CB} is increased sufficiently to cause the C-B depletion layer to extend right across the base to the E-B junction, the influence of the C-B junction field at the E-B junction causes injection to greatly increase and since the injected electrons are attracted out into the collector by the C-B junction field, a high current flow and failure results. The condition is known as *punch-through* and effectively creates a short circuit between emitter and collector.

Avalanche Multiplication
Reverse breakdown of a *pn* junction occurs due to avalanche multiplication when the electric field in the depletion layer is sufficiently high to enable a carrier to ionise a lattice atom causing carrier generation by collision (section 1.7). Depending on the doping levels at the junction, avalanche breakdown can occur at a bias voltage lower than that necessary to cause punch-through. If avalanche breakdown occurs at the C-B junction, I_C becomes greater than I_E whereby the static CB current gain α_{dc} ($-h_{FB}$) becomes greater than unity (figure 2.41). The rise in I_C causes increased power dissipation, temperature rise and failure if not restricted.

Rapid increase of I_C can occur, however, at a bias level well below that necessary for avalanche breakdown. Reduction in the effective base width due to an increase of C-B reverse bias reduces recombination in the base increasing the base transport factor B and the gain α_{dc}. From equation 2.36 it can be seen that for zero or positive I_B, I_C increases rapidly as α_{dc} approaches unity. Zero I_B corresponds to the base open-circuit in which case the collector current is the leakage component I_{CEO} (figure 2.43a) which increases rapidly

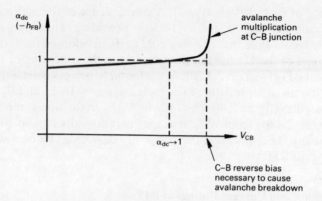

Figure 2.41 Increase of static CB current gain with C-B reverse bias

as α_{dc} approaches unity (equation 2.41). The voltage between collector and emitter at which this rapid increase in I_C occurs with the base open-circuit is the quoted rating V_{CEO}max shown in figure 2.42.

If the base is shorted to the emitter there are two parallel paths for the C-B leakage current: through the base contact or across the E-B junction (figure 2.43b). The negative base current component reduces the collector current compared with the case of open-circuit base, and the rapid increase in I_C corresponding to α_{dc} approaching unity does not occur until a voltage slightly

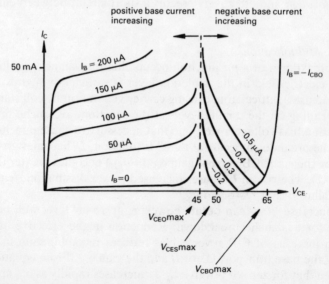

Figure 2.42 Voltage ratings for a typical low-power *npn* BJT showing characteristics for negative base current

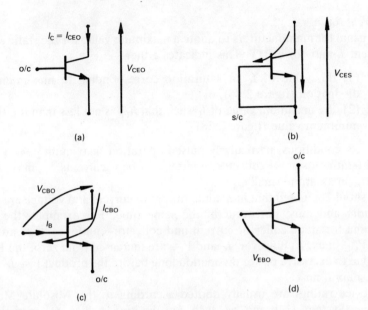

Figure 2.43 BJT voltage ratings

higher than V_{CEO}max, denoted by V_{CES}max as the base is shorted to the emitter. For practical operation with an external base resistance of finite value, the maximum value of V_{CE} that can be applied before I_C increases rapidly is between V_{CEO}max and V_{CES}max.

When avalanche multiplication at the C-B junction causes α_{dc} to be greater than unity, high positive collector currents can occur for *negative* values of base current (equation 2.36). A particular case of negative base current occurs when the emitter is open-circuit (figure 2.43c) in which case $I_B = -I_{CBO}$. C-B leakage I_{CBO} increases rapidly as avalanche breakdown approaches and the C-B reverse bias at which this occurs is the quoted rating V_{CBO}max (figure 2.42). Although significant negative base current does not occur during normal static operation for an *npn* BJT, the situation does occur in switching. When a reverse bias is applied to the B-E junction to drive the BJT from saturation to cut-off, a negative base current flows as stored base charge is removed and the B-E depletion region is re-established. Typical output characteristics for negative base current conditions are shown in figure 2.42. To avoid failure due to excess power dissipation during negative base current operation, care must be taken to restrict V_{CB}. This phenomenon is particularly important when operating at high power levels (section 2.2.23).

The breakdown voltage of the E-B junction for the collector open-circuit V_{EBO}max (figure 2.43d) is typically of the order of only 5 V due to the heavy doping of the emitter region.

Current Ratings

It is usual for manufacturers to quote a maximum value for the static collector current $I_C$max for a BJT. This indicates either

(1) the value of I_C at saturation corresponding to maximum power dissipation (figure 2.46), or

(2) the maximum value of I_C such that h_{FE} is not less than the specified minimum value (figure 2.38).

For a.c. conditions, particularly pulsed operation, maximum peak values of *total* (static + signal) collector, emitter and base currents I_{CM}max, I_{EM}max and I_{BM}max are normally quoted.

It should be noted that individual ratings of current and voltage are *isolated maxima* and must not occur at the same time. For example, the quoted maxima for static collector current and collector–emitter voltage are $I_C$max and V_{CEO}max. If, however, I_C and V_{CE} are increased from zero, the BJT will fail due to excessive power dissipation long before the product $V_{CE} I_C$ reaches V_{CEO}max $I_C$max.

Device ratings are usually quoted according to the Absolute Maximum Rating System [14] and as such are limiting values of operating and environmental conditions which *should not be exceeded under the worst probable operating conditions*. Values are chosen to provide acceptable serviceability taking no account of variations in device operating conditions due, for example, to the changes of device characteristics with time (ageing). Other rating systems [14] exist that are accepted by the International Electrotechnical Commission (IEC) and take account of normal variations in operating conditions (Design-centre Rating System) or changes in device characteristics in operation (Design-maximum Rating System).

2.2.20 Temperature Effects in BJTs

There are two aspects to temperature effects in devices

(1) the change of device characteristics with temperature which affects the performance of the device;

(2) operational limitations to avoid failure due to excessive temperature rise.

The terminal characteristics supplied by manufacturers are for a fixed device temperature, often 25 °C being a typical operating temperature at relatively low power levels.

Forward junction voltage and reverse leakage current are temperature-dependent due mainly to the variation of the intrinsic carrier density n_i with temperature (section 1.6.2). From equations 1.147 and 1.143, the changes of B-E forward bias V_{BE} and C-B leakage I_{CBO} with temperature for a silicon BJT are

$$\left.\frac{\partial V_{BE}}{\partial T}\right|_{I_E \text{ constant}} = \frac{V_{BE} - W_g/e}{T} - \frac{3k}{e} \qquad (2.59)$$

$$\approx -2 \text{ mV/°C} \qquad (2.60)$$

at 25 °C for I_E in the range 1–10 mA, and

$$\left.\frac{I_{CBO}(T_2)}{I_{CBO}(T_1)}\right|_{V_{CB} \text{ constant}} = \left(\frac{T_2}{T_1}\right)^{3/2} \exp\left[6.3 \times 10^3 \left(\frac{T_2-T_1}{T_1 T_2}\right)\right] \qquad (2.61)$$

$$\approx \exp[0.06(T_2 - T_1)] \qquad (2.62)$$

for small changes of temperature in the region of 25 °C. If I_{CBO} (25 °C) is the value of C-B leakage at 25 °C (298 K), typically 2–20 nA for a low-power BJT, the leakage current at a temperature T °C *near to 25 °C* is approximately given by

$$I_{CBO}(T) \approx I_{CBO}(25 \text{ °C}) \times \exp[0.06(T-25)] \qquad (2.63)$$

For temperature variation over a wider range, for example > 10 °C, the simplification introduced in equation 2.63 introduces large errors (> 10 per cent) and the more accurate relation (equation 2.61) must be used. Equation 2.61 leads to the useful results that for constant C-B reverse bias, I_{CBO} *approximately doubles per 8 °C rise in temperature near 25 °C and doubles per 20 °C rise in temperature near 150 °C*. It should be noted that the temperature referred to in this section is that of the semiconductor, or more specifically the junction temperature T_j, which may differ considerably from ambient temperature due to power dissipation within the device. The variations of B-E forward voltage V_{BE} and normalised C-B leakage I_{CBO} with temperature are shown in figure 2.44.

Similar consideration of germanium BJTs based on equations 1.147 and 1.141 shows that the rate of change of V_{BE} with temperature is similar to that for silicon BJTs although the value of V_{BE} (25 °C) is only 0.2 V. The fractional change of I_{CBO} for germanium BJTs is also similar to that of silicon BJTs but due to the fact that the 25 °C value for germanium is typically 1000 × that for silicon, a practical limit of about 90 °C is imposed on germanium BJTs.

An additional factor contributing to the temperature dependence of BJT characteristics is the variation of static current gain with temperature. As temperature increases, the lifetime of minority carriers in the base increases thereby causing an increase in h_{FE} (figure 2.44c).

Figure 2.45 shows the effect of an increase in temperature on the CE input and output characteristics. Decrease of V_{BE} with temperature causes the

(a)
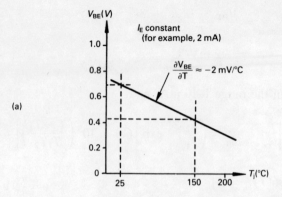

(b)
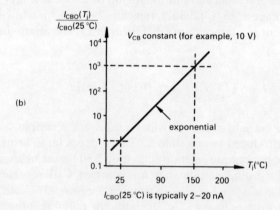

(c)
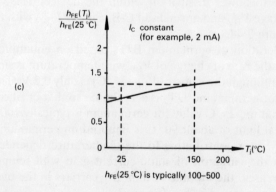

Figure 2.44 Change of V_{BE}, I_{CBO} and h_{FE} with the junction temperature for a typical low-power silicon BJT

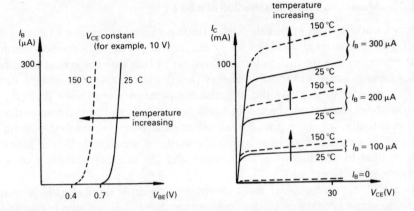

Figure 2.45 Effect of temperature on the CE input and output characteristics of a low-power silicon BJT

input characteristic to move to the left at a rate of approximately 2 mV/°C. Increase of h_{FE} with temperature causes I_C to increase for constant I_B and the output characteristic to move upwards as temperature increases, the spacing becoming wider at higher current levels. The movement of the $I_B = 0$ characteristic is due to the increase of I_{CEO} which depends on the increase in I_{CBO} from typically 10 nA at 25 °C to 10 μA at 150 °C and, to a lesser extent, on the increase of h_{FE} by a factor of typically 1.3 over this temperature range. Thus I_{CEO} increases typically from 3 μA at 25 °C to 4 mA at 150 °C for a BJT with h_{FE} (25 °C) = 300. The CB static output characteristic is only slightly affected by temperature as the CB static current gain h_{FB} varies only slightly with temperature. For silicon BJTs the CB characteristic corresponding to $I_B = 0$ is virtually unaffected by temperature relative to practical values of I_C, due to the extremely low value of I_{CBO} even at elevated temperatures.

Investigation of the thermal stability of practical circuits using silicon BJTs in the active (linear) mode in terms of stability factors [15] $\partial I_C/\partial V_{BE}$, $\partial I_C/\partial I_{CBO}$ and $\partial I_C/\partial h_{FE}$ shows that the change of V_{BE} with temperature is the most significant, followed by the change of h_{FE} and the temperature dependence of I_{CBO} is insignificant. For germanium BJTs, however, the much higher leakage current causes variation of leakage with temperature to be the most significant factor in thermal stability.

Variation of the device characteristics with temperature clearly affects the performance of the BJT in a particular application and circuit designs attempt to reduce the effect of these variations to provide good thermal stability. Although temperature increase may be due to ambient conditions, the power dissipated within the device is the most likely cause of massive temperature increases and the maximum allowable power dissipation of the device is the overriding limiting factor.

2.2.21 Maximum Power Dissipation of a BJT

When a static current I flows through a static potential difference V, electrical energy is dissipated at a rate of VI joules/second (watts). Dissipation means that energy is lost from the electrical system in that it is converted to some other form, usually into thermal energy (heat) which causes the temperature of the material to rise. As the material temperature rises above that of its surroundings, heat is conducted and radiated away from the material surface and eventually, for static power dissipation, steady-state is reached such that the rate of generation of heat within the material is balanced by the rate of loss of heat by conduction and radiation and the material attains a steady working temperature.

In a BJT operating under static conditions, power is dissipated at each of the junctions where potential differences exist due to the junction biases and, in the bulk emitter, base and collection regions due to the finite conductivity of the material. The emitter and substrate are heavily doped however and thus have low resistance. To a first-order approximation the total power dissipated P_{tot} within a BJT may be taken as the power dissipated at the two junctions, that is

$$\begin{aligned} P_{tot} &= V_{CB} I_C + V_{BE} I_E \\ &= V_{CB} I_C + V_{BE} (I_C + I_B) \\ &= V_{CE} I_C + V_{BE} I_B \end{aligned} \tag{2.64}$$

For active region operation $I_C \gg I_B$ and $V_{CE} \gg V_{BE}$, therefore

$$P_{tot} \approx V_{CE} I_C \tag{2.65}$$

Power dissipation in this mode of operation is thus concentrated at the C-B junction due to the relatively large reverse bias V_{CB}.

The power dissipation of a BJT is limited (P_{tot} max) either to ensure that specified properties such as h_{FE} min or I_{CBO} max are maintained or not exceeded, or to avoid excessive temperature rise which may lead to thermal runaway and eventual melting of the semiconductor.

Thermal runaway is a situation where the temperature rise within the device due to power dissipation causes the collector current to increase thereby further increasing the power dissipation and subsequent escalation results in failure.

The maximum allowable power dissipation of a device P_{tot} max is a fixed value for a particular chip/package arrangement in a fixed environment, the permitted variations of V_{CE} and I_C being limited by

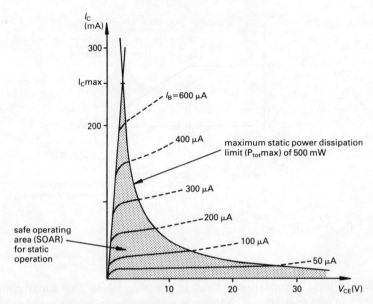

Figure 2.46 Safe operating area on the CE output characteristics is bounded by the maximum power dissipation limit

$$V_{CE} I_C \approx P_{tot} \text{ max} \qquad (2.66)$$

This maximum power limit imposes a boundary to the safe operating area (SOAR) on the I_C–V_{CE} BJT characteristics for static operation (figure 2.46). A transistor is safe from excessive power dissipation during active region operation if the load line and hence the Q-point is chosen to lie within the SOAR.

During switching, brief excursions beyond the maximum static power dissipation limit can be permitted as the operating point of the BJT moves from saturation to cut-off or vice versa. Device safety then depends upon the ratio of time spent above and below the static P_{tot} max limit and the frequency of the switching operation. Sufficient time must be allowed after each excursion beyond the P_{tot} max limit, relative to the thermal time constants involved, for the generated heat to disperse before the next excursion beyond P_{tot} max takes place.

2.2.22 Thermal Aspects

Heat generated within a BJT, notably at the C-B junction, is conducted to the package *case* or *mounting base* (mb) via the substrate of the chip and the

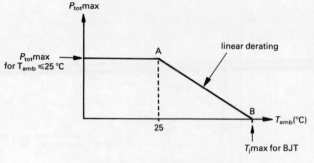

Figure 2.47 Power derating

bonding wires. From there, heat is transferred to the environment by conduction and radiation. The thermal system may be represented in terms of a thermal resistance $R_{th(j-amb)}$ as shown in the basic thermal model of figure 2.6 and the maximum static power dissipation, corresponding to the maximum permissible junction temperature T_j max, is given by equation 2.6.

T_j max depends on the semiconductor and the construction and material of the encapsulation. The maximum junction temperature for germanium BJTs is limited to about 90 °C by excessive increase in leakage current. For silicon devices the T_j limit is 175–200 °C unless plastic encapsulation is used, in which case the upper limit is about 150 °C.

The maximum static power dissipation P_{tot} max of low-power BJTs is normally quoted for an ambient temperature T_{amb} not exceeding 25 °C. If the transistor is required to operate with an ambient temperature > 25 °C, P_{tot} max must be *derated* (reduced) so that the junction temperature does not exceed T_j max. Figure 2.47 shows a typical derating curve, P_{tot} max being reduced linearly from the value at $T_{amb} = 25$ °C to zero at $T_{amb} = T_j$ max. Clearly if $T_{amb} = T_j$ max, no power can be dissipated in the device without the junction temperature rising above T_j max. The modulus of the slope of the derate line AB is termed the *derating factor*

$$\text{Derating factor} = \frac{P_{tot}\text{max (at } T_{amb})}{T_j\text{max} - T_{amb}} \qquad (2.67)$$

Comparison with equation 2.6 shows that

$$\text{derating factor} = \frac{1}{R_{th(j-amb)}} \qquad (2.68)$$

Example
A silicon BJT type BC107 (appendix H.3) has P_{tot} max $(T_{amb} \leq 25\,°C) = 300$ mW and $T_j$max = 175 °C. Thus

$$\text{derating factor} = 300\ mW/(175-25)°C = 2\ mW/°C$$

If this BJT is to be used in an environment where the highest probable ambient temperature is 55 °C, the maximum permissible power dissipation must be derated from 300 mW to

P_{tot} max (at $T_{amb} = 25\,°C$) − [derating factor × ΔT_{amb} above 25 °C]
= 300 mW − (2 mW/°C)(55−25 °C)
= 240 mW

The reduction of P_{tot} max for operation at $T_{amb} > 25\,°C$ means that the SOAR (figure 2.46) is reduced since the maximum power dissipation boundary moves towards the origin.

$R_{th(j-amb)}$ can be subdivided into $R_{th(j-case)}$ or $R_{th(j-mb)}$ between the junction and the package (described as the case or mounting base depending on the type of package) and $R_{th(case-amb)}$ or $R_{th(mb-amb)}$ between the package and the environment (figure 2.48a). $R_{th(j-case)}$ is of the order of 100–200 °C/W for low-power BJTs (< 1 W) and 1–10 °C/W for high-power types while $R_{th(case-amb)}$ is typically 100–300 °C/W and 30–100 °C/W respectively for the two power ranges.

The power dissipation capability of a transistor can be increased by improving the flow of heat from the package to the environment. This essentially means reducing $R_{th(j-amb)}$ so that P_{tot} can be increased without increasing the temperature difference $T_j - T_{amb}$ (equation 2.4). The device junction is inaccessible to the user but a thermal shunt or *heat sink* can be included between the case and ambient to improve the flow of heat from the case to the environment. The model of figure 2.48b is a first-order representation of the thermal system which is adequate for most practical purposes. R_{th} (hs) is a lumped thermal resistance describing heat flow within the bulk of the heat sink material together with conduction and radiation from its surface. R_{th} (i) represents a washer providing electrical insulation, if required, between the device case and the heat sink together with a soft metal washer and/or silicone grease used to reduce the effect of surface irregularities and produce a good thermal contact between the device package and the heat sink.

The selection of a heat sink to enable a BJT safely to dissipate a certain power level in an environment of known maximum temperature requires calculation of the thermal resistance of the heat sink R_{th} (hs) and then the determination of the size of material to provide that value of R_{th} (hs). From figure 2.48b the total thermal resistance between junction and ambient with the heat sink is

$$R_{th(j-case)} + \{R_{th(case-amb)}//[R_{th}(i) + R_{th}(hs)]\}$$

where // denotes 'in parallel with'. If the heat sink is to be really effective (R_{th} (i) + R_{th} (hs)) $\ll R_{th(case-amb)}$ and the junction to ambient values becomes approximately

$$R_{th(j-case)} + R_{th}(i) + R_{th}(hs)$$

When P_{tot} max is flowing through this thermal resistance, T_j must not exceed $T_j$max when T_{amb} is a maximum, thus

$$R_{th(j-case)} + R_{th}(i) + R_{th}(hs) \leq \frac{T_j\text{max} - T_{amb}\text{max}}{P_{tot}\text{max}} \qquad (2.69)$$

and the value of R_{th} (hs) can be determined. Calculation of the dimensions of material required to provide this value of thermal resistance is not simple as it requires the solution of a complex thermal field problem and depends on the type, thickness, orientation and surface finish of the material. However, manufacturers provide charts [16] based on practical measurements which allow the required area of heat sink to be selected to provide a certain thermal resistance taking account of the type of device package, the maximum power dissipation involved and the type (flat or finned extrusion), orientation (vertical or horizontal) and surface finish (blackened or bright) of the heat sink. A flat aluminium heat sink 100 mm × 100 mm × 2 mm thick having a

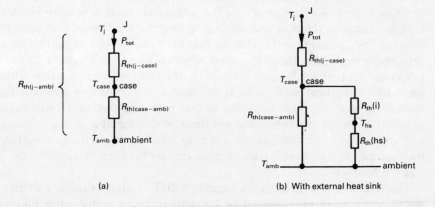

Figure 2.48 Thermal model of a BJT

Discrete Semiconductor Devices

bright surface has a thermal resistance of 5–10 °C/W depending on its orientation and the power dissipation. A similar area of finned aluminium extrusion would provide typically 1–3 °C/W.

When the dissipated power is substantial, that is, > 10 W, the heat radiated from the device causes the air surrounding the package to be well above the general ambient temperature. For this reason, manufacturers specify P_{tot} max for a maximum case or mounting base temperature rather than a maximum ambient temperature.

Example
A power transistor in a TO-3 style package has the following specification

P_{tot} max = 115 W for $T_{mb} \leq 60$ °C

T_j max = 200 °C

$R_{th(j-amb)} = 40$ °C/W

It is required to select a suitable heat sink which would allow safe dissipation of up to 45 W in an environment of up to 35 °C. The heat sink must be electrically insulated from the transistor.

From specification

$$R_{th(j-mb)} = \frac{T_j\text{max} - T_{mb}}{P_{tot}\text{max}}$$

$$= \frac{200-60}{115} = 1.2 \text{ °C/W}$$

Thus

$$R_{th(mb-amb)} = R_{th(j-amb)} - R_{th(j-mb)}$$

$$= 40 - 1.2 = 38.8 \text{ °C/W}$$

An insulating washer for a TO-3 style package with soft metal washer and silicone grease gives R_{th} (i) ≈ 1 °C/W. From the model of figure 2.48b, the thermal resistance between junction and ambient is therefore

$$1.2 + \left[\frac{38.8 \ (R_{th} \ (hs) + 1)}{R_{th} \ (hs) + 39.8}\right] \text{ °C/W}$$

In order to enable safe dissipation of 45 W at $T_{amb} = 35$ °C, this thermal resistance must not exceed

$$\frac{T_j\text{max} - T_{\text{amb}}\text{max}}{P_{\text{tot}}\text{max}} = \frac{200 - 35}{45} = 3.7\ °\text{C/W}$$

Thus

$$1.2 + \left\{\frac{38.8\ [R_{\text{th}}\ (\text{hs}) + 1]}{R_{\text{th}}\ (\text{hs}) + 39.8}\right\} \leq 3.7$$

From which R_{th} (hs) $\leq$ 1.7 °C/W. Manufacturer's data [16] shows that a thermal resistance of 1.7 °C/W can be provided by a finned heat sink approximately 100 mm square. Clearly a slightly larger physical size can be used to provide a safety margin.

The above consideration of thermal aspects is concerned with static conditions. When the power dissipation varies with time, each thermal resistance in the model of figure 2.48b must be shunted by a thermal capacitance to represent the thermal time constants. To a first-order approximation the static analysis can be used with P_{tot} replaced by the average power dissipated P_{av}. When the BJT is used as a switch, little power is dissipated during the ON and OFF periods as either v_{CE} or i_{C} is low. Power is dissipated as pulses during the transition from ON to OFF and vice versa (figure 2.50a). In such cases the thermal condition depends not only on the average power dissipated but also on the width of the power pulse t_p and the ratio of pulse width to the period T of the power waveform (termed the *duty cycle*, *d*). BJT data sheets often give the thermal impedance of the device as a function of pulse width for various values of duty cycle between 0.01 (single-shot) and unity (d.c.) (BC107 data, appendix H.3). Z_{th} for $d = 1$ is the same as the static value R_{th} since power is being dissipated continuously. For $d < 1$ the thermal impedance Z_{th} is less than R_{th} as the heat generated by a power pulse partially disperses before the next power pulse occurs. A duty cycle of 0.01 is regarded as a 'single-shot' situation where the interval between power pulses is sufficiently large to enable all the heat generated by a pulse to disperse before the arrival of the next pulse. In switching therefore, where power dissipation is intermittent, the analysis procedure is similar to that for the static case but using the average power dissipated and the appropriate values of thermal impedance.

2.2.23 The BJT at High Power Levels

A major problem in the operation of BJTs at high power levels is a phenomenon known as *second breakdown*. Progressive increase of V_{CE} in normal operation eventually leads to a rapid increase of I_{C} due to avalanche (first) breakdown. If the power dissipation is allowed to reach a certain level, V_{CE} falls abruptly to a few volts and I_{C} rises rapidly (figure 2.49a). This sudden change is termed second breakdown and invariably destroys the BJT.

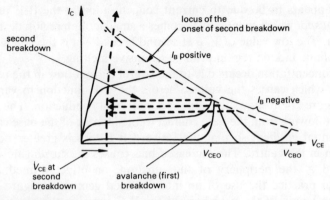

(a) Second breakdown

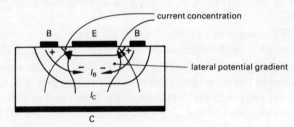

(b) *npn* power BJT during conduction

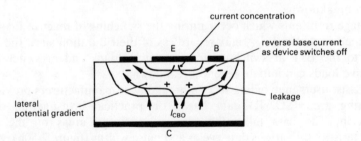

(c) *npn* power BJT at switch-off

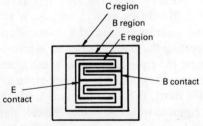

(d) Interdigitated geometry of a power BJT

Figure 2.49 BJT at high power levels

The effect appears to be due to current concentration at the B-E junction causing a *hot spot* which eventually becomes a small molten region destroying the junction. The low value of V_{CE} at second breakdown is the voltage drop across the ohmic molten region prior to extensive melting.

Current concentration occurs due to the lateral current flow in the resistive base region which causes the voltage across the B-E junction to vary with position. Figure 2.49b shows the situation in normal conduction. The lateral base current flowing from the base contact to the interior of the base causes a lateral potential gradient so that the forward B-E bias is greater near the contact than in the centre. The increased bias causes the current flow to be concentrated at the periphery of the junction, an effect termed *emitter crowding*. In practice the use of an interdigitated geometry (figure 2.49d) increases the length of the periphery of the B-E junction thereby reducing current concentration or shifting its significance to higher operating current levels.

BJTs operating at high current levels are particularly vunerable at switch-off. Application of reverse base drive to enable rapid removal of stored charge causes a negative base current and the resulting lateral potential gradient in the base causes current flow to be concentrated at the centre of the junction (figure 2.49c). The negative base current can also occur due to the leakage currents of the junctions when the device is OFF and at high voltage and/or high temperature. These currents can cause sufficient current concentration at the centre of the junction to create a hot spot and possible second breakdown.

Voltage transients which occur during the switching of reactive loads must be limited to within V_{CEO} max regardless of their duration since the energy input to the BJT at switch-on for capacitive loads and at switch-off for inductive loads can initiate second breakdown.

To assist users in avoiding second breakdown, manufacturers provide safe operating area (SOAR) data based on practical investigation of BJT operation. The data takes the form of enclosed areas on the output characteristics for various duty cycles and pulse widths (figure 2.50b). As long as the operating point of the BJT (v_{CE}, i_C) does not go outside the SOAR for the duty cycle and pulse width involved, the BJT is safe from second breakdown. Figure 2.50a shows typical variations of v_{CE} and i_C during a repetitive switching operation and the associated variation in power dissipation. By converting the power pulses to 'equivalent rectangular pulses' the user can determine the equivalent pulse width t_p and hence the duty cycle d ($= t_p/T$) for the operation. SOAR curves are usually provided for duty cycles of 0.01 (single shot), 0.1, 0.2, 0.5 and 1 (d.c.). The user selects the SOAR for d and t_p equal to or greater than the calculated values. By plotting the locus of the actual operating point during the power pulse on the SOAR the degree of safety can be estimated.

Discrete Semiconductor Devices 173

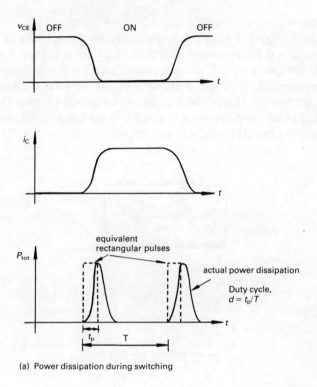

(a) Power dissipation during switching

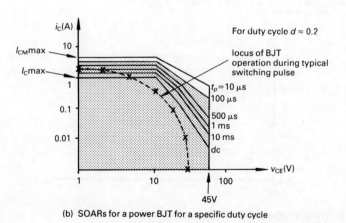

(b) SOARs for a power BJT for a specific duty cycle

Figure 2.50 Use of SOAR data to check the safety of operation of a BJT

Example

If t_p and T (figure 2.50a) are respectively 80 μs and 500 μs in a practical situation, $d = 0.16$. From the available data for $d = 0.01, 0.1, 0.2, 0.5, 1$ and $t_p = 10, 100, 500$ μs, 1, 10 ms, d.c., the SOAR for $d = 0.2$, $t_p = 100$ μs is the most appropriate, although it allows a margin of safety. The corresponding SOAR is then the shaded area in figure 2.50b and plotting the locus of the operating point using values of v_{CE} and i_C from figure 2.50a during the power pulse enables the BJT safety to be checked.

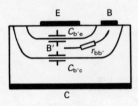

(a) Base resistance affects the charging times of $C_{b'e}$ and $C_{b'c}$

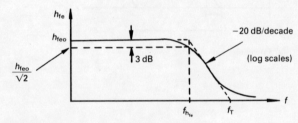

(b) Typical variation of small-signal CE current gain with frequency

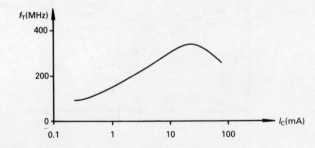

(c) Typical variation of transition frequency with static collector current for a low-power silicon BJT

Figure 2.51 Frequency response of a BJT

2.2.24 High-Frequency Properties of BJTs

For active region operation the high frequency response of BJTs is affected by

(1) depletion capacitance of the reverse-biased C-B junction;
(2) diffusion capacitance of the forward-biased B-E junction.

Low frequency operation refers to signal frequencies at which the time taken to charge and discharge device capacitances is negligible compared with the rate at which the input signal is varying. At such frequencies the delay between a change in B-E voltage and the corresponding change in collector current is thus negligible. As the frequency is increased these charging times become significant altering the current gain and the input and output impedances of the BJT.

Description of current flow within the base region of a real BJT is a complicated field problem but as a first-order representation, a hypothetical point B' is defined within the base region where current summation takes place. The junction capacitances can then be denoted as $C_{b'e}$ and $C_{b'c}$. At low current levels the main transistor action is in the central region beneath the emitter contact and B' can be taken as a point in this region (figure 2.51a). During dynamic operation, change in the state of charge of $C_{b'e}$ and $C_{b'c}$ via the base contact necessitates current flow through the bulk resistance $r_{bb'}$ introducing a time constant and hence slowing the response time. The base resistance $r_{bb'}$ is a function of V_{CB} as the depth of penetration of the C-B depletion region alters the width of the effective base region so altering $r_{bb'}$. Also $r_{bb'}$ is a function of I_C as the corresponding change of I_B alters the majority carrier density in the base causing its conductivity to vary. At higher current levels the variation of V_{BE} with position along the B-E junction (section 2.2.23) causes the active region to widen, reducing the length of the base current path and hence reducing $r_{bb'}$.

In normal operation, the C-B junction is reverse-biased and so $C_{b'c}$ is the depletion capacitance of the junction. For a planar diffused BJT the dopant profile of this junction is approximately linearly graded and $C_{b'c}$ is therefore proportional to $1/\sqrt[3]{(\psi_{CB} - V_{BC})}$ (section 1.6.2) or approximately $V_{CB}^{-1/3}$ for $V_{CB} \gg \psi_{CB}$. The variation of depletion layer capacitance with bias is shown in figure 1.27b. For a low-power silicon BJT, $C_{b'c}$, which is termed the *collector transition capacitance* C_{T_c} in manufacturers' data, is typically 5 pF at $V_{CB} = 0$ falling to 2 pF at $V_{CB} = 30$ V.

The B-E junction is forward-biased in normal operation and so its depletion capacitance is negligible. However, charge injection causes an appreciable diffusion capacitance due mainly to the stored charge in the base. If the minority carrier distribution in the base is assumed to be linear (section 2.2.7), the base charge for an *npn* BJT is (equation 2.23)

$$q_B = -\tfrac{1}{2}e\,\Delta n(0) w_B A$$

Since the base charge is negative in an *npn* BJT

$$C_{b'e} = \frac{dq_B}{dV_{EB}} = -\frac{dq_B}{dV_{BE}} = \frac{ew_B A}{2}\frac{d[\Delta n(0)]}{dV_{BE}} \qquad (2.70)$$

From equation 2.20, $\Delta n(0)$ can be related to the static collector current I_C which is approximately equal to the static emitter current I_E, thus

$$I_E \approx \frac{eD_n A \Delta n(0)}{w_B} \qquad (2.71)$$

Combining equations 2.70 and 2.71

$$C_{b'e} \approx \frac{w_B^2}{2D_n}\frac{dI_E}{dV_{BE}} \qquad (2.72)$$

The theoretical *I–V* relationship for the B-E junction (from equation 1.132) is

$$I_E = I_0\left[\exp\left(\frac{eV_{BE}}{kT}\right) - 1\right]$$

where I_0 is the theoretical reverse saturation current for the B-E junction. Thus theoretically

$$\frac{dI_E}{dV_{BE}} = \frac{e}{kT} I_0 \exp\left(\frac{eV_{BE}}{kT}\right)$$

$$= \frac{e}{kT}(I_E - I_0)$$

$$\approx \frac{eI_E}{kT}, \text{ as } I_E \gg I_0 \text{ in normal operation} \qquad (2.73)$$

dV_{BE}/dI_E is the *emitter slope resistance* r_e, therefore

$$r_e \approx \frac{kT}{eI_E} \approx \frac{1}{40 I_E} \quad \text{at } T = 25\,°C \qquad (2.74)$$

and $C_{b'e}$ from equation 2.72 can be seen to be directly related to the static emitter current

$$C_{b'e} \approx \frac{w_B^2}{2D_n}\frac{1}{r_e} \approx \frac{20 w_B^2 I_E}{D_n} \qquad (2.75)$$

As the base charge distribution is a consequence of charge movement across the region, $C_{b'e}$ takes account of the time taken for injected carriers to cross the base, in fact the term $w_B^2/2D_n$ in equation 2.75 is the *transit time* for injected electrons to cross the effective base width w_B. The dependence of $C_{b'e}$ on w_B shows that a reduction in base width improves the high frequency performance of the BJT. $C_{b'e}$ is typically 10–150 pF for a low-power silicon BJT.

At high frequency a component of base current is necessary to charge the device capacitances and so for a certain collector signal current i_c the base current i_b is larger than at low frequency causing the small-signal CE current gain to fall. Figure 2.51b shows the typical variation of $|h_{fe}|$ with frequency. The modulus is included as the phase relation between i_c and i_b is a function of frequency. At a frequency $f_{h_{fe}}$ (also termed f_β) known as the CE (h_{fe} or β) *cut-off frequency*, $|h_{fe}|$ has reduced to $h_{feo}/\sqrt{2}$, where h_{feo} is the low-frequency value of h_{fe}. The cut-off frequency describes the bandwidth of the BJT and a *figure of merit* for the device is its gain (at LF) — bandwidth product $h_{feo} f_{h_{fe}}$, often written $\beta_o f_\beta$.

If the high frequency roll-off of $|h_{fe}|$ is extrapolated to the frequency axis (figure 2.51b), the intercept is termed the *transition frequency* f_T of the BJT. Since at this intercept log $|h_{fe}| = 0$, that is, $|h_{fe}| = 1$, f_T may be defined as the *theoretical frequency at which the modulus of the small-signal CE current gain is equal to unity if the high frequency roll-off were continued at -20 dB/decade*. The value of f_T for a BJT is deduced by making gain measurements at frequencies around $f_{h_{fe}}$ and extrapolating linearly to the $|h_{fe}| = 1$ point. Consideration of high-frequency response in terms of the hybrid-π model (section 2.2.26, equations 2.147 and 2.148) shows that

$$f_T = h_{feo} f_{h_{fe}} = \beta_o f_\beta = \alpha_o f_\alpha \approx f_\alpha, \text{ as } \alpha_o \approx 1 \qquad (2.76)$$

where α_o is the LF small-signal current gain in CB and $f\alpha$ is the CB (α) *cut-off frequency*. Thus f_T is also termed the *gain-bandwidth product* of the BJT.

The transition frequency is a function of static current flow through the transistor (figure 2.51c) typically peaking at $I_C \approx 20$–30 mA for a low-power device. The variation of f_T with current can be linked to the variation of h_{fe} with current (section 2.2.17).

For a general-purpose low-power BJT f_T ($=f_\alpha$) is typically 300 MHz at $I_C =$ 10 mA falling to about 100 MHz at 1 mA. With h_{feo} of the order of 200, this gives CE cut-off frequencies of 1.5 MHz and 500 kHz at the two current levels respectively.

2.2.25 Noise in BJTs

Noise is generated in semiconductor devices [17,18] due to the discrete nature of charge carriers and the random nature of the electronic processes involving them. The most significant sources of noise are as follows.

(1) *Thermal (Johnson) noise.* In a resistive region, the random movement of charge carriers due to their random thermal energies and their collisions with the atoms of the material results in voltage fluctuations which are described in terms of the mean square thermal noise voltage across the region. Thermal noise is described as 'white' as it has a flat frequency spectrum throughout the normal operating range.

(2) *Shot noise* due to the discrete nature of charge carriers. Current flow through a region due to an applied voltage comprises a stream of discrete charges and so superimposed on the average flow is a fluctuating component. The degree of fluctuation is represented by the mean square shot noise current through the region.

(3) *Surface (1/f or flicker) noise* due to the random nature of free carrier generation and recombination. This is particularly apparent at a surface or interface where the crystal structure is not uniform. Surface noise can be significant at low frequency but diminishes as the operating frequency is increased.

In BJTs thermal noise is generated within the resistive bulk semiconductor regions, the base component being the most significant. Shot noise occurs at the B-E and C-B junctions since carriers entering the depletion region at a particular instant do not arrive at the other side of the region together causing a fluctuating current component. At very low frequencies surface noise due to interface effects tends to dominate.

The noise properties of a device are normally specified in terms of its *noise figure* (NF) which is the ratio in dB of the total noise power at the output to the output noise power due only to the source. If the *noise* powers at the input and output of the device are N_i and N_o respectively and the device provides a power gain A_p (figure 2.52a)

$$\text{NF} = 10 \log \frac{N_o}{A_p N_i} \text{ (dB)} \tag{2.77}$$

However, if the input and output *signal* powers are S_i and S_o, $A_p = S_o/S_i$ and

$$\text{NF} = 10 \log \frac{S_i}{N_i} \frac{N_o}{S_o} = 10 \log \frac{(S/N)_i}{(S/N)_o}$$

$$= (S/N)_i \text{ dB} - (S/N)_o \text{ dB} \tag{2.78}$$

Thus the noise figure can also be defined as the difference in the signal-to-noise power ratios in dB at the input and output of the device. As the signal and noise are developed across the same resistance, the S/N power ratio is equal to the square of the S/N voltage ratio and equation 2.78 can be written in terms of the voltage ratio, the only difference being that 10 log ( ) is replaced by 20 log ( ) due to the squared relationship.

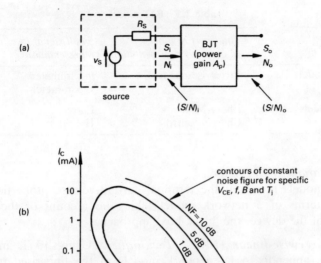

Figure 2.52   Noise performance of a BJT

The noise input to the device (N_i, figure 2.52a) depends on the degree of matching between the source and the device and so the noise figure is a function of the source resistance R_S. Noise figure is also a function of current flow through the BJT since the shot noise component particularly is dependent on current. Device data sheets often provide contours of constant noise figure for specific conditions of voltage, temperature, frequency and measurement bandwidth (BC107 data, appendix H.3) showing how NF varies with I_C and R_S (figure 2.52b) enabling the user to select I_C and R_S for low noise performance.

2.2.26   BJT Modelling

The representation of BJT performance is a wide topic due to the variety of models that are in widespread use. With the exception of the charge-control model, which is a mathematical model describing current flow in terms of the charge conditions within the BJT, all the models listed in table 2.5 are network models. Total performance models represent the performance of the BJT under both static (d.c.) and dynamic (a.c.) conditions whereas small-signal models describe only the a.c. properties at a particular operating point. If models take account of frequency-dependent properties such as charge storage they are termed high-frequency (HF) models, otherwise they are restricted to low-frequency (LF) representation.

Semiconductor Device Technology

Table 2.5 BJT models

	Total performance (d.c.+a.c.) representation	Small-signal (a.c. only) representation
Terminal models	Piecewise-linear	h–parameter y–parameter
Physical models	Ebers–Moll Charge–control	Early Hybrid–π

Terminal Parameter BJT models

Terminal models represent the terminal current-voltage properties of the device in terms of a network model taking no account of the internal operation of the device, the 'black-box' approach.

(1) *Piecewise-linear (PWL) Techniques.* Using PWL modelling techniques (appendix A.1), network models of the linearised input and output characteristics of the BJT (figure 2.53) can be produced and combined to form a model of BJT performance (figures 2.54a and b). Switching at the breakpoints is provided by the hypothetical voltage-controlled ideal switches which are represented by an unblanked diode symbol. Dependence of the breakpoint of the input characteristic on V_{CE} and that of the output characteristic on I_B is included in figure 2.54b by introducing device parameters $k_1 = \Delta V_{BE}/\Delta V_{CE}$ and $k_2 = \Delta I_C/\Delta I_B$. This model is labelled in terms of static voltages and currents and changes in these values. It describes the total performance comprising static components (I_B etc.) and signal

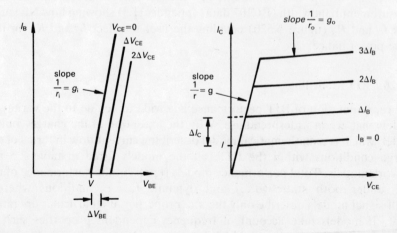

Figure 2.53 First-order piecewise linearisation of BJT CE input and output characteristics

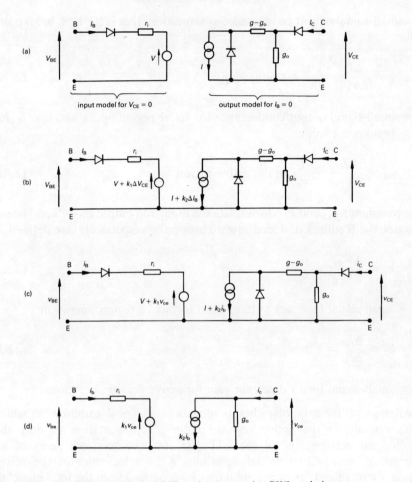

Figure 2.54 Terminal modelling of BJT performance using PWL techniques

components ($\Delta I_B = i_b$, etc.) and can therefore be labelled more generally as in figure 2.54c where, for example, i_B is the total base current $I_B + i_b$. The system of variable symbols is given in appendix B.2.

As the parameters g_i, g and g_o refer to the *slopes* of the characteristics they describe the relative *changes* of current and voltage and are therefore termed small-signal parameters. From figure 2.53,

r_i = small-signal input resistance for active region operation, that is, for V_{BE} > breakpoint value

$$r_i = \frac{\Delta V_{BE}}{\Delta I_B} = \frac{v_{be}}{i_b} \text{ in the active region} \tag{2.79}$$

g = small-signal output conductance at saturation, that is for $I_C <$ breakpoint value

$$g = \frac{\Delta I_C}{\Delta V_{CE}} = \frac{i_c}{v_{ce}} \text{ at saturation} \tag{2.80}$$

g_o = small-signal output conductance for active region operation, that is, for $I_C >$ breakpoint value

$$g_o = \frac{\Delta I_C}{\Delta V_{CE}} = \frac{i_c}{v_{ce}} \text{ in the active region} \tag{2.81}$$

The parameters k_1 and k_2 which relate the input and output models and hence describe the feedback and feedforward properties respectively, are defined as

$$k_1 = \frac{\Delta V_{BE}}{\Delta V_{CE}} = \frac{v_{be}}{v_{ce}} \tag{2.82}$$

= small-signal feedback voltage ratio for active region operation

$$k_2 = \frac{\Delta I_C}{\Delta I_B} = \frac{i_c}{i_b} \tag{2.83}$$

= small-signal forward current gain for active region operation.

Analysis of linear bipolar circuits such as small-signal amplifiers in which BJTs operate in the active region, requires representation of only the small-signal performance of the BJT, the static components being of no interest. A small-signal model describing a.c. performance in the active region about a fixed operating point may be produced from the total model of figure 2.54c by removal of the static components, the result being shown in figure 2.54d. By definition, the quiescent point lies in the active region of the characteristic and so both V_{BE} and I_C are greater than the breakpoint values. Thus the ideal switches and the static components of the breakpoint biases V and I in figure 2.54c can be deleted. In addition, slope conductance $g - g_o$, which is only applicable at saturation, that is below the breakpoint, is not involved in the active region model.

The models of figure 2.54 take no account of time lags between variations of voltage and the corresponding variations of current due to reactive properties and are therefore restricted to low-frequency representation.

(2) *Hybrid (h) Parameter Small-signal BJT Model.* Small-signal models may be developed by considering the BJT as a two-port system and relating the signal voltages and currents mathematically. By choosing the input current and output voltage as independent variables and forming

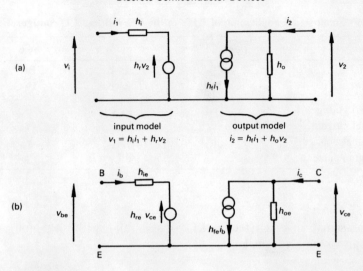

Figure 2.55 The h-parameter small-signal model of (a) a general two-port system (b) BJT performance in CE

mathematical relationships for the input voltage and output current, the following small-signal relationships are obtained

$$v_1 = h_i i_1 + h_r v_2 \tag{2.84}$$
$$i_2 = h_f i_1 + h_o v_2 \tag{2.85}$$

where suffixes 1 and 2 refer to the input and output ports respectively (appendix A.2). The four parameters relating the input and output variables are termed *hybrid* (h) parameters as they have mixed units, h_i being a resistance, h_r a voltage ratio, h_f a current ratio and h_o a conductance. The general h-parameter small-signal model of a two-port system corresponding to equations 2.84 and 2.85 is given in figure 2.55a. These equations lead to general definitions of the h-parameters and indicate the relevant measurement conditions

$$h_i = \left. \frac{v_1}{i_1} \right|_{v_2=0} \tag{2.86}$$

= small-signal input resistance with the output port short-circuit to the a.c. signal

$$h_r = \left. \frac{v_1}{v_2} \right|_{i_1=0} \tag{2.87}$$

Table 2.6 Small-signal application of BJTs in the CE, CB and CC configurations

Configuration	CE	CB	CC
Input port	B–E	E–B	B–C
Output port	C–E	C–B	E–C
Input signal voltage	v_{be}	v_{eb}	v_{bc}
Input signal current	i_b	i_e	i_b
Output signal voltage	v_{ce}	v_{cb}	v_{ec}
Output signal current	i_c	i_c	i_e

= small-signal reverse (feedback) voltage ratio with the input port open-circuit to the a.c. signal

$$h_f = \left.\frac{i_2}{i_1}\right|_{v_2=0} \quad (2.88)$$

= small-signal forward current ratio (gain) with the output port short-circuit to the a.c. signal

$$h_o = \left.\frac{i_2}{v_2}\right|_{i_1=0} \quad (2.89)$$

= small-signal output conductance with the input port open-circuit to the a.c. signal

It should be noted that practical measurement of these parameters requires either an a.c. open-circuit at the input or an a.c. short-circuit at the output. These conditions can be provided by a sufficiently large value inductor and capacitor respectively, the value being dependent on the signal (measurement) frequency. During measurement, the quiescent (d.c.) conditions must be maintained to provide operation at the required point (V_{CE}, I_C) in the active region and so the input and output ports can not be physically open or short-circuited.

Equations 2.84 to 2.89 are general as no reference is made to which terminals of the BJT constitute the input and output ports. Table 2.6 lists the input and output ports together with the input and output variables for the common-emitter (CE), common-base (CB) and common-collector (CC) configurations. As the numerical values of h_i, h_r, h_f and h_o are different for the three configurations, a second subscript is used to identify the particular mode of connection, e for CE, b for CB and c for CC. Thus the CE h-parameters for a BJT are h_{ie}, h_{re}, h_{fe} and h_{oe}, the CB parameters h_{ib}, h_{rb}, h_{fb}, h_{ob} and for CC h_{ic}, h_{rc}, h_{fc}, h_{oc}. The general model of figure 2.55a then

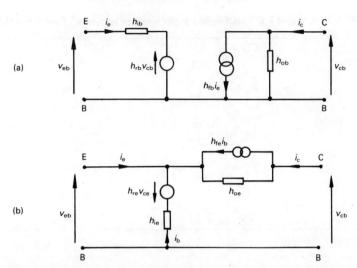

Figure 2.56 Small-signal h-parameter models of BJT performance in CB (a) in terms of CB parameters and (b) in terms of CE parameters

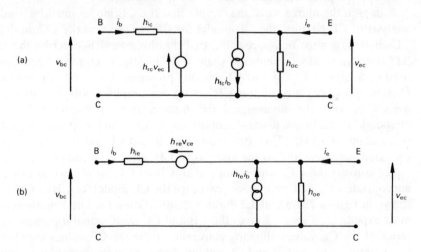

Figure 2.57 Small-signal h-parameter models of BJT performance in CC (a) in terms of CC parameters and (b) in terms of CE parameters

becomes that of figure 2.55b for CE, while those for CB and CC are given in figures 2.56a and 2.57a.

It should be noted that the CE h-parameter small-signal model (figure 2.55b) obtained from derivation of mathematical relationships between the BJT variables corresponds directly to the small-signal model of figure 2.54d

Table 2.7 CB and CC small-signal h-parameters in terms of the CE values

CE	CB	CC
h_{ie}	$h_{ib} \approx \dfrac{h_{ie}}{1 + h_{fe}}$	$h_{ic} = h_{ie}$
h_{re}	$h_{rb} \approx \dfrac{h_{ie}h_{oe}}{1 + h_{fe}} - h_{re}$	$h_{rc} = 1 - h_{re} \approx 1$
h_{fe} (β)	$h_{fb} \approx \dfrac{-h_{fe}}{1 + h_{fe}} = -\alpha$	$h_{fc} = -(1 + h_{fe})$
h_{oe}	$h_{ob} \approx \dfrac{h_{oe}}{1 + h_{fe}}$	$h_{oc} = h_{oe}$

which was derived from the BJT characteristics using PWL techniques. By comparison, $h_{ie} \equiv r_i$, $h_{re} \equiv k_1$, $h_{fe} \equiv k_2$ and $h_{oe} \equiv g_o$. These relationships help to clarify the practical significance of the h-parameters.

Although the above work may imply that the CE model must be used for analysis of CE circuits, the CB model for CB circuits and the CC model for CC circuits, it must be appreciated that the three models describe the same BJT properties, albeit measured slightly differently. Each model can thus be used to analyse CE, CB and CC circuits providing it is used appropriately. Due to the popularity of the CE circuit for amplification, manufacturers usually provide the small-signal CE h-parameters, particularly for BJTs intended for use in small-signal amplification at low and medium frequencies up to about 500 kHz. The corresponding CB and CC parameters are rarely provided however, and so for analysis of CB and CC circuits the user must either convert the CE values supplied to CB or CC form, and then use the appropriate CB or CC model, or rearrange the CE model for CB or CC use as shown in figures 2.56b and 2.57b respectively. Often the latter method is the most expedient. Table 2.7 gives the CB and CC small-signal h-parameters in terms of the CE values allowing conversion from the CE values supplied by manufacturers to the CB and CC values. The conversion expressions may be derived by comparison of the coefficients of the CE, CB and CC h-parameter equations corresponding to equations 2.84 and 2.85, noting that

$$\left. \begin{array}{l} i_e + i_b + i_c = 0 \\ v_{be} = -v_{eb} = v_{bc} - v_{ec} \\ v_{ce} = -v_{ec} = v_{cb} + v_{be} \end{array} \right\} \quad (2.90)$$

The approximate relationships in table 2.7 are simplifications based on the fact that for practical BJTs $h_{re} \ll 1$, $h_{rb} \ll 1$, $h_{ie} h_{oe} \ll 1$, $h_{ib} h_{ob} \ll 1$ and $h_{ic} h_{oc} \ll 1$. Detailed derivation of these conversion relationships together with exact relationships can be found in reference 19.

The BJTs intended for small-signal amplification at low and medium frequencies, manufacturers usually provide typical values of the CE h-parameters h_{ie}, h_{re}, h_{fe} and h_{oe} for a specific set of operating conditions such as $I_C = 2$ mA, $V_{CE} = 5$ V and $f = 1$ kHz together with indication of the spread of values. Typical values for a low-power BJT are given in table 2.8 and the corresponding values for CB and CC, as obtained using the conversion expressions in table 2.7, are listed in table 2.9.

Table 2.8 CE h–parameters for a typical low-power silicon planar epitaxial BJT at $I_C = 2$ mA, $V_{CE} = 5$ V and f = 1 kHz

	Minimum	Typical	Maximum
h_{ie} (kΩ)	1.5	2.5	4.5
h_{re}	10^{-4}	1.5×10^{-4}	2×10^{-4}
h_{fe}	120	220	300
h_{oe} (μS)	10	20	30

Table 2.9 Typical values for the CE, CB and CC h–parameters of a low-power silicon planar epitaxial BJT at $I_C = 2$ mA, $V_{CE} = 5$ V and f = 1 kHz

	CE	CB	CC
h_i (Ω)	2500	11	2500
h_r	1.5×10^{-4}	0.8×10^{-4}	1
h_f	220	-0.995	-221
h_o (μS)	20	0.09	20

Due to the non-linearity of the BJT characteristics in the active region, the h-parameters of a BJT are a function of the quiescent conditions V_{CE}, I_C. Analysis of a practical circuit first requires determination of the h-parameters corresponding to the quiescent conditions and manufacturers usually provide details of the variation of CE h-parameters with I_C and V_{CE} (BC107 data, appendix H.3). Typical variations normalised to the $I_C = 2$ mA and $V_{CE} = 5$ V values are shown in figure 2.58a and b. The h-parameters are also a function of device temperature (figure 2.58c).

It should be noted that the quoted values for BJT h-parameters are all real (in a mathematical sense) indicating that time lags due to finite charge transit time and charge storage effects are not represented. The model is thus only

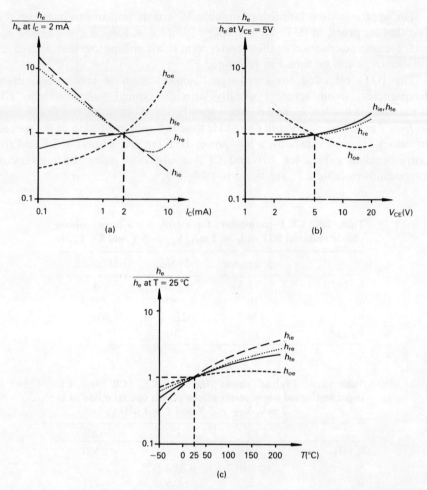

Figure 2.58 Typical variations of small-signal CE h-parameters for a low-power silicon BJT with I_C, V_{CE} and T at 1 kHz

applicable to low and medium frequency operation up to frequencies of the order of the device cut-off frequency, typically 500 kHz for CE and 1.5 MHz for CB at $I_C = 1$ mA. This limit is imposed by the errors introduced by neglecting reactive effects.

In theory the h-parameter model could represent the BJT at high frequency (HF) if the parameters were measured as complex quantities of the form $p + jq$ so that h_{ie} would become an impedance instead of a resistance, h_{re} and h_{fe} would then include phase as well as amplitude information and h_{oe} would become an admittance in place of a conductance. Equation 2.141 shows that h_{fe} is frequency dependent. The problem with this approach is the practical

Discrete Semiconductor Devices

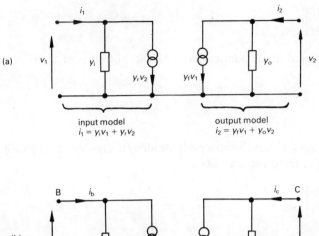

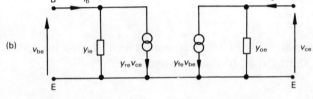

Figure 2.59 Small-signal y-parameter model of (a) a general two-port system (b) BJT performance in CE

difficulty of measurement of the quadrative components (jq) when a.c. open-circuit conditions are required (equations 2.87 and 2.89). Provision of a satisfactory a.c. open-circuit condition is difficult at HF and for this reason the y-parameter model is popular at HF as only a.c. short-circuit conditions are required in measurement of the parameter values. It is usual for manufacturers to provide only low frequency h-parameter information on device data sheets.

(3) *Admittance (y) Parameter Small-signal BJT Model.* By choosing the input and output signal voltages (v_1 and v_2) as the independent variables, mathematical relationships for the input and output signal currents (i_1 and i_2) may be written (appendix A.2)

$$i_1 = y_i v_1 + y_r v_2 \tag{2.91}$$

$$i_2 = y_f v_1 + y_o v_2 \tag{2.92}$$

The four parameters y_i, y_r, y_f and y_o that describe the properties of the device are all admittances, hence the name of the model. The corresponding network model is shown in figure 2.59a and from equations 2.91 and 2.92 the y-parameters can be defined as

$$y_i = \frac{i_1}{v_1}\bigg|_{v_2=0} \tag{2.93}$$

= small-signal input admittance with the output port short-circuit to the a.c. signal

$$y_r = \frac{i_1}{v_2}\bigg|_{v_1=0} \tag{2.94}$$

= small-signal reverse transfer admittance (transadmittance) with the input port short-circuit to the a.c. signal

$$y_f = \frac{i_2}{v_1}\bigg|_{v_2=0} \tag{2.95}$$

= small-signal forward transfer admittance (transadmittance) with the output port short-circuit to the a.c. signal

$$y_o = \frac{i_2}{v_2}\bigg|_{v_1=0} \tag{2.96}$$

= small-signal output admittance with the input port short-circuit to the a.c. signal

In each case, practical measurement requires a.c. short-circuit conditions at either the input or output ports; no a.c. open-circuit conditions are required. The y-parameter model is thus more suitable for BJT representation at high frequency than the h-parameter model as satisfactory a.c. short-circuit conditions are easier to achieve at HF than open-circuit conditions. The term *transfer admittance* or *transadmittance* is used in the case of y_r and y_f as these parameters describe the relationship between the current at one port of the device and the voltage at the other port, that is, a relationship between input and output or vice versa.

As in the h-parameter case, a second suffix is used to indicate the connection configuration to which the parameters refer. Figure 2.59b shows the common-emitter (CE) y-parameter BJT model labelled in terms of appropriate input and output variables (table 2.6) and CE y-parameters y_{ie}, y_{re}, y_{fe}, y_{oe}. In some texts numerical suffixes are used corresponding to the suffixes of i and v in equations 2.93–2.96, thus $11 \equiv i$, $12 \equiv r$, $21 \equiv f$, $22 \equiv o$ and the parameters are written $y_{11}, y_{12}, y_{21}, y_{22}$ although this does not indicate the connection configuration. It is possible to convert from the y-parameter values for one connection configuration to those for the other configurations as in the case of h-parameters although it is usually simpler to reconfigure the model in terms of the parameters supplied in an analogous fashion to the reconfigured CE h-parameter models of figures 2.56b and 2.57b.

In general, y-parameters are complex quantities of the form $y = g + jb$ where g is a small-signal conductance and b is a small-signal susceptance, describing the resistive and reactive properties respectively. BJT data sheets sometimes specify LF values of y-parameters, often at 1 kHz, in which case the values given are in effect the small-signal conductances g_i, g_r, g_f, g_o, together with terminal capacitances. Quoted h-parameters may be converted to y-parameter values (appendix A.2), [20] although since h-parameter values are normally given only for low-frequency operation, the corresponding y-parameter values are the conductances g_i, g_r, g_f, g_o. Typical values of these parameters for a low-power silicon planar epitaxial BJT in CE at $I_C = 2$ mA, $V_{CE} = 5$ V and $T = 25$ °C corresponding to the typical h-parameters in table 2.8 are $g_{ie} = 0.4$ mS, $g_{re} = 0.06$ μS, $g_{fe} = 88$ mS and $g_{oe} = 6.8$ μS.

BJTs intended for high frequency operation often have the real and quadrative components of the y-parameters (g and b) presented in graphical form as functions of frequency and since operation at HF is frequently in CB, these are usually the parameters provided.

Although the h and y-parameter sets are the most widely used in device representation, all the terminal parameter sets, [20] namely, z, y, g, h, a, b are useful in computer-aided analysis and design of electronic systems in which systems are considered as series/parallel/cascade connections between electronic subsystems.

Physical BJT Models
Physical models describe the current–voltage properties in terms of the physical mechanism of charge movement within the device.

(1) *Ebers–Moll BJT Model.* The Ebers–Moll model represents the BJT as two pn junction diodes (figure 2.60a) enabling representation of normal and inverse active region operation as well as saturation and cut-off. For normal operation (figure 2.60b), junction diode D_1, is forward-biased, D_2 is reverse-biased and the collector current comprises the component from the emitter $\alpha_N I_F$ together with reverse leakage across D_2. For inverse operation (figure 2.60c) D_1 is reverse-biased, D_2 forward-biased and the emitter current is made up of the component from the collector $\alpha_I I_R$ together with reverse leakage across D_1. In saturation, both D_1 and D_2 are forward-biased and flow is made up of all four components I_F, $\alpha_N I_F$, I_R, $\alpha_I I_R$ while at cut-off both diodes are reverse-biased and only leakage current flows. The parameters α_N and α_I are the CB static current gains for normal and inverted operation respectively.

Using the theoretical I–V relationship for a pn junction (equation 1.132), the diode currents I_F and I_R may be related to the biasing voltages V_{BE} and V_{BC}

$$I_F = I_{EBS} \left[\exp\left(\frac{eV_{BE}}{kT}\right) - 1 \right] \tag{2.97}$$

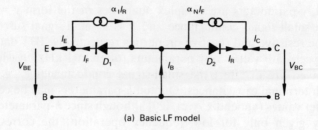
(a) Basic LF model

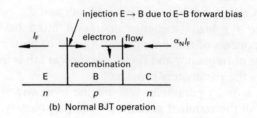
(b) Normal BJT operation

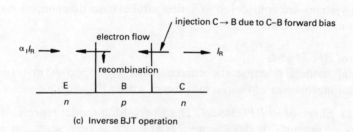
(c) Inverse BJT operation

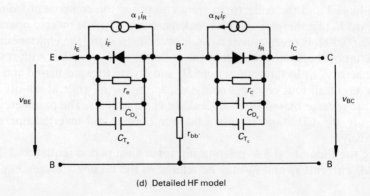
(d) Detailed HF model

Figure 2.60 Ebers–Moll BJT model

$$I_R = I_{CBS}\left[\exp\left(\frac{eV_{BC}}{kT}\right) - 1\right] \tag{2.98}$$

where I_{EBS} and I_{CBS} are the reverse leakage currents through the E-B and C-B diodes with the collector shorted to the base and the emitter shorted to the base respectively.

From the model of figure 2.60a

$$\left.\begin{aligned}I_E &= I_F - \alpha_I I_R \\ I_C &= \alpha_N I_F - I_R \\ I_B &= I_E - I_C \\ &= (1 - \alpha_N)I_F + (1 - \alpha_F)I_R\end{aligned}\right\} \tag{2.99}$$

Substituting for I_F and I_R from equations 2.97 and 2.98, the terminal currents become

$$I_E = I_{EBS}\left[\exp\left(\frac{eV_{BE}}{kT}\right) - 1\right] - \alpha_I I_{CBS}\left[\exp\left(\frac{eV_{BC}}{kT}\right) - 1\right] \tag{2.100}$$

$$I_C = \alpha_N I_{EBS}\left[\exp\left(\frac{eV_{BE}}{kT}\right) - 1\right] - I_{CBS}\left[\exp\left(\frac{eV_{BC}}{kT}\right) - 1\right] \tag{2.101}$$

$$I_B = (1 - \alpha_N)I_{EBS}\left[\exp\left(\frac{eV_{BE}}{kT}\right) - 1\right] + (1 - \alpha_F)I_{CBS}\left[\exp\left(\frac{eV_{BC}}{kT}\right) - 1\right] \tag{2.102}$$

Practical measurement of α_N, α_I, I_{EBS} and I_{CBS} enables these equations to be used to represent BJT performance in general since they describe operation at saturation and cut-off as well as in the normal and inverse active modes.

Multiplying equation 2.100 by α_N and subtracting from equation 2.101 gives

$$I_C = \alpha_N I_E - (1 - \alpha_N \alpha_I)I_{CBS}\left[\exp\left(\frac{eV_{BC}}{kT}\right) - 1\right] \tag{2.103}$$

Similarly

$$I_E = \alpha_I I_C + (1 - \alpha_N \alpha_I)I_{EBS}\left[\exp\left(\frac{eV_{BE}}{kT}\right) - 1\right] \tag{2.104}$$

With the emitter open-circuit, $I_E = 0$ and so the current $(1 - \alpha_N \alpha_I)I_{CBS} = I_{CBO}$, the collector–base leakage current with the emitter open-circuit. Similarly from equation 2.104, $(1 - \alpha_N \alpha_I)I_{EBS} = I_{EBO}$, the emitter–base leakage current with the collector open-circuit. Thus these equations, which were developed by J.J. Ebers and J.L. Moll [21] may be written

$$I_C = \alpha_N I_E - I_{CBO}\left[\exp\left(\frac{eV_{BC}}{kT}\right) - 1\right] \quad (2.105)$$

$$I_E = \alpha_I I_C + I_{EBO}\left[\exp\left(\frac{eV_{BE}}{kT}\right) - 1\right] \quad (2.106)$$

The basic model of figure 2.60a can be extended to include the diffusion and depletion (transition) capacitances C_D and C_T, the ohmic base resistance $r_{bb'}$ and base width modulation. For normal active region operation, the collector current increases (α_N increases) with collector–base reverse bias (section 2.2.6). This effect can be included in the model by a resistance r_c equal to $\Delta V_{CB}/\Delta I_C$. The corresponding effect for inverse operation is represented by r_e given by $\Delta V_{EB}/\Delta I_E$.

Due to the complexity of the terminal current equations (2.105 and 2.106), circuit analysis using the model is cumbersome if calculation is by hand. However, the use of digital computation removes the calculation difficulties introduced by the exponential terms and the model has become a popular method of representing BJT performance in computer-aided circuit analysis programs.

(2) *Charge-control BJT Model.* The charge-control model is a mathematical representation of the terminal currents of the BJT in terms of the charge stored in the base and the transit time of charge across the base. Figure 2.61a represents the distribution of minority charge in the base region of a BJT for normal active region operation (sections 2.2.7 and 2.2.8). The collector current is dependent on the rate of diffusion of minority carriers across the base and is thus related to the minority carrier gradient (equation 2.18) and hence to the stored charge in the base q_B. The collector current and the stored charge are given by equations 2.20 and 2.23 from which the total collector current i_C may be written

$$i_C \approx \frac{2D_n}{w_B^2}\left(\frac{e\Delta n(0)w_B A}{2}\right) = \frac{|q_B|}{w_B^2/2D_n} \quad (2.107)$$

where w_B is the effective base width (section 2.2.6). The term $w_B^2/2D_n$ is the average (transit) time τ_T taken by electrons to cross the base. Under static conditions (constant junction biases) the collector current is thus given by

$$I_C \approx \frac{Q_B}{\tau_T} \quad (2.108)$$

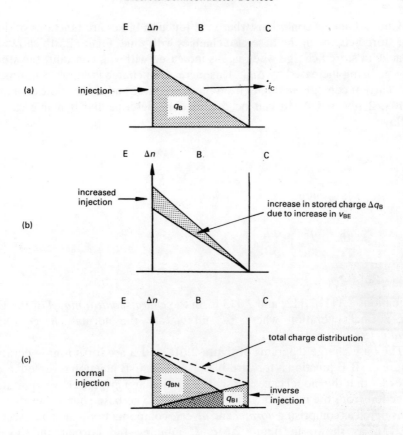

Figure 2.61 Stored base charge in a BJT

where Q_B is the magnitude of the static base charge.

During the diffusion of minority carriers across the base, some are lost due to recombination giving rise to a component of the base current. Thus in a first-order representation that ignores other components, the static base current is

$$I_B \approx \frac{Q_B}{\tau_B} \qquad (2.109)$$

where τ_B is the minority carrier lifetime (section 2.2.10) in the base. Combining equations 2.108 and 2.109

$$I_E \approx \frac{Q_B}{\tau_T} + \frac{Q_B}{\tau_B} \qquad (2.110)$$

Under dynamic conditions where the junction biases are functions of time, the stored charge in the base also changes with time. Figure 2.61b shows the change in stored charge when v_{BE} is increased with v_{CB} constant, the stored charge being increased by Δq_B. This increase in charge is supplied in time Δt as a current component dq_B/dt from the bias source and so the static equations 2.108, 2.109 and 2.110 can be modified to describe the dynamic case as follows

$$i_C \approx \frac{q_B}{\tau_T} \tag{2.111}$$

$$i_B \approx \frac{q_B}{\tau_B} + \frac{dq_B}{dt} \tag{2.112}$$

$$i_E \approx \frac{q_B}{\tau_T} + \frac{q_B}{\tau_B} + \frac{dq_B}{dt} \tag{2.113}$$

Equations 2.111, 2.112, and 2.113 form the *charge-control model* of the BJT for normal operation where τ_T and τ_B are the normal *charge-control parameters* of the device.

The base charge diagrams of figure 2.61a and b are for normal operation where the B-E junction is forward-biased and the C-B junction reverse-biased. In order that the model can represent saturation and inverse active operation, injection from the collector must be included. The base charge can then be considered as comprising *normal* and *inverse* components q_{BN} and q_{BI} (section 2.2.13), as shown in figure 2.61c. Combining the normal and inverse components, the *general charge-control model* of BJT performance becomes

$$i_B = i_{BN} + i_{BI} \tag{2.114}$$

$$= \frac{q_{BN}}{\tau_{BN}} + \frac{dq_{BN}}{dt} + \frac{q_{BI}}{\tau_{BI}} + \frac{dq_{BI}}{dt}$$

$$i_C = i_{CN} - i_{CI}$$

$$= \left(\frac{q_{BN}}{\tau_{TN}}\right) - \left(\frac{q_{BI}}{\tau_{TI}} + \frac{q_{BI}}{\tau_{BI}} + \frac{dq_{BI}}{dt}\right) \tag{2.115}$$

$$i_E = i_{EN} - i_{EI}$$

$$= \left(\frac{q_{BN}}{\tau_{TN}} + \frac{q_{BN}}{\tau_{BN}} + \frac{dq_{BN}}{dt}\right) - \left(\frac{q_{BI}}{\tau_{TI}}\right) \tag{2.116}$$

where suffixes N and I refer to the normal and inverse components respectively and the directions of i_C and i_E are as for normal operation. τ_{TN}, τ_{BN} and τ_{TI}, τ_{BI} are the *normal* and *inverse charge-control parameters* of the BJT. The static common-emitter current gains for normal and inverse operation, β_N and β_I, are given by

$$\beta_N = \frac{I_{CN}}{I_{BN}} = \frac{q_{BN}/\tau_{TN}}{q_{BN}/\tau_{BN}} = \frac{\tau_{BN}}{\tau_{TN}} \tag{2.117}$$

and

$$\beta_I = \frac{I_{EI}}{I_{BI}} = \frac{q_{BI}/\tau_{TI}}{q_{BI}/\tau_{BI}} = \frac{\tau_{BI}}{\tau_{TI}} \tag{2.118}$$

If the derivative terms in equations 2.114, 2.115 and 2.116 are neglected, these equations then describe the general static operation of the BJT and are a restatement of the Ebers–Moll equations 2.100, 2.101 and 2.102. The normal component of static base current I_{BN} is given by q_{BN}/τ_{BN} in terms of the charge-control model and, ignoring leakage, by $I_E - I_C$ with $V_{BC} = 0$ in terms of the Ebers-Moll model.

Thus, combining equations 2.100 and 2.101

$$\frac{q_{BN}}{\tau_{BN}} = I_{EBS}(1 - \alpha_N)\left[\exp\left(\frac{eV_{BE}}{kT}\right) - 1\right] \tag{2.119}$$

Similarly I_{BI} is given by q_{BI}/τ_{BI} and $I_E - I_C$ with $V_{BE} = 0$, from which

$$\frac{q_{BI}}{\tau_{BI}} = I_{CBS}(1 - \alpha_I)\left[\exp\left(\frac{eV_{BC}}{kT}\right) - 1\right] \tag{2.120}$$

Use of these link equations between the charge-control and Ebers–Moll models together with parameter relationships 2.117 and 2.118 shows that the two models are equivalent for static conditions.

The charge-control model is particularly useful in analysis of the large-signal (switching) performance of devices. Figure 2.62 shows the variations of base current, base charge and collector current during a typical switching cycle. Application of a positive source voltage V_s causes base current I_B, which is limited by R_s, resulting in charging of the base. As the C-B junction is held under reverse bias by V_{CC}, inverse components are zero and the growth of base charge is described by equation 2.114 with $i_B = I_B$, $q_{BI} = 0$ and $q_{BN} = q_B$. As the operating point of the BJT moves through the active region ($t_0 \rightarrow t_2$), the collector current rises until at t_2 saturation occurs and i_C is limited to the saturation value V_{CC}/R_L. The base charge continues to increase while i_B is positive, tending to the steady-state value $I_B\tau_{BN}$ (equation 2.114). At time t_3 the base drive is reversed and charge is removed from the base region resulting in a negative base current although the collector current cannot fall until the

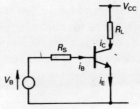
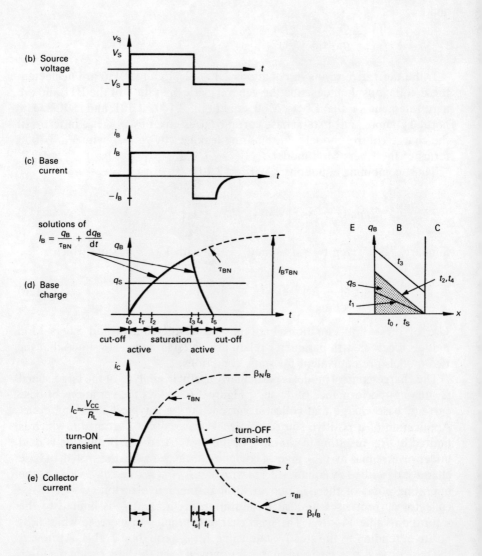

Figure 2.62 The BJT as a switch

BJT comes out of saturation (t_4). As the operating point moves back through the active region towards cut-off, the collector current falls until the base has been completely discharged (t_5). The performance of the device in switching the current flow through the load R_L is described in terms of the *rise time* (t_r), *storage time* (t_s) and *fall time* (t_f) of i_C (figure 2.62c).

As an example of the application of the charge-control model, consider investigation of the turn-ON transient to determine the rise time of the collector current. The turn-ON transient is described by

$$I_B = \frac{q_B}{\tau_{BN}} + \frac{dq_B}{dt}$$

Hence during this interval

$$q_B = I_B \tau_{BN} \left[1 - \exp\left(-\frac{t}{\tau_{BN}}\right) \right] \quad (2.121)$$

From equation 2.115, as the inverse component i_{CI} is zero in this case

$$i_C = \frac{q_B}{\tau_{TN}} = \frac{I_B \tau_{BN}}{\tau_{TN}} \left[1 - \exp\left(-\frac{t}{\tau_{BN}}\right) \right] \quad (2.122)$$

The rise time t_r in this simplified example is defined as the time taken for i_C to increase from zero to the saturation value I_C, thus from equation 2.122

$$I_C = \frac{I_B \tau_{BN}}{\tau_{TN}} \left[1 - \exp\left(-\frac{t_r}{\tau_{BN}}\right) \right]$$

From which

$$\begin{aligned} t_r &= \tau_{BN} \ln\left[\frac{1}{1 - \dfrac{I_C \tau_{TN}}{I_B \tau_{BN}}} \right] \\ &= \tau_{BN} \ln\left[\frac{1}{1 - \dfrac{I_C}{\beta_N I_B}} \right] \end{aligned} \quad (2.123)$$

using equation 2.117. Similar considerations allow description of the turn-OFF transient.

This simplified description of the switching performance of the BJT illustrates the use of the charge-control model. More detailed discussions together with explanation of the determination of the charge-control parameters for a BJT can be found in references 22 and 23.

(3) *Early Small-signal Physical BJT Model*. The most basic physical model of the small-signal performance of a BJT is that devised by J. Early (figure 2.63). As with other small-signal representations, this model applies to

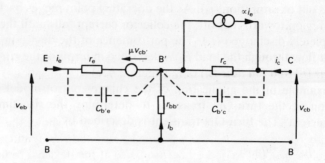

Figure 2.63 The Early physical model of small-signal BJT performance

small variations about a quiescent point in the active region which is fixed by external biasing. Unlike terminal models, however, performance is described in terms of the internal operation of the device.

Charge movement within the base region is not uniform but as a first-order approximation a hypothetical point B' is defined where the emitter, base and collector signal currents combine. The base resistance $r_{bb'}$ then describes the bulk resistance of the base region taking account of the non-uniformity of the current flow.

The emitter resistance r_e is the dynamic resistance of the forward-biased E-B junction and relates the change of injection into the base to the change of voltage across the junction. Using the theoretical I-V relationship for the junction, r_e can be related to the static emitter current (equation 2.74).

Due to base-width modulation, the emitter signal current is a function of the voltage across the collector–base junction (figure 2.30b). This internal feedback is represented in the Early model by the voltage source $\mu v_{cb'}$, where μ is a voltage feedback factor describing the degree of the effect. In operation, the signal at the collector causes the C-B voltage to vary which in turn modulates the effective width of the base. If the C-B voltage increases, the C-B depletion layer width increases, reducing the effective base width and increasing the emitter current. This effect is included in the model as the fedback voltage $\mu v_{cb'}$ augments the applied E-B voltage thus increasing the emitter current.

The portion of emitter signal current which flows into the collector is represented by the current source αi_e, the loss of carriers by recombination and injection from base to emitter being described by the small-signal common-base current gain α ($= -h_{fb}$, equation 2.35). For constant emitter current, the collector current is a function of C-B voltage as shown in the common-base output and transfer characteristics of figures 2.31b and 2.32b. This effect is included in the collector branch of the model by the collector resistance r_c. The gain term α is defined for constant C-B voltage (equation

2.35) and the resistance r_c takes account of the slight variation of gain due to base-width modulation causing a finite collector conductance.

As the frequency of operation is increased, the diffusion capacitance of the forward-biased E-B junction $C_{b'e}$ and the depletion capacitance of the reverse-biased C-B junction $C_{b'c}$ become progressively more significant and inclusion of these effects produces a high-frequency model.

Comparison of the low-frequency Early model (figure 2.63 omitting $C_{b'e}$ and $C_{b'c}$) with the CB h-parameter model (figure 2.56a) shows the degree of correlation between r_e and h_{ib}, $\mu v_{cb'}$ and $h_{rb}v_{cb}$, αi_e and $-h_{fb}i_e$ and r_c and h_{ob}. The h-parameter model, based on performance at the device terminals, does not attempt to represent the voltage drop within the bulk material of the base and so a base resistance equivalent to $r_{bb'}$ is not included.

The following values are typical for the Early parameters of a low-power silicon planar epitaxial BJT: $r_{bb'} = 50\text{--}500\ \Omega$, $r_e = 5\text{--}150\ \Omega$, $r_c = 200\ \text{k}\Omega\text{--}10\ \text{M}\Omega$, $\mu \approx 10^{-4}$, $\alpha \approx 0.995$, $C_{b'e} = 10\text{--}500\ \text{pF}$, $C_{b'c} = 1\text{--}10\ \text{pF}$. The ohmic resistances of the emitter and collector neutral regions are negligible compared with r_e and r_c respectively.

The operating frequency above which the high-frequency model must be used cannot be specified directly since it depends on the parameters of the particular BJT at the chosen operating point and the degree of accuracy required. Clearly $C_{b'e}$ and $C_{b'c}$ must be included when they have a significant effect in the model, particularly in comparison with r_e and r_c.

(4) *Hybrid-π Small-signal Physical BJT Model.* The Early model discussed above relates directly to the physical operation of the BJT but it is inconvenient for BJT circuit analysis as it contains two dependent sources $\mu v_{cb'}$ and αi_1 (figure 2.64a). Manipulation of the high-frequency Early model yields a single generator model known as the *hybrid-π model* which is particularly useful as its parameters are independent of frequency over a wide range.

The first stage in the manipulation is the application of network theory to replace the feedback voltage source $\mu v_{cb'}$. This results in a C-E feedback resistance r_e/μ and a C-B' resistance $r_e/\mu(1-\alpha)$ as shown in figure 2.64b. The current i_1 is given by

$$i_1 = -\frac{v_{b'e}}{r_e} \tag{2.124}$$

and so the current source becomes

$$-\alpha i_1 = -\alpha\left(-\frac{v_{b'e}}{r_e}\right) = g_m v_{b'e} \tag{2.125}$$

where the parameter $g_m = \alpha/r_e$ is termed the *mutual conductance* or *transconductance* of the BJT and defines the LF relationship between output current and input voltage. Using the relationship for r_e (equation 2.74) derived from the theoretical I–V relationship for the forward-biased B-E junction, g_m may be written

$$g_m = \frac{\alpha}{r_e} \approx \frac{e\alpha I_E}{kT} = \frac{eI_C}{kT} \approx 40 I_C \text{ at } T = 25\ °C \tag{2.126}$$

Equation 2.126 shows that theoretically g_m is directly related to the *static* collector current having values of 40 mS at $I_C = 1$ mA and 200 mS at 5 mA for example. Since the actual B-E I–V relationship deviates from the theoretical exponential form, however, equation 2.126 is only approximate becoming less realistic at higher current levels.

The most popular connection configuration for BJTs is common-emitter and the hybrid-π model is normally presented in this form. Figure 2.64c shows the CB model of figure 2.64b rearranged into CE form. In order to retain the $g_m v_{b'e}$ source conveniently at the output, the source in figure 2.64c between B' and C is replaced by two sources between B' and E and C and E. Inspection of figure 2.64d shows that a current $g_m v_{b'e}$ still flows away from node C and into node B'. The current source between B' and E has a voltage $v_{b'e}$ applied across it and so this source behaves as a resistance of value $-r_e/\alpha$ which, when combined with r_e, provides the resistance $r_{b'e} = r_e/(1-\alpha)$ (figure 2.64d). The final form of the high-frequency CE hybrid-π BJT model is given

Table 2.10 Approximate hybrid-π parameter relationships and values

	In terms of Early model parameters	In terms of h-parameters	Typical value†
$r_{b'e}$	$\dfrac{r_e}{1-\alpha}$	$\dfrac{h_{fe}}{g_m}$	2 – 10 kΩ
$r_{bb'}$		$h_{ie} - r_{b'e}$	50–500 Ω
$r_{b'c}$	$\dfrac{r_e}{\mu(1-\alpha)} = \dfrac{r_{b'e}}{\mu}$	$\dfrac{r_{b'e}}{h_{re}} \approx \dfrac{h_{fe}}{h_{re}g_m}$	10–20 MΩ
r_{ce}	$\dfrac{r_e}{\mu}$	$\dfrac{1}{h_{oe}}$	20–100 kΩ
g_m		$\dfrac{eI_C}{kT} \approx 40 I_C$ at T = 25 °C	40 mS
$C_{b'c}$		C_{T_C}	1–10 pF
$C_{b'e}$		$\dfrac{g_m}{2\pi f_T} - C_{b'c}$	10–500 pF

†For a low-power silicon BJT at $I_C = 1$ mA, $V_{CE} = 5$ V, T = 25 °C

in figure 2.64e, the resistances $r_e/\mu(1-\alpha)$ and r_e/μ being labelled $r_{b'c}$ and r_{ce} respectively. The name of the model derives from the π topology formed by the B'-E, B'-C and C-E branches.

It has been found [24] that the parameters of the hybrid-π model may be considered to be independent of frequency up to approximately one third the transition frequency of the BJT. For a modern general-purpose low-power silicon BJT, f_T is typically > 100 MHz at $I_C = 1$ mA (figure 2.51c) in which case the frequency independent properties of the model parameters can be assumed at least up to operating frequencies of 30 MHz, making the model popular for BJT circuit analysis at moderately high frequencies.

The major disadvantage of the model, as with all physical models, is that many of the parameters cannot be measured directly due to the inaccessibility of the hypothetical point B'. For this reason the parameters have to be derived; they are rarely given in manufacturers' device data. The model resistances can be derived either from the expressions derived above in terms of r_e, μ and α from the Early model (table 2.10) or more conveniently by comparison of the low-frequency hybrid-π model (figure 2.65a) with the corresponding h-parameter model (figure 2.65b) for which low-frequency parameters are normally provided by manufacturers.

As $r_{b'c}$ is high

$$g_m v_{b'e} \approx h_{fe} i_b \tag{2.127}$$

but

$$v_{b'e} \approx r_{b'e} i_b \text{ as } r_{b'c} \text{ is high}$$

thus

$$r_{b'e} \approx \frac{h_{fe}}{g_m} = \frac{r_e}{1-\alpha} \tag{2.128}$$

as $g_m = \alpha/r_e$ (equation 2.125) and $h_{fe} = \beta = \alpha/(1-\alpha)$.

Internal feedback is described by $r_{b'c}$ and h_{re} in the two models. Considering the potential divider effect of $r_{b'e}$ and $r_{b'c}$, the fedback voltage is

$$v_{ce} \frac{r_{b'c}}{r_{b'e} + r_{b'c}} \approx h_{re} v_{ce} \tag{2.129}$$

from which

$$r_{b'c} \approx \frac{r_{b'e}}{h_{re}} \approx \frac{h_{fe}}{h_{re} g_m} \text{ or } \frac{r_e}{\mu(1-\alpha)} \tag{2.130}$$

using equation 2.128 and noting that $h_{re} \approx \mu \ll 1$.

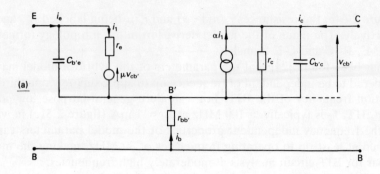

(a)

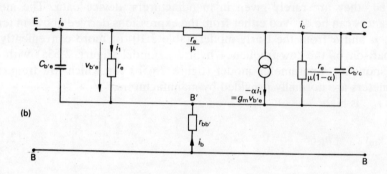

(b)

Figure 2.64 Development of the high-frequency CE hybrid-π BJT model (continued opposite)

Again since $r_{b'c}$ is high

$$r_{ce} \approx \frac{1}{h_{oe}} \qquad (2.131)$$

and from the input loop

$$r_{bb'} \approx h_{ie} - r_{b'e} \qquad (2.132)$$

Equation 2.132 is unreliable since h_{ie} and $r_{b'e}$ are of similar order. In practice, however, $r_{bb'}$ is usually of little importance due to its small value in comparison with $r_{b'e}$.

Capacitance $C_{b'c}$ is the B-C depletion (transition) layer capacitance and is normally quoted by the manufacturer as C_{T_c}. The capacitance $C_{b'e}$ is not normally provided in device data and has to be determined from knowledge

Discrete Semiconductor Devices

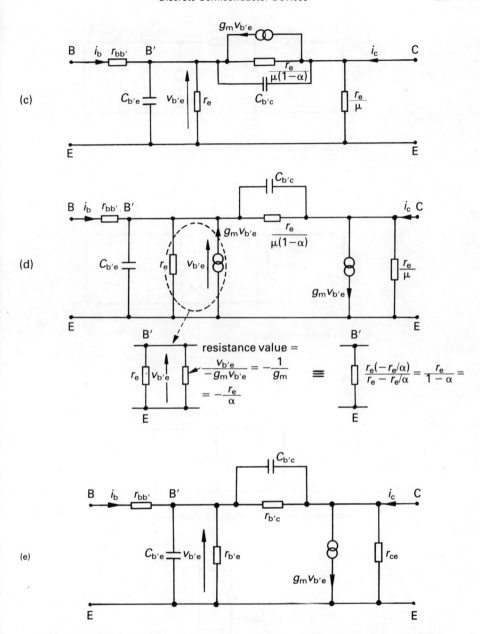

of the transition frequency f_T at the particular operating current. Equation 2.75 gives an expression for $C_{b'e}$ but it is in terms of device parameters w_B or τ_T that are not readily obtainable. By definition the small-signal CE current gain with the output an a.c. short-circuit is unity at the transition frequency. For high-frequency operation with the output short-circuit, the hybrid-π

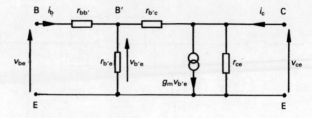

(a) Low-frequency CE hybrid – π BJT model

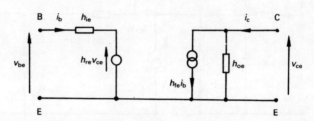

(b) CE h-parameter BJT model

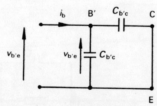

(c) Approximate HF input model with C-E short-circuit

(d) Approximate output model with C-E short-circuit

Figure 2.65 Derivation of hybrid-π model parameters

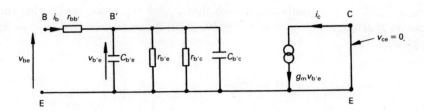

Figure 2.66 Hybrid-π BJT model for C-E a.c. short-circuit

input model reduces to that of figure 2.65c, the relatively high values of $r_{b'c}$ and $r_{b'e}$ being insignificant. Thus

$$i_b = j\omega v_{b'e}(C_{b'e} + C_{b'c}) \qquad (2.133)$$

and from the corresponding output model of figure 2.65d

$$i_c = g_m v_{b'e} \qquad (2.134)$$

The modulus of the CE current gain with the output short-circuit is therefore

$$\left|\frac{i_c}{i_b}\right| = \frac{g_m}{2\pi f(C_{b'e} + C_{b'c})} \qquad (2.135)$$

since $\omega = 2\pi f$. At the frequency f_T this gain is theoretically unity and so

$$C_{b'e} = \frac{g_m}{2\pi f_T} - C_{b'c} \qquad (2.136)$$

$C_{b'e}$ takes account of the time taken for injected carriers to cross the base (section 2.2.24) and thus describes the change of gain and phase shift with frequency.

Table 2.10 lists the expressions that can be used to determine the hybrid-π parameters together with typical values for a low-power silicon BJT. In general the parameters are a function of I_C, V_{CE} and T (reference 24).

The high-frequency hybrid-π model can be used to describe conveniently the variation of BJT current gain with frequency. The small-signal common-emitter current gain of a BJT, β or h_{fe}, is defined as i_c/i_b with $v_{ce} = 0$ (equation 2.57). With $v_{ce} = 0$ (C-E a.c. short-circuit) the hybrid-π model is shown in figure 2.66. As a first-order approximation that is certainly valid up to and beyond the cut-off frequency, f_β, $r_{bb'}$ may be neglected in comparison with the

$r_{b'e}$, $r_{b'c}$, $C_{b'e}$, $C_{b'c}$ combination such that $v_{b'e} \approx v_{be}$. Also, comparing the relative magnitudes of $r_{b'c}$ and $r_{b'e}$ (table 2.10) it can be seen that $r_{b'c}$ can be neglected, thus

$$i_b \approx v_{be} \left[\frac{1}{r_{b'e}} + j\omega (C_{b'e} + C_{b'c}) \right] \tag{2.137}$$

As $i_c = g_m v_{b'e} \approx g_m v_{be}$, the small-signal CE current gain of the BJT is

$$\beta = h_{fe} = \frac{g_m}{\frac{1}{r_{b'e}} + j\omega (C_{b'e} + C_{b'c})}$$

$$= \frac{g_m r_{b'e}}{1 + j\omega r_{b'e} (C_{b'e} + C_{b'c})} \tag{2.138}$$

At low frequency, capacitive effects within the device are negligible and the *low-frequency small-signal CE current gain* of the BJT h_{feo} or β_o is equal to $g_m r_{b'e}$ as the term $\omega r_{b'e} (C_{b'e} + C_{b'c})$ is negligible ($\omega \to 0$). Thus equation 2.138 may be written

$$h_{fe} = \frac{h_{feo}}{1 + j\omega r_{b'e} (C_{b'e} + C_{b'c})} \tag{2.139}$$

The high frequency performance of the BJT is described by the *CE* (h_{fe} or β) *cut-off frequency* $f_{h_{fe}}$ or f_β which is the frequency at which $|h_{fe}|$ has reduced to $h_{feo}/\sqrt{2}$. From equation 2.139 the cut-off frequency is therefore given by

$$2\pi f_{h_{fe}} r_{b'e} (C_{b'e} + C_{b'c}) = 1$$

or

$$f_{h_{fe}} = f_\beta = \frac{1}{2\pi r_{b'e} (C_{b'e} + C_{b'c})} \tag{2.140}$$

The general gain equation (2.139) can then be written

$$h_{fe} = \frac{h_{feo}}{1 + j\frac{f}{f_{h_{fe}}}}$$

or

$$\beta = \frac{\beta_o}{1 + j\frac{f}{f_\beta}}$$

$$\tag{2.141}$$

where $h_{fe} = \beta$ = small-signal CE current gain at frequency f and $h_{feo} = \beta_o$ = small-signal CE current gain at low frequency.

Equation 2.141 is clearly complex, describing the magnitude and phase relationship between the collector and base signal currents i_c and i_b, and may be expressed in the form

$$h_{fe} = |h_{fe}| \angle h_{fe}$$

$$= \frac{h_{feo}}{\sqrt{\left[1 + \left(\frac{f}{f_{h_{fe}}}\right)^2\right]}} \angle -\tan^{-1}\frac{f}{f_{h_{fe}}} \qquad (2.142)$$

The variation of the magnitude of the gain $|h_{fe}|$ is illustrated in figure 2.51b. The phase term varies from zero at low frequency ($f \ll f_{h_{fe}}$) to $-\pi/2$ at high frequency ($f \gg f_{h_{fe}}$) indicating that the variation of collector current is in phase with the variation of base current at low frequency but lags by a quarter cycle at very high frequency.

Using the relationship between CB and CE small-signal current gains $h_{fb} = -h_{fe}/(1 + h_{fe})$ or $\alpha = \beta/(1 + \beta)$ (table 2.7) and substituting for h_{fe} or β from equation 2.141 gives corresponding expressions for the variation of the small-signal common-base current gain of a BJT with frequency. Thus

$$h_{fb} = \frac{-h_{feo}/(1 + jf/f_{h_{fe}})}{1 + h_{feo}/(1 + jf/f_{h_{fe}})} = \frac{-h_{feo}}{1 + h_{feo} + jf/f_{h_{fe}}}$$

$$= \frac{-h_{feo}}{1 + h_{feo}} \left[\frac{1}{1 + j\frac{f}{(1+h_{feo})f_{h_{fe}}}} \right]$$

$$= \frac{h_{fbo}}{1 + j\, f/f_{h_{fb}}} \qquad (2.143)$$

or

$$\alpha = \frac{\alpha_o}{1 + j\, f/f_\alpha}$$

where $h_{fb} = -\alpha$ = small-signal CB current gain at frequency f, $h_{fbo} = -\alpha_o$ = small-signal CB current gain at low frequency and $f_{h_{fb}} = f_\alpha$ = CB (h_{fb} or α) cut-off frequency of the BJT at which $|h_{fb}| = h_{fbo}/\sqrt{2}$. Note from equation 2.143

$$\left. \begin{array}{l} f_{h_{fb}} = (1 + h_{feo}) f_{h_{fe}} \\ \text{or} \\ f_\alpha = (1 + \beta_o) f_\beta \end{array} \right\} \quad (2.144)$$

As $1 + h_{feo} = 1/(1 + h_{fbo})$ or $1 + \beta_o = 1/(1 - \alpha_o)$, from the expression for h_{fb} in terms of h_{fe} (table 2.7)

$$\left. \begin{array}{l} f_{h_{fe}} = (1 + h_{fbo}) f_{h_{fb}} \\ \text{or} \\ f_\beta = (1 - \alpha_o) f_\alpha \end{array} \right\} \quad (2.145)$$

A figure of merit for a BJT is its gain-bandwidth product (at LF) $h_{feo} f_{h_{fe}}$ or $\beta_o f_\beta$. But $\beta_o = \alpha_o/(1 - \alpha_o)$ and $f_\beta = (1 - \alpha_o) f_\alpha$, therefore

$$\beta_o f_\beta = \frac{\alpha_o}{1 - \alpha_o} (1 - \alpha_o) f_\alpha = \alpha_o f_\alpha \quad (2.146)$$

so that the CE and CB gain-bandwidth products are numerically the same.

The *transition frequency* f_T of a BJT is the theoretical frequency at which the magnitude of the small-signal CE current gain would be unity if the -20 dB/decade high-frequency roll-off were maintained (figure 2.51b). Thus from equation 2.142

$$\frac{h_{feo}}{\sqrt{\left[1 + \left(\dfrac{f_T}{f_{h_{fe}}}\right)^2\right]}} = 1$$

and as

$$f_T \gg f_{h_{fe}}$$
$$f_T \approx h_{feo} f_{h_{fe}} = \beta_o f_\beta \quad (2.147)$$

Combining equations 2.146 and 2.147

$$f_T \approx \beta_o f_\beta = \alpha_o f_\alpha \approx f_\alpha \text{ as } \alpha_o \approx 1 \quad (2.148)$$

The transition frequency is thus also termed the *gain-bandwidth product* of the BJT.

A useful simplification is to isolate the input and output loops of the hybrid-π model. Figure 2.67a gives an approximate HF representation of the

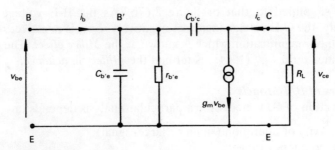

(a) Approximate HF representation of loaded BJT using the hybrid – π model

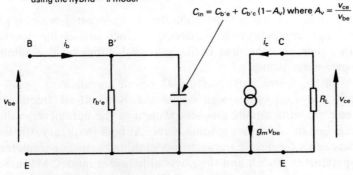

(b) Simplified model of loaded BJT

Figure 2.67 Miller capacitance

BJT with its external load R_L using the hybrid-π model. Resistance $r_{b'c}$ is omitted due to its high value in comparison with the effect of $C_{b'c}$, and r_{ce} is neglected in comparison with typical values of R_L of a few kilohms. In addition, except at very high frequency, $r_{bb'}$ has only a minor effect in comparison with the $r_{b'e}$, $C_{b'e}$ combination and can therefore be omitted, whereby $v_{b'e} \approx v_{be}$. Thus

$$i_b = \frac{v_{be}}{r_{b'e}} + j\omega C_{b'e} v_{be} + j\omega C_{b'c}(v_{be} - v_{ce}) \qquad (2.149)$$

But v_{ce}/v_{be} is the voltage gain A_v of the arrangement and so

$$\begin{aligned} i_b &= \frac{v_{be}}{r_{b'e}} + j\omega [C_{b'e} + C_{b'c}(1 - A_v)]v_{be} \\ &= \frac{v_{be}}{r_{b'e}} + j\omega C_{in} v_{be} \end{aligned} \qquad (2.150)$$

where C_{in} is the effective input capacitance. From equation 2.150 the model can be simplified to that of figure 2.67b. As the B-E voltage increases positively, the C-E voltage increases negatively so A_v is negative, resulting in capacitance magnification which is known as the *Miller effect*. The magnified capacitance term $C_{b'c}(1-A_v)$ is termed the *Miller capacitance*.

Summary of BJT Models

The selection of BJT model for a particular analysis depends on

(1) type of operation (small or large-signal);
(2) frequency range;
(3) availability of model parameter values.

For large-signal (switching) operation the charge-control model is popular although many computer-aided analysis routines utilise the Ebers–Moll model. In either case the user is likely to find difficulty in obtaining the relevant parameter values.

In the case of small-signal (linear) operation such as small-signal amplification at frequencies well below the BJT cut-off frequency, the *h*-parameter model is usually most convenient as the parameter, values are usually supplied by the device manufacturer. At high frequency, the user can choose between a terminal parameter model using complex parameters such as the *y*-parameter model, and the physical hybrid-π model. Manufacturers often supply *y*-parameter details for BJTs intended for operation at high frequency such as in the UHF range. Application can be cumbersome, however, unless computer assistance is available. For hand calculation the hybrid-π model is popular as the parameter values can be assumed to be constant up to about $f_T/3$.

It should be noted that a high-frequency terminal parameter model with complex parameters of the form $p + jq$ is equivalent to a high-frequency physical model which is essentially the low-frequency model comprising mathematically real parameters with the frequency dependent properties included in the form of model capacitances.

2.3 FIELD-EFFECT TRANSISTORS (FETs)

A field-effect transistor is essentially a semiconductor resistor, the value of resistance being controlled by the potential applied to a control electrode. The conducting region is known as the *channel* which may be either *p* or *n*-type. Terminals at either end of the channel are referred to as the *source* (*S*) and *drain* (*D*). Application of a drain–source voltage V_{DS} causes a drift of charge carriers along the channel resulting in a drain current I_D. Since the current is due almost entirely to the flow of one type of carrier, the majority carrier, (electrons in *n*-channel FETs and holes in *p*-channel FETs) these devices are described as *unipolar* transistors or majority-carrier devices.

Channel conductivity is altered by varying the number of charge carriers able to contribute to conduction along the channel. The potential applied to the control electrode, known as the *gate* (G), creates an electric field which alters the number of free carriers in the channel, hence the term *field-effect*. This control may be created in two ways resulting in two distinct types of FET

(1) *Junction FET* (JFET) The cross-sectional area of the conducting channel is controlled by the depth of penetration of the depletion region formed at the gate–channel *pn* junction by application of a reverse bias.
(2) *Insulated gate FET* (IGFET) The controlling electric field is induced capacitively across a thin insulating layer between the gate electrode and the surface of the channel. The most common IGFET structure comprises a metal gate electrode which is insulated from the semiconductor channel by a layer of silicon dioxide resulting in a metal–oxide–semiconductor structure referred to as a MOSFET or MOST. There are two types of IGFET

(1) *Enchancement-type* that does not conduct for $V_{GS} = 0$. Channel conductivity is increased (enhanced) by application of V_{GS}.
(2) *Depletion-type* that conducts for $V_{GS} = 0$ and channel conductivity may be increased or decreased by V_{GS}.

Circuit symbols for the various types of FETs are shown in figure 2.68. The gate is shown in line with the source terminal to avoid confusion in the case of an unlabelled symbol although a symmetrical symbol is also used with the gate midway between source and drain. Channel type is shown by the direction of the gate arrow in the case of a JFET and by the substrate arrow for an IGFET. For a JFET, the gate arrow shows the conventional direction of gate current if the gate–channel junction were to be forward-biased, a condition which does not occur in normal operation. A similar representation is used for the IGFET with reference to the substrate–channel junction.

Non-conduction of the channel for $V_{GS} = 0$ (normally OFF) is indicated by the dashed channel symbol in the case of an enhancement-type IGFET (figure 2.68b) while the normally ON (conduction for $V_{GS} = 0$) property of the depletion-type IGFET is represented by a full line (figure 2.68c).

The IGFET symbols of figure 2.68b and c indicate four terminal devices, but it is normal for the substrate to be connected to the source during manufacture (figure 2.68d). Multigate IGFETs are also produced (figure 2.68e), the drain current being a function of the separate gate signals. JFETs, depletion-type MOSTS and enhancement-type MOSTS are designated type A, B, and C FETs respectively by the U.S. standards organisation JEDEC.

Since electrons have higher mobility than holes, *n*-channel FETs have a better frequency response and switching speed than corresponding *p*-channel types and therefore have become more popular. For this reason subsequent consideration of structure and performance refers to *n*-channel versions but,

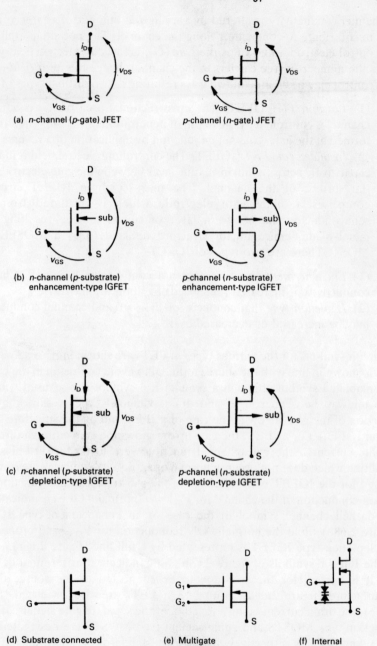

Figure 2.68 FET circuit symbols

if the roles of electrons and holes are exchanged and voltage polarities and current directions are reversed, the work applies equally to *p*-channel FETs.

2.3.1 Junction Field-effect Transistors (JFETs)

Structure

Figure 2.69 shows the essential features of a low-power lateral *n*-channel JFET. The term *lateral* refers to the direction of current flow in the channel relative to the chip surface and is in contrast to the vertical flow in a V-JFET used at higher power levels (section 2.3.7).

An *n*-type epitaxial layer typically 5 μm thick is grown on a 100 μm thick *p*-type substrate and a 4–4.5 μm deep p^+ diffusion into the surface forms the gate region, leaving the channel region only 0.5–1 μm thick. Shallow n^+ diffusions form source and drain contact regions at either end of the channel and surface metallisation completes the structure. JFETs are packaged using plastic encapsulation or hermetically sealed metal cans having styles similar to the low-power BJT packages of figure 2.10.

Operation

For simplicity, discussion of operation is based on the uniform channel region shown dashed in figure 2.69. As the gate region is heavily doped relative to the channel, the depletion region formed at the gate–channel junction

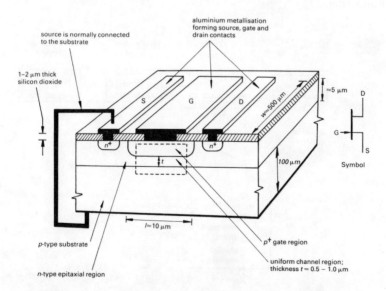

Figure 2.69 Structure of a low-power lateral *n*-channel JFET

extends mainly into the channel. Normally the substrate is not as heavily doped as the gate region and so the depletion region does not penetrate as far into the channel at the substrate–channel junction as at the gate–channel junction. When there is no current flow through the channel these depletion regions are uniform (figure 2.70a).

Effect of V_{DS}
With V_{GS} constant (for example, $V_{GS} = 0$), an increase in V_{DS} causes current flow I_D through the resistive channel and an approximately linear drop in potential along the channel results. The bias across the gate–channel and substrate–channel junctions therefore varies with position, being greatest at the drain end. Consequently the depletion layers become non-uniform, penetrating deeper into the channel at the drain end (figure 2.70b). Increase of I_D with V_{DS} is approximately linear for small values of V_{DS} (typically < 2 V). Progressive increase of V_{DS} eventually causes the depletion regions almost to meet (figure 2.70c) and further increase of I_D with V_{DS} is greatly reduced. This is described as the *pinch-off effect* and the value of V_{DS} at the start of the pinch-off region defines the pinch-off voltage V_P. For convenience, V_P is defined as $-V_{DS}$ at the start of the pinch-off region for $V_{GS} = 0$. The value of drain current at the pinch-off point for $V_{GS} = 0$, that is, for G-S short-circuit, is designated I_{DSS}. It should be noted that the depletion regions cannot actually meet since this would reduce I_D to zero. The cross-sectional area of the undepleted channel becomes very small and the lateral electric field, which maintains the high current density, increases preventing the depletion regions from meeting.

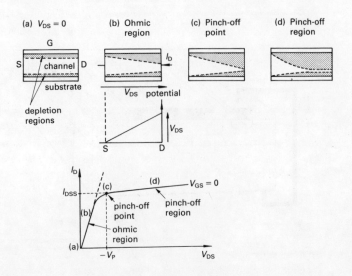

Figure 2.70 Stages of operation of a JFET for increasing V_{DS} with $V_{GS} = 0$

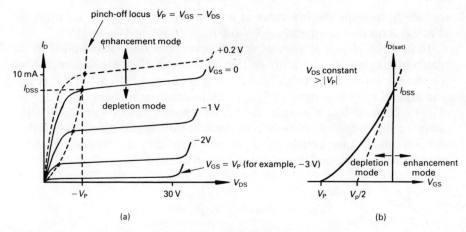

Figure 2.71 Output and transfer characteristics for an n-channel JFET

Effect of V_{GS}

As V_{GS} alters the width of the gate–channel depletion region, it follows that the onset of pinch-off depends on V_{GS} and that I_D is a function of V_{GS} as well as V_{DS}. An increase in the gate–channel reverse bias (negative V_{GS} for an n-channel JFET) causes a wider depletion region than for $V_{GS} = 0$ and pinch-off therefore occurs at a lower value of V_{DS}. Application of a negative V_{GS} depletes the channel of free carriers and this is therefore referred to as the *depletion mode* of operation. Positive V_{GS} increases (enhances) channel conductance by narrowing the gate–channel depletion region resulting in an increased drain current for the same V_{DS}, termed the *enhancement mode*. However, enhancement-mode operation entails forward biasing the gate–channel junction resulting in a greatly increased gate current and therefore it is not usual or advisable to operate a JFET in this region. Figure 2.71a shows typical output characteristics for an n-channel JFET.

If V_{GS} is increased sufficiently in the depletion mode the channel becomes pinched-off, whatever the value of V_{DS}, giving an alternative definition of pinch-off voltage as the value of V_{GS} to pinch off the channel when $V_{DS} = 0$. When neither V_{GS} nor V_{DS} is zero, both contribute to pinch-off and the locus of the pinch-off point (figure 2.71a) is described by the equation

$$V_P = V_{GS} - V_{DS} \qquad (2.151)$$

The definitions of V_P in terms of V_{DS} when $V_{GS} = 0$ and in terms of V_{GS} when $V_{DS} = 0$ are both difficult to use in practice and it is common to avoid the term pinch-off voltage and to define $V_{GS(off)}$ as the voltage required to reduce I_D to a specific low value at a particular value of V_{DS}, for example, $V_{GS(off)}$ is quoted as typically -3 V for $I_D = 10$ nA at $V_{DS} = 15$V.

In linear applications such as amplification, the JFET is operated in the pinch-off region (figure 2.70) so that changes in the gate–source voltage produce approximately proportional changes in drain current. In this region the drain current is designated $I_{D(sat)}$ and the relationship between $I_{D(sat)}$ and V_{GS} for constant V_{DS} is given by the transfer characteristic (figure 2.71b). The slope of the output characteristic in the pinch-off region gives the effective output conductance of the JFET for low frequency a.c. operation and is termed the small-signal common-source conductance (g_{os})

$$g_{os} = \frac{1}{r_{ds}} = \left.\frac{\Delta I_{D(sat)}}{\Delta V_{DS}}\right|_{V_{GS} \text{ constant}} = \left.\frac{i_d}{v_{ds}}\right|_{V_{gs}=0} \qquad (2.152)$$

where r_{ds} is the small-signal drain-source resistance or drain-slope resistance.

Change of output current $\Delta I_{D(sat)}$ with input voltage ΔV_{GS} is called the small-signal common-source forward transconductance g_{fs} (formally mutual conductance g_m) of the device

$$g_{fs} = g_m = \left.\frac{\Delta I_{D(sat)}}{\Delta V_{GS}}\right|_{V_{DS} \text{ constant}} = \left.\frac{i_d}{v_{gs}}\right|_{v_{ds}=0} \qquad (2.153)$$

The parameters g_{fs} and g_{os} are the real parts, that is, the effective low frequency values, of the common-source y-parameters y_{fs} and y_{os} which are described generally in appendix A.2 as y_{21} and y_{22}.

For normal depletion-mode operation, input to a JFET is via the reverse-biased gate–channel junction and so the static gate current I_G is very low, typically 2–20 nA at 25 °C, and $R_{GS} = V_{GS}/I_G$ is correspondingly very high, typically $> 10^8$ Ω. Therefore, as far as static and low-frequency operation is concerned, the gate–source can usually be regarded as an open-circuit.

If V_{DS} is increased sufficiently, breakdown of the gate–channel junction causes I_D to increase rapidly.

Theoretical Analysis of JFET Operation

Consider a uniform section of the channel region as shown in figure 2.72. From equation 1.137 the depletion layer width of a p^+n junction ($N_a \gg N_d$) is given by

$$d = \sqrt{\left[\frac{2\epsilon}{eN_d}(\psi - v)\right]} \qquad (2.154)$$

where V is the bias applied to the p region relative to the n. Due to the heavy

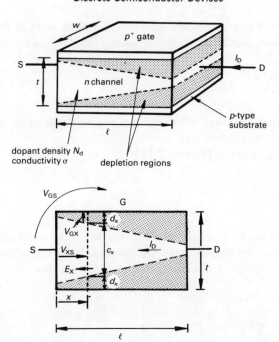

Figure 2.72 Analysis of the channel region of an n-channel JFET

doping of the p side, the depletion region lies almost entirely on the n side of the junction.

Figure 2.72 shows a first-order representation of the geometry of the depletion regions for small values of V_{DS} below pinch-off. It is assumed that the change of depletion layer width d with position x along the channel due to the voltage drop V_{XS} is linear and for simplicity the gate–channel and substrate–channel depletion regions are taken to be identical.

The bias across the depletion layer V varies from V_{GS} at the source end to $V_{GS} - V_{DS}$ at the drain end and, at a distance x along the channel from the source end, the bias $V_{GX} = V_{GS} - V_{XS}$.

If the bias is considerably greater than the junction contact potential ψ, equation 2.154 reduces to

$$d = \left(-\frac{2\epsilon V}{eN_d}\right)^{1/2} \tag{2.155}$$

and the depletion layer width at a position x along the channel

$$d_x = \left(-\frac{2\epsilon V_{GX}}{eN_d}\right)^{1/2} = \left[-\frac{2\epsilon(V_{GS}-V_{XS})}{eN_d}\right]^{1/2} \tag{2.156}$$

At the pinch-off point, the depletion layers almost meet at the drain end, that is, at $x = l$, $d_x \approx t/2$, and the bias voltage $V_{GS} - V_{DS} = V_P$, thus

$$\frac{t}{2} \approx \left(-\frac{2\epsilon V_P}{eN_d}\right)^{1/2} \tag{2.157}$$

where t is the total thickness of the channel region between the gate–channel and substrate–channel junctions.

Combining equations 2.156 and 2.157

$$d_x \approx \frac{t}{2}\left(\frac{V_{GS} - V_{XS}}{V_P}\right)^{1/2} \tag{2.158}$$

The drain current I_D is a drift of electrons due to a lateral electric field E which causes a current density J_D in the channel. If the lateral field is defined as being in the negative x direction, that is, the same direction as I_D (figure 2.72)

$$J_D = \frac{I_D}{A} = \sigma E \tag{2.159}$$

where A is the cross-sectional area of the channel and σ is the channel conductivity. Tapering of the conducting channel causes A, J_D and E to vary with x and, at a particular position x along the channel

$$I_D = \sigma A_x E_x \tag{2.160}$$

From figure 2.72, using equation 2.158 for d_x

$$\begin{aligned}A_x &= wc_x \\ &= w(t - 2d_x) \\ &\approx wt\left[1 - \left(\frac{V_{GS} - V_{XS}}{V_P}\right)^{1/2}\right]\end{aligned} \tag{2.161}$$

where w is the channel width.

The lateral electric field at x which causes $I_D(x)$ is

$$E_x = \frac{dV_{XS}}{dx} \tag{2.162}$$

using the definition of electric field (equation 1.37) and noting that E_x has been defined in the negative x direction (figure 2.72). Therefore from equation 2.160

$$I_D \approx \sigma wt \left[1 - \left(\frac{V_{GS} - V_{XS}}{V_P}\right)^{1/2}\right] \frac{dV_{XS}}{dx} \qquad (2.163)$$

At $x = 0$, $V_{XS} = 0$ and at $x = l$, $V_{XS} = V_{DS}$, thus

$$\int_0^l I_D \, dx \approx \sigma wt \int_0^{V_{DS}} \left[1 - \left(\frac{V_{GS} - V_{XS}}{V_P}\right)^{1/2}\right] dV_{XS}$$

from which,

$$I_D \approx \frac{\sigma wt}{l} \left[V_{XS} + \frac{2}{3} V_P \left(\frac{V_{GS} - V_{XS}}{V_P}\right)^{3/2}\right]_0^{V_{DS}}$$

$$= \frac{\sigma wt}{l} \left[V_{DS} + \frac{2}{3} V_P \left(\frac{V_{GS} - V_{DS}}{V_P}\right)^{3/2} - \frac{2}{3} V_P \left(\frac{V_{GS}}{V_P}\right)^{1/2}\right] \qquad (2.164)$$

Equation 2.164 gives I_D as a function of V_{GS} and V_{DS} in the ohmic region below pinch-off, that is, for values of V_{GS} and V_{DS} not sufficient to cause the depletion regions to meet at the drain end.

At pinch-off $V_{GS} - V_{DS} = V_P$, therefore from equation 2.164 the drain current at pinch-off $I_{D(sat)}$ is

$$I_{D(sat)} \approx \frac{\sigma wt}{l} \left[V_{DS} + \frac{2}{3} V_P - \frac{2}{3} V_P \left(\frac{V_{GS}}{V_P}\right)^{3/2}\right]$$

$$= \frac{\sigma wt}{l} V_P \left[\frac{V_{DS}}{V_P} + \frac{2}{3} - \frac{2}{3} \left(\frac{V_{GS}}{V_P}\right)^{3/2}\right]$$

But from equation 2.151, at pinch-off, $V_{DS}/V_P = (V_{GS}/V_P) - 1$ therefore

$$I_{D(sat)} \approx \frac{\sigma wt}{l} V_P \left[\frac{V_{GS}}{V_P} - \frac{1}{3} - \frac{2}{3} \left(\frac{V_{GS}}{V_P}\right)^{3/2}\right]$$

When $V_{GS} = 0$, $I_{D(sat)}$ is I_{DSS} and so

$$-\frac{\sigma wt}{3l} V_P = I_{DSS}$$

Note that the numerical values of V_P and I_{DSS} are of opposite sign (table 2.11). Thus

Table 2.11 Signs of numerical values of variables for n and p–channel FETs

	n–channel	p–channel
JFETs (JEDEC type A FETs)		
Drain–source voltage V_{DS}	+	−
Drain current I_D	+	−
Gate–source voltage V_{GS} (for normal depletion-mode operation)	−	+
Pinch-off voltage V_P or $V_{GS(off)}$	−[1]	+[1]
Depletion-type MOSTs (JEDEC type B FETs)		
Drain–source voltage V_{DS}	+	−
Drain current I_D	+	−
Gate–source voltage V_{GS}	−[2] or +[3]	+[2] or −[3]
Pinch-off voltage V_P or $V_{GS(off)}$	−[1]	+[1]
Enhancement-type MOSTs (JEDEC type C FETs)		
Drain–source voltage V_{DS}	+	−
Drain current I_D	+	−
Gate–source voltage V_{GS}	+[4]	−[4]
Threshold voltage V_T	+[5]	−[5]

Directions are as defined in figure 2.68.
1. Gate–source cut-off voltage
2. Depletion-mode operation (usual application)
3. Enhancement-mode operation
4. Enhancement-mode operation only
5. V_{GS} to establish channel (turn MOST ON)

$$I_{D(sat)} \approx I_{DSS}\left[1 - 3\frac{V_{GS}}{V_P} + 2\left(\frac{V_{GS}}{V_P}\right)^{3/2}\right] \tag{2.165}$$

Equation 2.165 gives the theoretical value of drain current at pinch-off based on the first-order approximation of conditions in a uniform channel as shown in figure 2.72.

In practice, JFETs have an asymmetrical structure and the relation between $I_{D(sat)}$ and V_{GS} (the transfer characteristic of figure 2.71b) is found to be approximately

$$I_{D(sat)} = I_{DSS}\left(1 - \frac{V_{GS}}{V_P}\right)^2 \tag{2.166}$$

Based on this practical relation, the small-signal common-source forward transconductance g_{fs} of a JFET (equation 2.153) is

$$g_{fs} = \frac{\partial I_{D(sat)}}{\partial V_{GS}} = I_{DSS}\frac{\partial}{\partial V_{GS}}\left(1 - \frac{V_{GS}}{V_P}\right)^2 = -\frac{2I_{DSS}}{V_P}\left(1 - \frac{V_{GS}}{V_P}\right)$$

$$= -\frac{2I_{DSS}}{V_P}\left(\frac{I_{D(sat)}}{I_{DSS}}\right)^{1/2} = -\frac{2}{V_P}(I_{D(sat)} I_{DSS})^{1/2} \qquad (2.167)$$

It should be noted that g_{fs} is a function of $I_{D(sat)}$ as shown by the non-linear nature of the transfer characteristic of figure 2.71b.

The data sheet for a typical low-power silicon planar epitaxial JFET (type 2N 5457) is given in appendix H.4.

Although the above consideration is for an n-channel JFET, the results apply equally to a p-channel device. The signs of the numerical values of the various variables for the two types are listed in table 2.11.

2.3.2 Enhancement-type MOSTs (e-MOSTs)

Structure

Figure 2.73 shows the main structural features of a low-power lateral enhancement-type MOST. It shows an n-channel device that uses a lightly doped p-type substrate ($\sigma \approx 100$ S/m) into which shallow heavily doped n-type source and drain regions are diffused. Surface metallisation, usually of aluminium, forms the source and drain contacts together with the gate electrode. The insulating silicon dioxide layer is typically 1 μm thick except

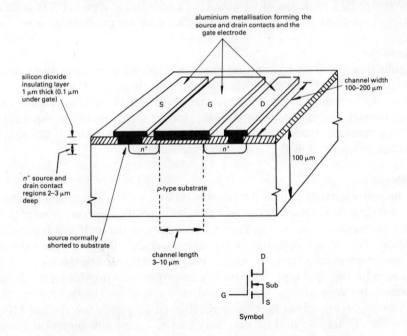

Figure 2.73 Structure of a low-power lateral n-channel enhancement-type MOST

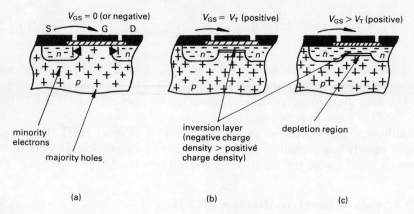

Figure 2.74 Formation of an inversion layer in an n-channel enhancement-type MOST allowing drain–source conduction

under the gate electrode where the thickness is of the order of 0.1 μm. MOSTs are normally produced as three terminal devices by shorting the source to the substrate. This is done by extending the source metallisation to make contact with the p-type substrate. They are packaged using plastic encapsulation or hermetically sealed metal cans having styles similar to the low-power BJT packages of figure 2.10. Enhancement-type MOSTs have the simplest structure of all transistors resulting in fewer processing steps.

Operation
Conduction between source and drain is controlled by conditions at the surface of the substrate directly under the gate electrode. For zero gate bias no conducting channel exists, the source and drain being connected via two back-to-back pn^+ diodes (figure 2.74a). Under these conditions, application of V_{DS} causes a drain current I_{DSS} of less than 20 nA which is the reverse leakage component at the drain–substrate junction.

Effect of V_{GS}
If the gate electrode is biased positively relative to the source and substrate, an electric field is established which attracts minority electrons present in the p-type substrate to the surface region under the gate oxide. If V_{GS} is increased sufficiently, the net negative charge density at the surface of the substrate becomes just greater than the positive charge density causing the surface layer to *invert* becoming n-type. This surface *inversion layer* provides a conducting channel between the n^+ source and drain regions (figure 2.74b). The *enhancement* of current flow by application of V_{GS} gives this type of MOST its name. The value of V_{GS} necessary just to create the inversion layer is known as the *threshold voltage* V_T or $V_{GS(th)}$ and is typically in the range of

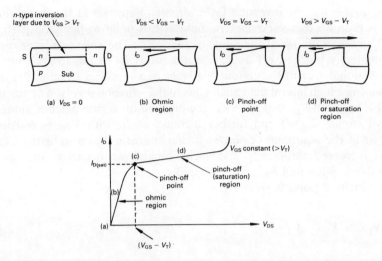

Figure 2.75 Stages of operation of an enhancement-type MOST for increasing V_{DS} with V_{GS} ($>V_T$) constant

1–5 V. V_T is dependent on the thickness of the insulating oxide layer under the gate and the substrate doping, although when shorted to the substrate, the source provides extra electrons which may be attracted to the channel region and the substrate doping is then less significant. V_T has been reduced to a conveniently low value by reduction of the thickness of the gate oxide so that the required electric field can be established for a low value of V_{GS}. A minimum oxide thickness of about 0.1 μm is imposed by breakdown due to minute holes which form in the layer during growth.

Application of a negative gate bias reverses the direction of the transverse electric field at the surface of the substrate causing attraction of holes to the region which maintains the back-to-back diode structure and no channel exists.

As there is no conducting channel for $V_{GS} = 0$, the enhancement-type MOST is described as being normally OFF. Conduction occurs only when $V_{GS} > V_T$, termed the enhancement mode of operation.

Effect of V_{DS}

For $V_{GS} > V_T$ the depth of the inversion layer for $V_{DS} = 0$ is determined by $V_{GS} - V_T$. When $V_{DS} > 0$ the lateral current flow I_D through the resistive inversion layer causes a voltage drop along the layer so that the transverse electric field and hence the degree of inversion varies with position along the channel (figure 2.75b). For $V_{DS} < V_{GS} - V_T$, the distortion of the inversion layer is only slight and the $I_D - V_{DS}$ relationship is approximately linear resulting in the *ohmic* region of the output characteristic (figure 2.75). When

$V_{DS} = V_{GS} - V_T$, the inversion layer almost disappears at the drain end as there is then no transverse electric field at this point to maintain it (figure 2.75c) and operation is at the verge of *pinch-off*. The inversion layer cannot be completely pinched off as this would cause I_D to fall to zero but the cross-sectional area of the conducting layer becomes very small at the drain end causing a high lateral field analogous to the situation involved at pinch-off in a JFET. When $V_{DS} > V_{GS} - V_T$, the pinched-off region extends along the channel (figure 2.75d) and further increase of I_D with V_{DS} is restricted, resulting in the *saturation* or *pinch-off* characteristic shown in figure 2.75. If V_{DS} is increased sufficiently, avalanche breakdown occurs in the narrow pinched-off region and I_D increases rapidly.

The pinch-off point is given by

$$V_{DS} = V_{GS} - V_T \qquad (2.168)$$

which is known as the *saturation* or *pinch-off line*. If V_{GS} is increased, pinch-off does not occur until a higher value of V_{DS} and consequently $I_{D(sat)}$ is also increased. Variation of V_{GS} thus results in a family of output characteristics as shown in figure 2.76a. Application as an amplifier involves operation in the saturation region so that changes in gate–source voltage produce approximately proportional changes in drain current. In this region the relationship between $I_{D(sat)}$ and V_{GS} for constant V_{DS} is given by the transfer characteristic (figure 2.76b). The slope of the output characteristic in the saturation region is the small-signal common-source output conductance g_{os} and the slope of the transfer characteristic is the small-signal

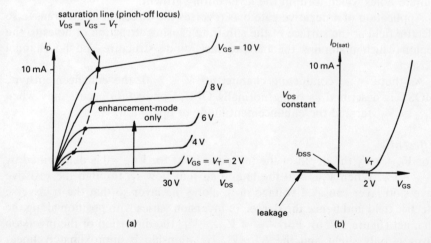

Figure 2.76 Output and transfer characteristics for an *n*-channel enhancement-type MOST

common-source forward transconductance g_{fs} given, as for the JFET, by equations 2.152 and 2.153.

As the gate terminal is insulated from the channel by the oxide layer, the static gate current is extremely low, typically I_{GSS} (the gate current with $V_{DS} = 0$) is less than 10 pA for V_{GS} up to ± 30 V, resulting in $R_{GS} > 10^{12}$ Ω. In fact values of R_{GS} up to 10^{15} Ω are not uncommon. For static or low-frequency operation the gate–source can therefore be regarded as an open-circuit which is one of the most useful properties of the MOST and enables systems with very high input resistance to be produced which draw negligible current from the signal source.

The high input resistance introduces a safety problem however since it is possible for sufficient static charge to build up on the well-insulated gate electrode to cause breakdown of the thin oxide layer even with no voltages applied to the device. Manufacturers' data sheets usually list precautions that should be followed to avoid the build up of static charge, namely

(1) MOST leads should be shorted together during storage and handling;
(2) handle MOSTs by the case instead of by the leads;
(3) do not insert or remove MOSTs from circuits with the power applied as transient voltages can damage the insulating layer.

MOSTs are normally supplied with the leads shorted together with metal foil or with conducting foam to provide a leakage path for gate charge. In addition it is common for MOSTs to have back-to-back breakdown diodes fabricated within the chip to provide gate protection (figure 2.68f). The doping levels of the diodes are chosen to provide a breakdown voltage less than that of the oxide layer which is typically of the order of 20 V.

Replacement of the metal gate electrode by a grown silicon gate introduces a degree of flexibility since the threshold voltage can be controlled by the doping of the silicon gate.

Theoretical Analysis of the Operation of an n-Channel Enhancement-type MOST

Figure 2.77 shows the essential features of the MOST where x is the distance along the channel from the source and y is the direction perpendicular to the channel.

The drain current I_D is a drift of electrons along the channel due to a lateral electric field E which causes a current density J_D. If the lateral field is defined as being in the negative x direction, that is, in the same direction as I_D (figure 2.77)

$$J_D = \frac{I_D}{A} = \sigma E \qquad (2.169)$$

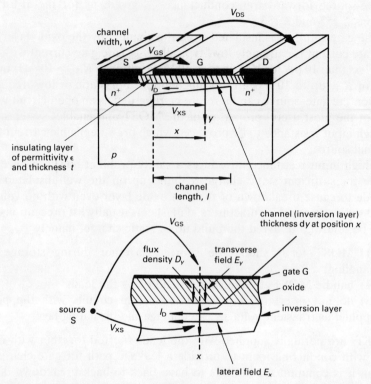

Figure 2.77 Analysis of the channel region of an *n*-channel enhancement-type MOST

where A is the cross-sectional area of the channel and σ is the channel conductivity. Tapering of the channel due to I_D causes A, J_D and E to vary with position along the channel and at a particular point distance x from the source

$$I_D = \sigma A_x E_x \tag{2.170}$$

From figure 2.77, the channel thickness at x is dy and so the cross-sectional area of the channel at x is

$$A_x = w \, dy \tag{2.171}$$

where w is the channel width.

The lateral electric field at x in the direction shown in figure 2.77, which causes I_D, is

$$E_x = \frac{dV_{XS}}{dx} \tag{2.172}$$

from the definition of electric field (equation 1.37) and noting that E_x has been defined in the negative x direction.

Channel conductivity depends on the charge density created in the inversion layer by the transverse electric field E_y. The voltage across the insulating layer at x, which provides the transverse electric field E_y, is $(V_{GS} - V_{XS})$, thus

$$E_y = \frac{V_{GS} - V_{XS}}{t} \tag{2.173}$$

This field creates a surface charge density (C/m^2) at the surface of the channel region which is given by the electric flux density D_y, hence the surface charge density is

$$D_y = \epsilon E_y = \frac{\epsilon}{t}(V_{GS} - V_{XS}) \tag{2.174}$$

where ϵ and t are the permittivity and thickness of the insulating layer.

The conducting channel is not formed however until inversion takes place at $(V_{GS} - V_T)$ and so the effective surface charge density at x providing channel conductivity σ for $V_{GS} > V_T$ is $(\epsilon/t)(V_{GS} - V_T - V_{XS})$.

As the thickness of the channel at x is dy, the volume charge density at this point ρ_x (C/m^3) is

$$\rho_x = \frac{\epsilon}{t\,dy}(V_{GS} - V_T - V_{XS}) \tag{2.175}$$

and by comparison with equation 1.41, the channel conductivity is then

$$\sigma = \text{charge density} \times \text{carrier mobility}$$

$$= \rho_x \mu_n$$

$$= \frac{\epsilon \mu_n}{t\,dy}(V_{GS} - V_T - V_{XS}) \tag{2.176}$$

Substituting for A_x, E_x and σ from equation 2.171, 2.172 and 2.176 into equation 2.170, the drain current is given by

$$I_D = \frac{\epsilon \mu_n}{t\,dy}(V_{GS} - V_T - V_{XS})\, w\, dy\, \frac{dV_{XS}}{dx}$$

$$= \frac{\epsilon \mu_n w}{t}(V_{GS} - V_T - V_{XS})\frac{dV_{XS}}{dx} \tag{2.177}$$

At $x = 0$, $V_{XS} = 0$ and at $x = l$, $V_{XS} = V_{DS}$, thus

$$\int_0^l I_D \, dx = \frac{\epsilon \mu_n w}{t} \int_0^{V_{DS}} (V_{GS} - V_T - V_{XS}) \, dV_{XS}$$

from which

$$I_D = \frac{\epsilon \mu_n w}{tl} \left[(V_{GS} - V_T) V_{XS} - \tfrac{1}{2} V_{XS}^2 \right]_0^{V_{DS}}$$

$$= \frac{\epsilon \mu_n w}{tl} \left[(V_{GS} - V_T) V_{DS} - \tfrac{1}{2} V_{DS}^2 \right] \qquad (2.178)$$

Equation 2.178 gives I_D as a function of V_{GS} and V_{DS} in the *ohmic* region below pinch-off (saturation) for $V_{GS} > V_T$. If V_{DS} is increased sufficiently, the pinch-off condition is reached when the inversion layer at the drain end almost disappears (figure 2.75c). The pinch-off point occurs when $V_{DS} = V_{GS} - V_T$ (equation 2.168) and the corresponding drain current $I_{D(sat)}$ from equation 2.178 is

$$I_{D(sat)} = \frac{\epsilon \mu_n w}{2tl} (V_{GS} - V_T)^2$$

$$= \frac{\epsilon \mu_n w V_T}{2tl} \left(\frac{V_{GS}}{V_T} - 1 \right)^2 \qquad (2.179)$$

$$\equiv K \left(\frac{V_{GS}}{V_T} - 1 \right)^2$$

where K is a parameter for the MOST which can be obtained from the value of $I_{D(sat)}$ at a particular value of V_{GS} (transfer characteristic, figure 2.76b). Typically $V_T = 2$ V and $I_{D(sat)} = 10$ mA at $V_{GS} = 5$ V giving K = 4.4 mA. Note that the value of I_{DSS} is not particularly useful for the enhancement-type MOST since the device is switched OFF when $V_{GS} = 0$ and I_{DSS} is the leakage current.

The capacitance C of the gate–oxide–channel capacitor is $\epsilon lw/t$ and equations 2.178 and 2.179 can therefore be written in terms of C.

The small-signal common-source forward transconductance g_{fs} of the enhancement-type MOST in the pinch-off (saturation) region can be determined from equation 2.179

$$g_{fs} = \frac{\partial I_{D(sat)}}{\partial V_{GS}} = \frac{\epsilon \mu_n w}{tl} (V_{GS} - V_T) \qquad (2.180)$$

The forward transconductance is a gain term for the MOST describing the change in output current caused by a change in input voltage. Equation 2.180

shows that this gain can be increased by reducing the channel length l and increasing the channel width w.

Enhancement-type MOSTs are widely used as voltage-controlled switches while in the linear mode they are used in signal amplification being operated in the saturation region above pinch-off. The dependence of channel resistance on input voltage and the isolated nature of the input, also makes the device useful as a voltage-controlled resistor in the ohmic region below pinch-off. Equation 2.178 gives the drain current in this region from which the static drain-source conductance G_{DS} is

$$G_{DS} = \frac{I_D}{V_{DS}} = \frac{\epsilon \mu_n w}{tl}(V_{GS} - V_T - \tfrac{1}{2}V_{DS}) \tag{2.181}$$

G_{DS} can thus be controlled by V_{GS}.

For small-signal operation in this region, the output slope conductance g_{os} is given by

$$g_{os} = \frac{\partial I_D}{\partial V_{GS}} = \frac{\epsilon \mu_n w}{tl}(V_{GS} - V_T - V_{DS}) \tag{2.182}$$

$$\approx \frac{\epsilon \mu_n w}{tl}(V_{GS} - V_T) \text{ for small values of } V_{DS}$$

The above consideration of the n-channel enhancement-type MOST also applies to the p-channel device except that channel conduction is then due to *hole* drift and hence hole mobility μ_p must be used. The signs of the numerical values of the various variables for the two types as defined in figure 2.68 are listed in table 2.11.

Manufacturer's data for a typical low-power enhancement-type MOST (type 3N163) is given in appendix H.5.

2.3.3 Depletion-type MOSTs (d-MOSTs)

Structure
The depletion-type MOST has a similar structure to that of the enhancement-type device except that a conducting channel region is diffused into the substrate surface so that the device is normally ON, that is, a conducting channel exists between source and drain for $V_{GS} = 0$. Usually an offset gate geometry is used as shown in figure 2.78. This reduces the gate–drain capacitance C_{GD} hence improving the high-frequency performance making the device useful as an HF amplifier.

Operation
The operation of a depletion-type MOST is very similar to that of a JFET.

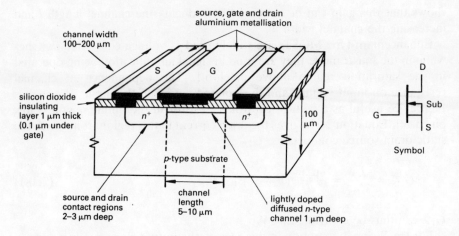

Figure 2.78 Structure of a low-power lateral n-channel depletion-type MOST

For $V_{GS} = 0$, application of V_{DS} causes a drain current to flow through the conducting channel. As V_{DS} is increased, the lateral voltage drop along the channel causes pinch-off to occur at the drain end and a saturating output characteristic results. Considering the n-channel device (figure 2.79), a negative gate bias causes electrons to be repelled from the surface of the channel reducing the conductivity and hence reducing the drain current. This is known as the *depletion mode* of operation since V_{GS} *depletes* the channel of free carriers (figure 2.79a and b). As this is the normal mode of operation for this type of device, it is called a *depletion-type* MOST. If the negative V_{GS} bias is increased sufficiently, the channel region is pinched-off even for $V_{DS} = 0$. As in the case of the JET, the value of V_{GS} necessary to pinch-off or cut-off the channel with $V_{DS} = 0$ is termed the pinch-off voltage V_P. However, since it is difficult to specify V_P in practice, device data sheets usually define $V_{GS(off)}$ as the value of V_{GS} necessary to reduce $I_{D(sat)}$ to a specific low value such as 10 μA at a mid-range value of V_{DS} such as 15 V. Typically $V_{GS(off)}$ is in the range of -1 to -6 V for an n-channel device.

If the gate is biased positively with respect to the source and substrate, electrons are attracted to the channel region thus increasing the conductivity and extending the range of operation. This is described as the *enhancement mode* of operation (figure 2.79a and b).

As with the enhancement-type MOST, the input resistance of the depletion-type device is very high (typically $R_{GS} > 10^{12}$ Ω) due to the gate–channel insulating layer and the same precautions must be taken regarding the accumulation of static charge on the gate to avoid breakdown.

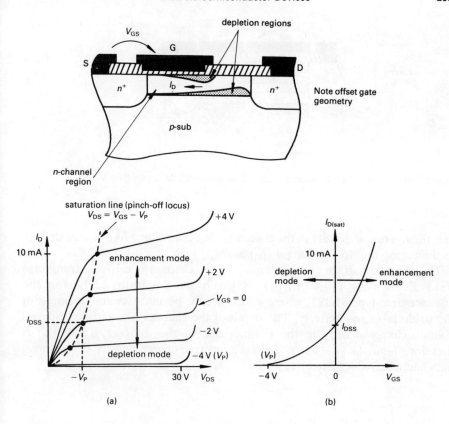

Figure 2.79 Output and transfer characteristics for an n-channel depletion-type MOST

Theoretical Analysis of the Operation of an n-channel Depletion-type MOST
The drain current at particular values of V_{DS} and V_{GS} depends on the conductivity of the channel and hence on the density of free carriers in the channel. For an enhancement-type MOST the free carrier density depends on the net voltage across the oxide layer above the threshold voltage, that is, $(V_{GS} - V_T - V_{XS})$ at a point distance x along the channel from the source, where V_{XS} is the lateral voltage drop along the channel due to I_D (equation 2.175). In the case of a depletion-type MOST, the free carrier density in the channel can be considered to comprise the free charge present due to the channel doping *and* the free charge induced by virtue of the field across the oxide layer.

Figure 2.80 shows the essential features of an n-channel depletion-type MOST and forms the basis of a first-order analysis of the performance of the device.

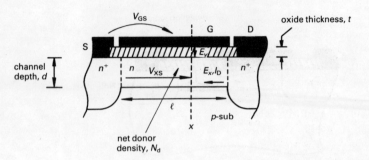

Figure 2.80 Analysis of the channel region of an *n*-channel depletion-type MOST

If the net donor density in the channel is N_d, the volume free charge density in the region due to the dopant (ρ_d) is eN_d.

The induced surface charge density in the channel due to the transverse field E_y is $(\epsilon/t)(V_{GS}-V_{XS})$ as given by equation 2.174 for the enhancement-type MOST, where ϵ and t are the permittivity and thickness of the oxide layer respectively. The depth of the channel is d and the induced volume charge density in the channel due to the transverse field (ρ_f) is therefore $(\epsilon/td)(V_{GS}-V_{XS})$. Thus the net free-charge density determining the channel conductivity at x, ρ_x is

$$\rho_x = \rho_d + \rho_f$$
$$= \rho_d + \frac{\epsilon}{td}(V_{GS}-V_{XS}) \tag{2.183}$$

When $V_{DS}=0$, $I_D=0$ and $V_{XS}=0$ and under these conditions the channel becomes pinched-off when $V_{GS}=V_P$. In a first-order analysis, pinch-off occurs when the channel becomes totally depleted. Thus for $V_{XS}=0$, $\rho_x=0$ when $V_{GS}=V_P$. From equation 2.183 therefore

$$0 = \rho_d + \frac{\epsilon V_P}{td}$$
$$\rho_d = -\frac{\epsilon V_P}{td} \tag{2.184}$$

Substituting for ρ_d in equation 2.183

$$\rho_x = \frac{\epsilon}{td}(V_{GS}-V_P-V_{XS}) \tag{2.185}$$

and therefore the channel conductivity σ is

$$\sigma = \frac{\epsilon\mu_n}{td}(V_{GS} - V_P - V_{XS}) \tag{2.186}$$

By comparison with the analysis of the enhancement-type MOST (section 2.3.2), the drain current at x is given by $\sigma A_x E_x$ (equation 2.170) where A_x is the cross-sectional area of the channel at x and E_x is the lateral field at x.

For the depletion-type MOST, $A_x = wd$ where w is the channel width and if the lateral field is defined in the negative x direction, that is, the same direction as I_D (figure 2.80), $E_x = dV_{XS}/dx$ from which

$$I_D = \frac{\epsilon\mu_n}{td}(V_{GS} - V_P - V_{XS})\,wd\,\frac{dV_{XS}}{dx} \tag{2.187}$$

Integration over the channel length l, that is, $x = 0 \to l$, $V_{XS} = 0 \to V_{DS}$ as in the enhancement-type MOST analysis yields

$$I_D = \frac{\epsilon\mu_n w}{tl}\left[(V_{GS} - V_P)V_{DS} - \tfrac{1}{2}V_{DS}^2\right] \tag{2.188}$$

Equation 2.188 gives I_D as a function of V_{GS} and V_{DS} below pinch-off. At pinch-off, $V_{GS} - V_{DS} = V_P$ as in the JFET case (equation 2.151). Therefore from equation 2.188 the drain current at pinch-off or saturation $I_{D(sat)}$ is

$$I_{D(sat)} = \frac{\epsilon\mu_n w}{2tl}(V_{GS} - V_P)^2 \tag{2.189}$$

When $V_{GS} = 0$, $I_{D(sat)} = I_{DSS}$ and so

$$I_{DSS} = \frac{\epsilon\mu_n w V_P^2}{2tl} \tag{2.190}$$

thus

$$\begin{aligned}I_{D(sat)} &= I_{DSS}\left(\frac{V_{GS}}{V_P} - 1\right)^2 \\ &= I_{DSS}\left(1 - \frac{V_{GS}}{V_P}\right)^2\end{aligned} \tag{2.191}$$

Equation 2.191 is the same as the practical square law relation for a JFET (equation 2.166) and hence leads to equation 2.167 for the small-signal common-source forward transconductance g_{fs} of a depletion-type MOST.

The above analysis has considered the free-charge density in the channel to be made up of two components, ρ_d due to the channel doping and ρ_f due to the transverse field. Depletion-mode operation corresponds to the case where ρ_f and ρ_d are of opposite sign so that the net charge density is reduced. This occurs when the numerical values of V_{GS} and V_P have the same sign (equation 2.185), that is, for negative V_{GS} for an n-channel device. When V_{GS} is of opposite sign to V_P, ρ_f and ρ_d are of the same sign and the net charge density is increased; this is enhancement-mode operation. From equation 2.184 it can be seen that the pinch-off voltage V_P is a function of ρ_d and hence of the net donor density in the channel N_d.

The above discussion of the n-channel depletion-type MOST also applies to a p-channel device if the roles of electrons and holes are interchanged and voltage polarities and current directions are reversed. The signs of the numerical values of the various variables for the two types as defined in figure 2.68 are listed in table 2.11.

Manufacturer's data for a typical low-power depletion-type MOST (type BFR 29) is given in appendix H.6.

Although depletion-mode MOSTs and JFETs exhibit similar characteristics there are two important differences that should be noted.

(1) The MOST has a gate–source resistance R_{GS} several orders of magnitude greater than the JFET. Typically $R_{GS} > 10^{12}$ Ω for the MOST resulting in a static gate current of a few picoamps, whereas for the JFET $R_{GS} > 10^8$ Ω and I_G is typically a few nanoamps.

(2) Although enhancement-mode operation for a JFET is restricted to V_{GS} less than about 0.5 V (figure 2.71a) due to forward conduction of the gate–channel junction and is thus of little practical use, the depletion-type MOST can be operated in the enhancement-mode with practical values of V_{GS} (figure 2.79). This means that the device can be operated in a linear application without a gate bias.

2.3.4 FETs in Practice

The 'gain' parameter of a FET is the forward transadmittance y_f which is the relation between output current and input voltage. For common-source operation, the gain at low frequency is given by the CS forward transconductance g_{fs}, the real part of y_{fs}. Analysis of the three types of FET shows that high g_{fs} requires a short, wide channel. In practice the minimum channel length in lateral FETs is limited to 1–5 μm by fabrication limitations such as mask alignment. Channel width is increased by the use of annular or serpentine layouts as shown in figure 2.81, the structural diagrams of figures 2.69, 2.73 and 2.78 being portions of practical discrete FET structures.

The performance of MOSTs depends critically on the properties of the insulating layer. In the early years of production manufacturers experienced considerable difficulty in producing MOSTs with stable characteristics. Although silicon diode is an excellent insulator, certain ions, particularly

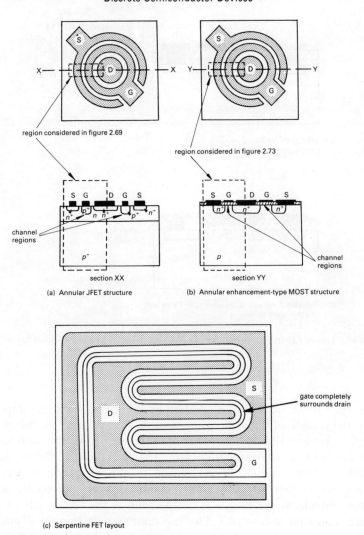

Figure 2.81 Practical FET layouts to increase channel width

sodium, can migrate through the layer under the influence of the applied transverse electric field. Typically $|V_{GS}|$ is in the range 2–4 V, and for a gate oxide thickness of 0.1 μm, the transverse field is 2×10^7–4×10^7 V/m; the breakdown field for silicon dioxide being approximately 5×10^8 V/m. The presence of these ions near the channel caused the device characteristics to drift, the threshold and pinch-off voltages changing considerably with time. Stability was improved initially by scrupulous control of the cleanliness of the

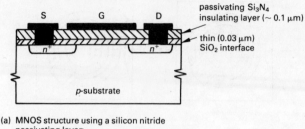

(a) MNOS structure using a silicon nitride passivating layer

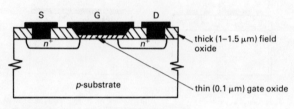

(b) Threshold voltage is reduced to 2–4 V by use of a thin gate oxide

Figure 2.82 Use of (a) silicon nitride passivation and (b) a thin gate insulating layer

fabrication environment and in addition by the formation of an impervious layer at the surface of the oxide using phosphorous prior to the metallisation stage. A more recent development has been the use of a silicon dioxide/silicon nitride sandwich as the insulating layer. Ion migration in silicon nitride is negligible and the use of silicon nitride as the insulating surface layer produces an excellent barrier to contaminants. However, direct replacement of silicon dioxide by the nitride is not possible as the silicon–silicon nitride interface is difficult to control and for this reason a very thin ($\approx$ 30 nm) oxide layer is used as an interface between the silicon and the nitride layer forming an MNOS structure (figure 2.82a). Protection of the chip surface from contamination is described as *surface passivation* and the use of silicon nitride provides such excellent protection that the technique is termed *sealed-junction* technology. Improved passivation has provided not only improved reliability and stability in MOSTs, but has enabled the degree of hermeticity required of the package to be relaxed enabling packaging costs to be reduced by the use of plastic encapsulation in place of hermetically sealed metal packages.

To enable enhancement-type MOSTs to be used with low power supply voltages, for example 5 V or less, the threshold voltage must be less than 4 V. This has been achieved by reduction of the oxide thickness under the gate so that the required transverse field can be created by a small value of gate–source voltage. Figure 2.82b shows a typical cross-sectional structure. It is usual to deposit the thick (1–1.5 μm) *field* oxide by thermal oxidation of silane (SiH_4) at about 400 °C. The oxide covering the gate area is then etched away and the thin ($\approx$ 100 nm) *gate* oxide is grown slowly by thermal oxidation of the silicon surface at 1200 °C. A lower limit of 100 nm is set by the formation of voids (*pin-holes*) in the oxide layer which reduce the breakdown field of the layer. Silicon nitride has also been useful in the reduction of threshold voltage. The permittivity of silicon nitride is higher than that of the dioxide (ϵ_r = 7.5 and 4 respectively) and so, for the same thickness of insulating layer, the required transverse field is established at a lower value of V_{GS} in the case of a mainly nitride layer, providing a lower threshold voltage.

When enhancement-type MOSTs were first introduced during the early 1960s they were almost exclusively of the *p*-channel type. Although it was appreciated that *n*-channel MOSTs have a performance advantage over the *p*-channel type due to μ_n being greater than μ_p, reliable enhancement-type *n*-channel MOSTS could not be produced due to the presence of positive ions in the oxide layer. These ions caused the surface of the *p*-type substrate to invert even with no applied gate bias producing depletion-type device properties. *p*-channel enhancement-type MOSTs did not suffer from this problem because positive ions in the oxide layer made the surface of the *n*-type substrate more heavily *n*-type, simply increasing the threshold voltage. Improvements in cleanliness and surface passivation enabled reliable *n*-channel enhancement-type MOSTs to become readily available during the 1970s.

From the user's point of view, a major difference between FETs and BJTs is that performance stability depends on different factors. With silicon BJTs, the variation of V_{BE} with temperature is the most significant factor and provided that this variation is swamped, most simply by inclusion of an emitter resistor, reasonable quiescent point stability can be obtained. In the case of FETs, the effect of temperature is far less of a problem than the spread of characteristics for devices of the same type number. FET characteristics as described by I_{DSS}, V_P, K and V_T (equations 2.166, 2.179 and 2.191) are very sensitive to channel geometry and also to the thickness of the gate oxide in the case of MOSTs. Examination of low-power FET data sheets shows that typical spreads are I_{DSS} = 2–9 mA, $|V_P|$ = 1–7 V, K = 0.2–2 mA and V_T = 1–5 V. Unless FETs are to be individually selected, circuit design must cater for such spreads while providing reasonable quiescent point stability. This can be achieved by the use of both fixed and self bias.[25]

Pinch-off and threshold voltages vary only slightly with temperature while I_{DSS} and K decrease by about 0.5%/°C rise in temperature due to the increase in channel resistance. As drain current decreases as the temperature increases, thermal runaway does not occur in FETs. In the case of JFETs, the gate current is the reverse leakage across the gate–channel *pn* junction and I_G

therefore increases with temperature as given by equations 2.1 and 2.2 and figure 2.5b for the *pn* junction diode.

Although the planar JFET and metal–oxide–semiconductor IGFET structures based on silicon are the most widely available, new structures and materials have recently been introduced aimed at improving the high-frequency and/or high-power performance of FETs (section 2.3.6 and 2.3.7). A high-frequency JFET structure has been developed using a Schottky (metal–semiconductor) junction to control channel conductivity in place of the conventional *pn* junction. The device is termed a Schottky–barrier FET or MESFET (*me*tal-*s*emiconductor FET) and is based on a GaAs substrate. In the case of IGFETs, the use of materials other than silicon dioxide to provide gate–channel insulation has led to the introduction of the acronym MISFET (*m*etal–*i*nsulator–*s*emiconductor FET) to describe generally the device structure.

2.3.5 FETs: Ratings and Power Considerations

Current flow through the resistive channel of a FET causes power dissipation and the temperature of the semiconductor rises. As with other devices, the maximum permissible temperature of the semiconductor together with the thermal properties of the package sets a limit to the maximum power dissipation.

For time varying operation, the average power dissipated in the channel is

$$P_{\text{tot(av)}} = \frac{1}{T} \int_0^T v_{\text{DS}} i_{\text{D}} \, dt \tag{2.192}$$

Under static conditions

$$P_{\text{tot}} = V_{\text{DS}} I_{\text{D}} \tag{2.193}$$

The hyperbola $V_{\text{DS}} I_{\text{D}} = P_{\text{tot}}\text{max}$ plotted on the static output characteristics bounds the safe operating area for the device (figure 2.83) and establishes the maximum drain current $I_{\text{D}}\text{max}$.

Avalanche breakdown of the gate–channel or substrate–channel depletion regions sets the drain–source voltage limit $V_{\text{DS}}\text{max}$. The maximum gate–source voltage $V_{\text{GS}}\text{max}$ is limited by breakdown of the reverse-biased gate–channel junction in a JFET and by breakdown of the oxide insulating layer in a MOST.

2.3.6 High-frequency Performance of FETs

The high frequency response of FETs is limited by

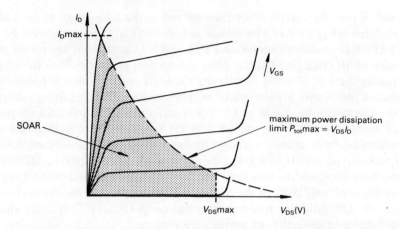

Figure 2.83 Safe operating area on the static common-source output characteristics of a FET

(1) capacitance: mostly gate–channel capacitance;
(2) channel length;
(3) mobility of carriers in the channel.

The change of channel conductivity due to a variation in gate–source voltage requires the state of charge of the gate–channel capacitance to change and the speed of response therefore depends on the associated time constant. In addition, the transit time of carriers between source and drain depends on channel length and carrier mobility. Both capacitance and channel length depend on device geometry and dimensions whereas carrier mobility depends on the semiconductor material.

There have been two main approaches to the improvement of the HF performance of FETs

(1) the use of gallium arsenide GaAs which has an electron mobility μ_n approximately six times that of silicon;
(2) modifications to the device geometry to reduce channel length and capacitance.

Manufacturers have encountered considerable difficulties in fabricating reliable devices with repeatable properties using high electron mobility group III–V compound semiconductors such as gallium arsenide. High performance n-channel GaAs JFETs are produced however in which the gate electrode forms a Schottky (metal–semiconductor) junction with the channel region. Control of channel conductivity depends on the depth of penetration of the junction depletion region as in a conventional pn JFET but the simpler structure allows very short channel lengths of the order of 1 μm to be obtained (figure 2.84a). In addition, use of the Schottky barrier contact avoids some of the difficulties encountered in the fabrication of *pn* junction structures using GaAs. The device is known as a *Schottky-barrier FET* or MESFET (*me*tal–*s*emiconductor FET) and is intended for operation in the region of 10 GHz.[26] Source–drain conduction is restricted to the channel layer by the use of a semi-insulating GaAs substrate which is produced by using a mid-gap dopant such as chromium. The resulting substrate conductivity can be as low as 10^{-10} S/m using this technique. An aluminium gate forms the rectifying contact with GaAs while ohmic source and drain contacts are formed using a gold–germanium alloy. Such devices are normally depletion-type (d-MESFETs) although enhancement-types (e-MESFETS) have been produced using a very thin (≈ 0.1 μm) implanted channel layer so that the depletion layer of the Schottky junction pinches the channel off for $V_{GS} = 0$. Operation at reduced temperature (COLDFET) reduces channel resistance and so further improves performance.

The frequency response of conventional lateral enhancement-type MOSTs can be improved by reduction of the overlaps between the gate metallisation and the n^+ source and drain diffusions shown in figure 2.73, so reducing capacitances C_{gs} and C_{gd}. Normal n^+ source and drain diffusions are performed and gate metallisation positioned so as not to overlap these regions (figure 2.84b). Using the metallisation as a mask, shallow n^+ 'extension' regions are then implanted through the insulating layer. The directional property of ion-implantation ensures that the n^+ regions do not extend under the gate metallisation. Speed improvements of up to × 5 have been reported using this technique.

The minimum channel length of the basic enhancement-type MOST (figure 2.73) is limited to 2–3 μm by tolerances in the photolithographic process. During the 1970s two structural innovations to the basic structure enabled devices with sub-micron channel lengths to be produced with corresponding improvements in performance. The two devices, known as double-diffused MOS (DMOS) and V-groove MOS (VMOS) transistors, are the most advanced silicon enhancement-type MOSTs.

The DMOST is a lateral device in which the channel length is determined by the spacing between two diffusions performed through the same window. Using a lightly doped *n* or *p*-type substrate, a *p*-type region is diffused through the source window followed by shallower n^+ diffusions through the same source window and the drain window. An *n*-channel device is thus created (figure 2.84c) with a channel length of 0.5–2 μm which depends not on tolerances in the photolithographic process but on dopant source concentration, diffusion time and process temperature analogous to the control of base width in a BJT. The substrate can be either *p* or *n*-type provided that it is lightly doped as the surface will be n^- in either case. If the substrate is p^- or π-type (from pi, *p*-intrinsic) the effect of the net positive charge at the silicon–silicon dioxide interface causes the surface to invert to n^-. Since the substrate doping is typically less than 10 per cent of that of the

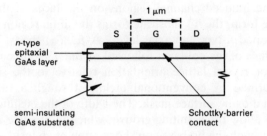

(a) GaAs Schottky-barrier FET (MESFET)

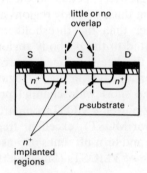

(b) Reduction of C_{gs} and C_{gd} using ion-implantation

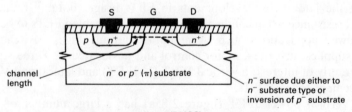

(c) n-channel double-diffused (DMOS) structure

Figure 2.84 FET developments to improve HF performance

channel, a sufficiently high breakdown limit V_{DS}max of the order of 40 V is maintained.

The VMOST structure utilises the same principle involved in DMOS technology whereby channel length is controlled by diffusion instead of photolithographic tolerances as in the basic MOST. Figure 2.85c shows the cross-section of a discrete n-channel VMOST. The structure departs from the normal lateral arrangement of MOSTs, current flow between source and drain being along the induced channel regions on the faces of the etched V-groove. In discrete form, the n^+ substrate forms the drain region on which is grown a lightly doped n-type epitaxial layer. A p^- (π) 'body' region is diffused into the surface of the epi-layer followed by the n^+ source diffusion. By suitable choice of crystal lattice orientation relative to the surface, a V-groove can be formed by conventional etching through a rectangular window in the silicon dioxide surface mask. The width of the etching window must be sufficient to allow the resulting groove, which in practice is slightly asymmetric, to cut through the body region. Formation of an insulating layer over the groove surface and aluminium metallisation to form the gate and source contacts completes the structure. Conduction between source and drain along the faces of the groove depends on the formation of inversion layers across the typically 1 µm thick p^- body regions adjacent to the groove by application of V_{GS} greater than the threshold voltage of the device. Usually the n^+ source is shorted to the p^- body region by extension of the source metallisation so that the heavily doped source region can provide electrons for the channel inversion layer in the p^- region and hence improve the turn-on performance. The short wide channel provides high gain (g_{fs} typically > 100 mS for a low-power VMOST compared with 1–5 mS for a conventional low-power lateral MOST), excellent frequency response and switching speeds with turn-on and turn-off times of 5 ns or less for low-power types. Data for a typical range of VMOSTs (Siliconix types VN 46/66/88AF) is given in appendix H.7.

2.3.7 High-power FETs

During the late 1970s developments in FET design, notably reduction of channel resistance when conducting $R_{DS(on)}$, have enabled FETs to be used in high-power applications such as amplifier output stages, switched-mode power supplies, inverters, motor control and pulse circuits. Three structures are currently of interest: a vertical JFET (V-JFET) and lateral and V-groove enhancement-type MOSTs.

The vertical power JFET (figure 2.85a) has a large number of parallel channels between the bars of a gate 'grid'. Application of a negative V_{GS} bias causes depletion regions to spread out from the bars of the grid reducing drain–source conductance. The n^+ substrate forms the drain which connects directly to the device mounting base providing excellent heat sinking. V-JFETs are currently available with $P_{tot}\text{max} = 100$ W, $I_D\text{max} = 10$ A, $V_{DS}\text{max} = 200$ V and complementary p and n-channel types are used in amplifier output stages. Their exceptionally linear operation in this application provides very low harmonic distortion (< 0.05 per cent) with a frequency response extending to 100 kHz.

The conventional low-power lateral MOST has been developed into a high-power device with typically $P_{tot}\text{max} = 100$ W, $I_D\text{max} = 7$ A and $V_{DS}\text{max}$

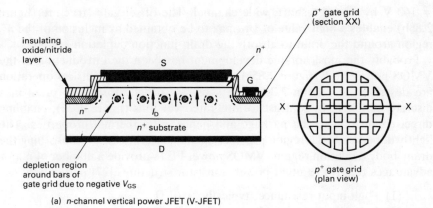

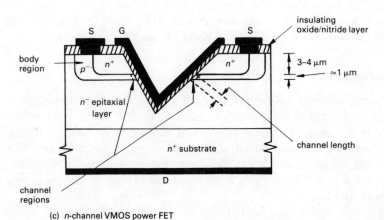

Figure 2.85 Structural details of power FETs

= 160 V by use of a short, wide channel. The offset gate structure (figure 2.85b) enables a high value of V_{DS}max to be obtained by inclusion of the n^- region around the drain to absorb the drain junction depletion region.

Probably the most notable development has been the introduction of the VMOS power FET (figure 2.85c). Device structure and electrical operation are described in section 2.3.6. The substrate usually forms the drain in a discrete VMOST, in contrast to integrated versions (figure 4.9d), enabling direct connection to the package and hence good thermal properties. The lightly doped epitaxial region provides an adequate V_{DS}max by absorbing the drain–body depletion region. VMOS power FETs provide a number of major advantages relative to other power transistor structures,[27] particularly

(1) High input resistance, typically $>10^7$ Ω.
(2) Low ON resistance $R_{DS(on)}$, typically 0.05–5 Ω.
(3) Inherent thermal stability. As the channel resistance increases with temperature, problems of thermal runaway, secondary breakdown and current-hogging do not occur.
(4) Very fast switching speed and excellent frequency response due to the short channel length and the absence of minority carrier storage which restricts the speed of power BJTs. Typical turn-on and turn-off times are 5–10 ns at 2 A and 50–100 ns at 10 A which are considerable improvements compared with switching times for corresponding BJTs (typically 100–500 ns).
(5) Very high gain (typically g_{fs} up to 3000 mS at $I_D > 5A$) enables replacement of BJT Darlington arrangements

A particularly useful feature is that since I_D reduces with increase in temperature due to the increase of channel resistance, higher power capability can be achieved by simple parallelling of devices without the problem of *current-hogging* which occurs with BJTs. Data for a typical range of VMOSTs (Siliconix types VN46/66/88AF) is given in appendix H.7.

2.3.8 Noise in FETs

Noise in semiconductor devices [17,18] can be characterised as thermal noise, shot noise and surface noise (section 2.2.25). In JFETs, thermal noise is generated in the resistive channel region and in the source and drain diffused regions. Shot noise is introduced at the gate–channel junction by reverse leakage, while surface noise due to the generation–recombination mechanism in the junction depletion regions is unimportant except at low frequencies below about 100 Hz.

As the active region of a MOST lies close to the silicon–silicon dioxide interface, surface noise is particularly significant and dominates thermal and shot noise effects typically up to 10 kHz. Above this frequency, thermal noise generated in the channel dominates.

Shot noise is largely insignificant in FETs as forward-biased junctions do not occur in MOSTs, or in JFETs for normal operation and consequently noise in FETs is less than in BJTs. The noise properties of FETs, which are specified in terms of noise figure in the same way as for BJTs (equation 2.78), are considered in detail in references 18 and 28.

2.3.9 FET Modelling

Apart from operation as a variable resistor in the ohmic region, for which FET performance can be represented by an approximately linear resistance of value dependent on V_{GS}, the small-signal operation of FETs is normally in the saturation region beyond pinch-off where the gain is greatest. The common-source y-parameter model of figure 2.86a (from appendix A.2) provides a general representation of the small-signal performance of a FET in the saturation region.

In general, each admittance parameter y is of the form $g+jb$ describing the in-phase (resistive) and quadrature (reactive) relationships between the signal currents and voltages. At low frequency the reactive components representing energy storage properties, and described by the susceptance (b) components, are negligible and the parameters are entirely real (conductances g).

The high resistance between gate and source provided in the JFET by the reverse-biased gate–channel junction and in the MOST by the insulating oxide layer, means that g_{is} is very low. Also at low frequency, internal feedback due to coupling between drain and gate is negligible and so $g_{rs} \approx 0$. The gate signal current is therefore very low at low frequency and the gate–source port may then be represented by an open-circuit. The low frequency performance is thus conveniently represented by the model of figure 2.86b.

At high frequency the reactive components of the y-parameters (b_{is}, b_{rs}, b_{fs} and b_{os}) become significant and must be included in the model. Data supplied by manufacturers for FETs intended for use at high frequency usually give g_{is}, b_{is}, g_{rs}, b_{rs}, g_{fs}, b_{fs}, g_{os} and b_{os} as functions of frequency for typical quiescent conditions of V_{DS}, V_{GS} and I_D (for example, 2N5457 JFET data, appendix H.4 and BFR 29 d-MOST data, appendix H.6). Often, however, device data sheets provide capacitance information in place of susceptance in which case the high frequency model of figure 2.84c is more applicable. This model is approximately equivalent to the y-parameter model of figure 2.84a where $r_{gs}//C_{gs} \equiv y_{is}$, $r_{ds}//C_{ds} \equiv y_{os}$ and the $r_{gd}//C_{gd}$ combination describes the internal feedback represented by $y_{rs} v_{ds}$. In many cases, however, at frequencies at which it is necessary to use the high-frequency model, the resistive components r_{gs} and r_{gd} are sufficiently large compared with the respective reactances of C_{gs} and C_{gd} that they can be neglected in which case the high-frequency model reduces to that of figure 2.86d.

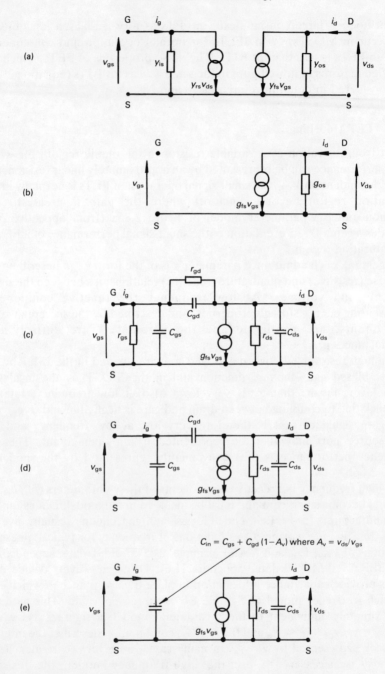

Figure 2:86 Network models of the small-signal performance of FETs (a) general common-source y-parameter model, (b) low-frequency model, (c) general high-frequency model, (d and e) simplified high-frequency models

Manufacturers usually quote C_{iss} and C_{rss} in place of the terminal capacitances C_{gs}, C_{gd} and C_{ds} where

C_{iss} = common-source input capacitance with the output (D-S) port short-circuit to a.c. signals (that is, for $v_{ds} = 0$)

$$= C_{gs} + C_{gd} \qquad (2.194)$$

C_{rss} = common-source reverse transfer capacitance with the input (G-S) port short-circuit to a.c. signals (that is, for $v_{gs} = 0$)

$$= C_{gd} + C_{ds} \qquad (2.195)$$

For many FETs, C_{ds}, which is mainly the package capacitance, is much less than C_{gd} in which case

$$C_{gd} \approx C_{rss} \qquad (2.196)$$
$$C_{gs} \approx C_{iss} - C_{rss} \qquad (2.197)$$

In practice when $v_{ds} \neq 0$, the voltage across C_{gd} in figure 2.86d is $(v_{gs} - v_{ds})$ and the input signal current is given by

$$i_g = j\omega C_{gs} v_{gs} + j\omega C_{gd} (v_{gs} - v_{ds})$$

But v_{ds}/v_{gs} is the voltage gain A_v and so

$$i_g = j\omega [C_{gs} + C_{gd}(1-A_v)] v_{gs}$$
$$= j\omega C_{in} v_{gs} \qquad (2.198)$$

where C_{in} is the input capacitance of the FET. This leads to the useful high-frequency model of figure 2.86e. As an increase of v_{gs} causes a decrease of v_{ds} in application as an amplifier, A_v is negative resulting in capacitance magnification (Miller effect) and the term $C_{gd}(1-A_v)$ is normally dominant.

Typical values of the small-signal parameters for low-power FETs are listed in table 2.12.

Further aspects of high-frequency FET operation may be introduced leading to a more precise model. [29] An important feature that must be considered as the period of v_{gs} is reduced to the order of the transit time of carriers in the channel, is that instead of the variation of i_D following closely the variation of v_{GS}, i_D tends to follow the average value $v_{GS(AV)}$. This is due

Table 2.12 Typical small-signal parameter values for low-power FETs

	JFET	MOST
g_{fs} at $V_{GS} = 0$, T = 25° C	1–10 mS	1–20 mS
r_{gs}	$>10^8 \, \Omega$	$>10^{12} \, \Omega$
r_{gd}	$>10^8 \, \Omega$	$>10^{12} \, \Omega$
$r_{ds} = {}^1/g_{os}$	10–500 kΩ	1–50 kΩ
C_{iss}	2–20 pF	2–10 pF
C_{rss}	1–5 pF	0.3–1 pF
C_{gs}	1–15 pF	2–10 pF
C_{gd}	1–5 pF	2–5 pF
C_{ds}	0.1–0.5 pF	0.1–0.5 pF

to the fact that at such high frequencies, the motion of a single carrier through the channel is due to a varying v_{GS} and the net flow is the average of the resultant motion leading to a reduction in gain. This effect may be introduced into the model in terms of distributed resistance and capacitance between the gate electrode and the channel region.

2.4 SWITCHING DEVICES

Electrical switching using semiconductor devices can be divided into *signal* switching, as in digital circuits, and *power* switching, as in applications such as motor speed control. Although *pn* junction diodes, BJTs and FETs are used extensively in both these general switching areas, there are certain

Table 2.13 Semiconductor switching devices

General-purpose   *pn* junction diode
                          BJT
                          FET

Special-purpose
    (1) *Signal switching*
            fast switching *pn* junction diode
             Schottky–barrier diode
    (2) *Power switching*
             semiconductor controlled rectifier (SCR)
             diode a.c. switch (Diac)†
             triode a.c. switch (Triac)
             unijunction transistor (UJT)†

†used primarily as triggering devices

special-purpose devices which have been developed specifically for switching applications (table 2.13) and it is these special-purpose devices that are considered in this section.

2.4.1 Fast-switching pn Junction Diode

Instantaneous switching of a diode from the conducting to non-conducting state or vice versa is not possible due to the diode capacitance. Under steady-state forward conduction the injection of carriers into each region causes a build up of minority charge either side of the junction which is described in terms of the diffusion capacitance C_D of the junction (section

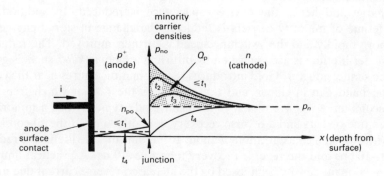

(a) Minority carrier distributions

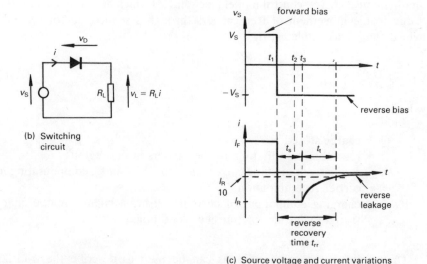

(b) Switching circuit

(c) Source voltage and current variations

Figure 2.87 The diode as a switch

1.6.2). After switching, a transient period exists during which the diode *recovers* to its non-conducting state, the stored minority charge flowing from the junction area discharging the diffusion capacitance. Also during this transient period the junction depletion region forms, consistent with the value of applied reverse bias, which may be considered as charging of the depletion capacitance C_T. The length of this transient period is known as the *reverse recovery time* t_{rr} (figure 2.87). Similarly at switch ON, the depletion region shrinks (C_T discharges) and the minority charge distribution is re-established (C_D charges) resulting in a *forward recovery time* t_{fr}. These recovery times limit maximum speed at which the diode can be switched.

The switching speed may be improved by reducing the junction area and the thickness of the semiconductor regions so that the amount of stored charge and hence the device capacitance is reduced. In addition, if the lifetime of minority carriers is reduced, the change in stored charge occurs more rapidly and the switching speed is further improved. The reduction in carrier lifetime is accomplished by diffusing a *lifetime killer* such as gold into the semiconductor. Gold introduces recombination centres near the centre of the band gap in silicon and thus increases the rate at which free carriers recombine. A conventional p^+n silicon diode typically has a minority hole lifetime of 1 μs and a reverse recovery time of 100 ns. If the silicon is doped with gold at a concentration of about 10^{20} atoms/m^3, the lifetime is reduced to 50–100 ns and the reverse recovery time to 5–10 ns. A practical limit to the gold dopant density is imposed by the increased reverse current due to carrier generation in the depletion region.

The most useful model of the switching performance of a diode is the mathematical charge-control model (section 1.8) which describes current flow i due to the movement of free carriers through a semiconductor region in which charge accumulates, by

$$i = \frac{Q}{\tau} + \frac{dQ}{dt} \tag{1.163}$$

where

Q = charge stored
τ = recombination lifetime of the carriers in the region
$Q/\tau = I$, quiescent or static current flow to maintain Q, compensating for carrier recombination.
dQ/dt = charging component of current causing a change in the charge stored under a.c. or transient conditions.

This model of charge conditions can be used to describe the transient behaviour of a diode during forward and reverse recovery. In the practical case of a p^+n diode, the forward current is almost entirely due to hole injection from the anode into the cathode, electron injection from cathode to anode being insignificant due to the relative doping levels of the anode and cathode. The relevant charge stored is thus the concentration of *excess* holes Q_p just inside the cathode, indicated by the shaded area in figure 2.87a.

For steady-state forward bias this stored charge is constant and so, from equation 1.163, the forward current is given by

$$I_F = \frac{Q_p}{\tau_p} \quad (2.199)$$

where τ_p is the lifetime of holes in the cathode region.

Following the sudden reversal of v_S this excess stored charge is removed which accounts for the *storage* period t_s (figure 2.87c) during which time a reverse current flows of magnitude I_R. Use of the charge-control model during this interval enables the time variation of stored charge and subsequently the storage time t_s to be determined. From equation 1.163

$$-I_R = \frac{Q_p}{\tau_p} + \frac{dQ_p}{dt} \quad (2.200)$$

Hence

$$Q_p = -I_R \tau_p + (I_F + I_R) \tau_p \exp\left(-\frac{t}{\tau_p}\right) \quad (2.201)$$

since $Q_p(0) = I_F \tau_p$ from equation 2.199, where $t = 0$ has been taken as the switching instant shown as t_1 in figure 2.87c. At a time t_s following this instant, the stored charge has been removed and so $Q_p(t_s) = 0$ can be used in conjunction with equation 2.201 to determine the storage time t_s

$$0 = -I_R \tau_p + (I_F + I_R) \tau_p \exp\left(-\frac{t}{\tau_p}\right)$$

hence

$$t_s = \tau_p \ln\left(\frac{I_F}{I_R} + 1\right) \quad (2.202)$$

Storage time and hence reverse recovery time t_{rr} can therefore be shortened by reduction of the minority carrier lifetime consistent with previous comments regarding the use of a lifetime killer to improve switching speed. This analysis illustrates the use of the charge-control model in the consideration of one aspect of the transient behaviour of a diode.

After t_3 (figure 2.87c) the reverse current falls exponentially as the depletion layer capacitance charges via R_L and at the same time the diode voltage v_D rises exponentially to the applied reverse bias $-V_S$. Eventually the diode recovers its reverse-biased state, the depletion layer capacitance being

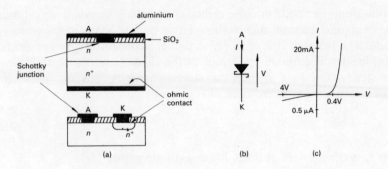

Figure 2.88 Schottky-barrier diode: typical structures, circuit symbol and *I–V* characteristic

fully charged to $-V_S$, and the reverse current then being the reverse leakage value due to thermally generated holes in the cathode near the junction. If a time t_4 is defined as the instant at which the reverse current has fallen to 10 per cent of I_R, the time interval $t_3 \rightarrow t_4$ is termed the *transition* time t_t (another name for depletion layer being transition layer) and the reverse recovery time t_{rr} is then the sum of the storage and transition times. The minority carrier distributions near the junction during this interval ($t_1 \rightarrow t_4$) are shown in figure 2.87a, the growth of the depletion layer not being shown.

2.4.2 Schottky-barrier Diode

A Schottky-barrier diode comprises a rectifying metal–semiconductor junction formed between aluminium, gold or platinum metallisation and lightly doped *n*-type silicon or gallium arsenide (figure 2.88a) encapsulated in a low capacitance package. The conduction properties of a rectifying metal–semiconductor junction are discussed in section 1.6.1 in terms of the effect of an applied bias on the energy of free carriers. An alternative way of considering the rectification properties of an aluminium/*n*-type silicon junction is to appreciate that the aluminium acts as a *p*-type dopant at the surface of the *n*-type silicon resulting in *pn* junction properties. If the silicon is heavily doped, however, the *p*-type dopant due to the aluminium is swamped by the *n*-type dopant and an ohmic contact results. The name Schottky-barrier derives from the *Schottky effect* whereby the potential barrier at the junction is slightly reduced by the charge at the metal surface.

The rectification properties of Schottky diodes depend critically upon conditions at the metal–semiconductor interface. It was not until the late 1960s that adequate control of the surface in terms of extreme cleanliness to avoid surface states due to impurities and defects, together with protection by surface passivation, made commercial production viable.

The lack of holes on the metal side of the Schottky junction means that charge transfer under forward bias conditions (metal positive with respect to the n-type semiconductor) is due almost entirely to the injection of electrons from the semiconductor into the metal. For this reason, conduction is said to be by majority carriers in contrast to the case of a *pn* semiconductor junction where forward current is due to the injection of minority carriers into each region from the opposite side of the junction. The potential barrier or contact potential is lower for a Schottky junction than for a typical *pn* junction, the forward voltage drop being typically 0.4 V (figure 2.88c) compared with 0.7 V for a *pn* junction. For electrons to be injected into the metal from the semiconductor under forward bias, they must have sufficient energy to cross the potential barrier at the junction and on entering the metal they have energies considerably greater than the mean electron energy or Fermi level in the metal. These injected electrons are thus termed 'hot' electrons and the device is often described as a *hot carrier diode*.

As forward current flow is by the injection of electrons into the metal which already contains a high density of free electrons, the storage effect which limits the switching speed of *pn* junction diodes does not occur in Schottky diodes. In addition, by making the junction area (surface metallisation) small, the depletion capacitance due to space charge can also be kept small. The over-all small capacitance provides an extremely fast switching speed of typically a few picoseconds.

Due to the high electron concentration in the metal the drift of electrons across the junction into the semiconductor is considerably higher than in the *pn* junction case resulting in a reverse leakage current at least an order of magnitude greater than for a corresponding silicon *pn* junction (figure 2.88c). More detailed consideration of Schottky diodes can be found in reference 30.

The static performance of a Schottky diode can be modelled using the same techniques described in section 2.1.4 for the *pn* junction diode. As a high-speed switch, however, the switching delay is usually of most interest and, if this delay is not negligible in comparison with other transients within the system, it can be introduced mathematically in the system analysis as a finite time interval.

In addition to use as a high-speed switch, the good rectification properties of the Schottky diode at frequencies up to 10 GHz and beyond make it useful as a UHF signal detector. It also finds application as a mixer at UHF and due to the low forward voltage drop it is used in integrated form as a voltage clamp to prevent saturation in BJTs and so improve the switching speed of bipolar logic circuits (section 4.7.5).

2.4.3 Semiconductor-controlled Rectifier (SCR)

The semiconductor (or silicon)-controlled rectifier or *thyristor* is a multilayer *pn* diode structure having the property that forward conduction is controllable by the signal applied to a third terminal known as the *gate*. Details of typical structure and characteristics are given in figure 2.89. Low-current SCRs ($I_A \leq 1$ A) are produced in low-power 'transistor-style' packages such as the TO-39 BTX 18 in figure 2.89a. The majority of SCRs are in the medium to high-current range (1–200 A) and are produced in either flat plastic encapsulation or stud-mounted metal packages (figure 2.89b). Very high-power devices that are capable of handling average forward currents up to 1.5 kA and reverse voltages approaching 4 kV, are normally encapsulated in a 'hockey puck' (as in ice hockey) package typically up to 100 mm diameter $\times$ 35 mm thick. In this type of encapsulation the silicon slice is sandwiched between heat-sink electrodes forming the upper and lower surfaces of the 'puck'. These electrodes are water-cooled in the highest current types.

Operation of the SCR may be considered with reference to the basic structure shown in figure 2.89e. With the gate terminal open-circuit ($I_G = 0$), application of a small positive anode–cathode voltage V_{AK} causes junction J_2 to be reverse-biased and the only current flow is due to leakage across this junction, this being the *forward-blocking state*. If V_{AK} is increased, eventually, at $V_{AK} = V_{BO}$ the *forward breakover voltage* avalanche breakdown of J_2 enables forward current flow and for further increase of V_{AK} the SCR behaves as a forward-biased diode due to junctions J_1 and J_3 (figure 2.89f). When the anode current I_A is reduced, V_{AK} remains low (typically of the order of 1 V) until I_A has been reduced to below the *holding current* I_H which is typically 10 mA for a low-power device up to approximately 1 A for a high-power type. This latching situation, whereby forward conduction cannot be halted until I_A has been reduced below a certain threshold, is due to the fact that the flow of charge above this threshold is sufficient to maintain

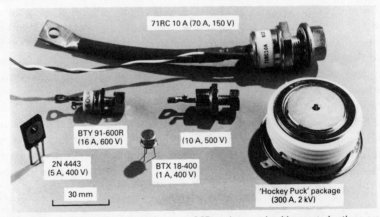

(a) Various low, medium and high power SCR package styles (the quoted ratings are the maximum average ON-state current $I_{T(AV)}$max and the maximum repetitive peak reverse voltage V_{RRM}max)

Figure 2.89 Semiconductor (silicon) controlled rectifier (SCR) (continued opposite)

Discrete Semiconductor Devices

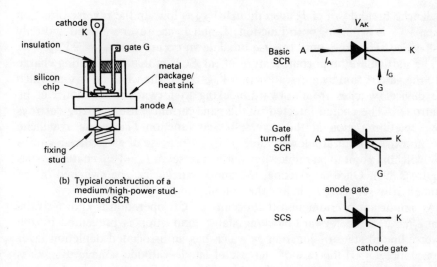

(b) Typical construction of a medium/high-power stud-mounted SCR

(c) Circuit symbols

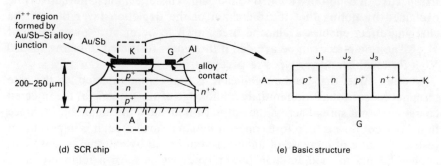

(d) SCR chip

(e) Basic structure

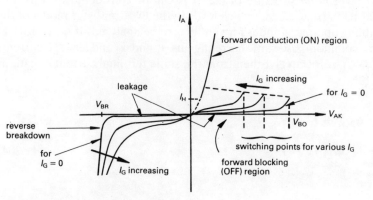

(f) I_A–V_{AK} characteristics for various I_G

avalanche breakdown of J_2 even though V_{AK} is low. In the reverse direction when V_{AK} is negative, both junctions J_1 and J_3 are reverse-biased and only leakage current flows until reverse breakdown occurs at $V_{AK} = -V_{BR}$.

The particularly important feature of the SCR is that by injecting charge into the gate region by application of a V_{GK} pulse, the value of V_{AK} at which the device switches from forward-blocking to forward-conduction can be controlled. The charge injected by the gate current pulse partially destroys the depletion region at the reverse-biased junction J_2 causing avalanche breakdown to occur at a lower value of V_{AK}. Increase of gate current causes the switching point to be reduced, resulting in a set of $I_A - V_{AK}$ characteristics (figure 2.89f). Once conducting, the anode current can be switched off by momentarily reducing I_A below the holding value.

An important phenomenon that occurs in SCR operation is known as the *rate effect*. In the forward-blocking state, conduction is prevented by the reverse-biased state of junction J_2 which has an associated depletion layer capacitance C_T. If the rate of increase of anode–cathode voltage dv_{AK}/dt is fast, the charging current $C_T\, dv_{AK}/dt$ may be sufficiently large to initiate avalanche breakdown of J_2 and hence cause a switch to the conducting state irrespective of conditions applied to the gate. This effect may alternatively be explained by noting that the reduction of the depletion layer due to the charging effect enables avalanche breakdown to occur at a lower value of V_{AK}. The rate effect can be avoided if dv_{AK}/dt is restricted to below 10–300 V/μs for low-high power types respectively.

Another effect that can be significant, is that charge injected via the gate terminal causes conduction initially at the edge of the cathode as the injected charge does not spread across the entire cathode area instantaneously. Since initial current flow is therefore concentrated in a small area, it is important to restrict the rate of increase of anode current di_A/dt by external means. This applies mainly to medium-high power types which have a relatively large junction area, the limit being typically in the range 50–400 A/μs.

The four layer *pnpn* structure of the SCR can be considered to consist of two merged BJTs (figure 2.90) which leads to the two-transistor model of the device enabling operation under forward bias to be analysed. If transistors T_1 and T_2 have common-base static current gains α_1 and α_2 and leakage currents I_{CBO_1} and I_{CBO_2} respectively, then from the static terminal current equations 2.9 and 2.10

$$I_{C_2} = \alpha_2 I_K + I_{CBO_2} \tag{2.203}$$

$$I_{B_1} = (1-\alpha_1) I_A - I_{CBO_1} \tag{2.204}$$

but

$$I_{B_1} = I_{C_2} \text{ and } I_K = I_A + I_G$$

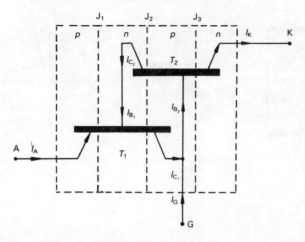

Figure 2.90 Two-transistor model of an SCR

thus

$$(1-\alpha_1) I_A - I_{CBO_1} = \alpha_2(I_A + I_G) + I_{CBO_2} \qquad (2.205)$$

The static anode current under forward bias is therefore

$$I_A = \frac{\alpha_2 I_G + I_{CBO_1} + I_{CBO_2}}{1 - (\alpha_1 - \alpha_2)} \qquad (2.206)$$

For values of V_{AK} well below the switching point V_{BO}, $\alpha_1 + \alpha_2 \ll 1$ and the anode current is due mainly to leakage. As V_{AK} increases, the gains α_1 and α_2 increase due to multiplication at the 'collector' junction J_2, and at $\alpha_1 + \alpha_2 = 1$ switching occurs at which I_A is limited by the external circuit and equation 2.206 no longer applies.

Some commercially available SCRs can be turned off by applying a negative pulse to the gate. They are termed *gate turn-off* (GTO) SCRs, the circuit symbol being given in figure 2.89c. The negative pulse applied to the gate, typically −5 V for ≈1 μs, removes holes from the *p*-type gate region. This effectively reduces the gain of the *pnp* transistor T_1 (two-transistor model, figure 2.90) and if the doping levels and dimensions are such that the removal of holes is sufficient to bring the *npn* transistor T_2 out of saturation, the SCR switches off.

Other SCRs have external connection to both the internal *p* and *n* regions of the *pnpn* structure giving an anode gate (*n* region) and a cathode gate (*p* region). This two-gate device is termed a *semiconductor-controlled switch* (SCS) and gives the added flexibility compared with the basic SCR of

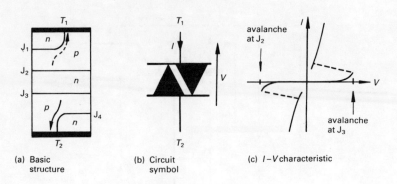

Figure 2.91 Diac

switch-on either by a positive pulse at the cathode gate or by a negative pulse at the anode gate. In addition, unwanted switching due to the rate effect can be reduced by capacitively coupling the gate not used for triggering to either A in the case of the anode gate or K for the cathode gate. This provides a charging path for junction J_2 during the voltage change so that the device current is not sufficiently high to initiate breakdown. The circuit symbol for a SCS is given in figure 2.89c.

SCRs are widely used in current-switching applications such as d.c. and a.c. motor speed control, choppers for electric vehicle control (d.c. ON–OFF control), d.c.–a.c. power supplies (inverters) used for standby power supplies and switched control of a.c.–d.c. power supplies.

2.4.4 Diac

The $I_A - V_{AK}$ characteristic of the SCR is not symmetrical since for negative V_{AK} there are two reverse-biased junctions in series. In order to cause reverse breakdown, V_{AK} must be sufficiently large to cause both junctions to be at breakdown and the sustaining effect due to the current flow, that occurs for positive V_{AK}, does not occur. The *diode a.c.* switch (diac) or bilateral diode switch is effectively two *pnpn* structures in parallel giving a symmetrical switching characteristic (figure 2.91).

With the applied voltage V small and positive, junction J_2 is forward-biased but J_3 is reverse-biased and so only leakage current flows. If V is increased, eventually avalanche breakdown of J_3 occurs and current flows along junction J_4 as shown in figure 2.91a. The voltage drop along the current path on the p-side of J_4 causes it to become partially forward-biased and the injected carriers then provide the sustaining effect causing the switching action. Junction J_1 has no effect during forward conduction. As the structure is symmetrical, a similar switching action occurs for negative V initiated by

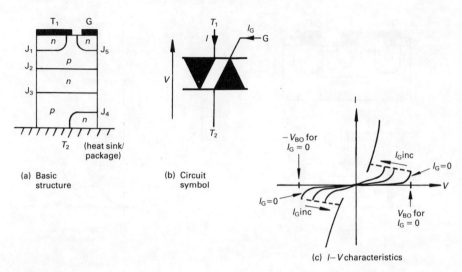

Figure 2.92 Triac

breakdown of J_2 and sustained by injection created by the lateral current along J_1, junction J_4 then having no effect.

Diacs are usually encapsulated in a cylindrical plastic package with axial leads and are used primarily for triggering triacs (section 2.4.5).

2.4.5 Triac

The *triode a.c.* switch or *triac* may be considered as a diac with the addition of a gate terminal allowing control of the switching point (figure 2.92a). As it may be triggered by an applied gate current pulse for either polarity of applied voltage, it behaves as an 'a.c. SCR'.

For $I_G = 0$ the operation for $\pm V$ is identical to that of the diac (section 2.4.4). The switching point for positive V may be controlled by supplying electrons to the gate region by application of a negative gate current pulse. The injected electrons are swept across junctions J_5 and J_2 and reduce the depletion region of the reverse-biased J_3 junction, thus reducing the switching voltage. For negative V, the switching point is determined by J_2 and a supply of holes via the gate by application of a positive I_G pulse, reduces the depletion region at J_2 and hence the switching point.

Triacs are packaged similarly to SCRs and are used in a.c. switching and phase control applications such as a.c. motor speed control, remote switching and dimming of lighting and a.c. power control. Triacs and SCRs enable control of high currents by a low-power control source and since they provide 'static switching', that is, not involving mechanical movement of contacts, problems such as contact bounce and arcing are avoided.

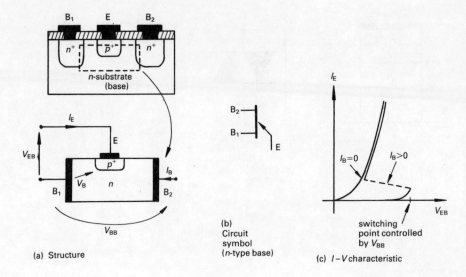

Figure 2.93 Unijunction transistor

2.4.6 Unijunction Transistor

The unijunction transistor (UJT) is essentially a junction diode in which forward conduction below the switching point can be prevented by the voltage developed on one side of the junction by a bias current flowing through the resistive semiconductor (figure 2.93a).

For zero bias current I_B, the $I_E - V_{EB}$ characteristic is the forward characteristic of the p^+n junction which has a lower slope than a conventional pn junction diode due to the light doping of the n-type 'base' region. With V_{BB} applied causing current flow I_B through the base region, a bias voltage V_B is established between the B_1 base contact and the junction so that the junction cannot become forward-biased until $V_{EB} > V_B$. The device thus behaves as a switch (figure 2.93c) with the switching point controlled by the bias voltage V_B and hence by the applied control voltage V_{BB}. The voltage V_{EB} falls after switching as carriers injected from the p^+ emitter region cause the conductivity of the n-type base to increase causing V_B to fall. This increases injection resulting in a regenerative switching action. UJTs are used in triggering and timing circuits.

2.5 DEVICES USED IN SIGNAL GENERATION

This section is concerned with special-purpose devices that are used mainly in the generation of microwave signals (nominally 1–300 GHz, appendix D), notably the tunnel diode, Gunn diode, varactor diode, step-recovery diode and the range of avalanche transit-time (ATT) devices. These high-frequency devices which are fabricated using silicon, germanium or group III–V semiconductors such as gallium arsenide are commonly packaged in low-capacitance, low-inductance assemblies of the type shown in figure 2.96c.

2.5.1 Tunnel Diode

When the doping levels of a silicon *pn* junction exceed about 10^{25} atoms/m^3, the electric field created in the narrow depletion region of a reverse-biased junction is sufficiently high to cause field emission (ionisation of lattice atoms by electrostatic attraction) and hence reverse current flow (Zener breakdown) at low reverse bias voltages of only a few volts (section 1.7). Increase in the doping concentration reduces the breakdown voltage and if the doping is sufficiently heavy ($\approx 10^{27}$ atoms/m^3) to cause *degeneracy* (section 1.2.3) the junction is at breakdown even for small *forward* bias. At higher forward bias the depletion region disappears and with it the high electric field, causing the forward characteristic to return to the normal form for a forward-biased *pn* junction. This change of mechanism under forward bias results in a region of *negative slope resistance* (figure 2.94a) which causes a degree of instability and provides the basis of use as an oscillator. Typically the peak current I_P lies in the range 50 μA–10 A, the peak to valley current ratio I_P/I_V is 5–20, the peak voltage V_P is 50–200 mV and the valley voltage V_V 0.5–1 V depending on the semiconductor, doping levels and geometry of the junction.

This unusual operation can be explained by reference to the energy diagram of the junction (figure 2.95). The Fermi level of a degenerate

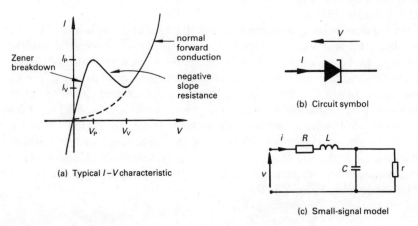

Figure 2.94 Tunnel diode

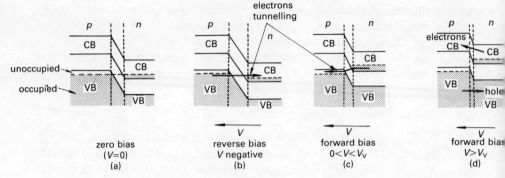

Figure 2.95 Energy diagrams describing the conduction mechanism for a tunnel diode

semiconductor is within the valence band (VB) for *p*-type material and within the conduction band (CB) for *n*-type so that there is a range of levels at the top of the VB in *p*-type material that have a low probability of being occupied and a range of levels at the bottom of the CB in *n*-type material that have a high probability of being occupied (figure 2.95a). Under conditions of reverse bias or small forward bias, the increase of energy of the electrons on the *p* or *n*-side of the junction respectively, allows current flow across the narrow depletion region to vacant levels on the other side of the junction (figure 2.95b,c). This current flow mechanism is different from the drift and diffusion components associated with normal conduction across a *pn* junction (section 1.6.2) and is described as electrons 'tunnelling' under or through the potential barrier; hence the name of the device. At higher forward bias ($V > V_V$) the energy of CB electrons on the *n*-side of the junction does not correspond to that of vacant levels in the VB on the *p*-side (figure 2.95d) and tunnelling does not occur. Current flow is then mainly due to diffusion of CB electrons and VB holes as in normal forward conduction. The tunnelling effect was first described by Esaki in 1958 and early devices utilising this effect were called Esaki diodes. [31]

In application as a microwave source, the tunnel diode is biased to operate between V_P and V_V, the negative slope properties causing an increase in voltage to be accompanied by a decrease in current and vice versa. By including the device in a loop having a gain greater than unity, the circuit oscillates. In practice the feedback loop contains a tuned circuit to define the frequency of oscillation. Figure 2.94c gives a small-signal model representing tunnel diode behaviour in the negative slope resistance region where r is the slope resistance (typically $-100\ \Omega$), C is the capacitance of the junction and package (≈ 5 pF) and R and L are the series resistance and inductance of the bulk semiconductor, package and leads (typically 1 Ω and 5 nH respectively). The diode impedance Z is thus

$$Z = R + j\omega L + \frac{r/j\omega C}{r + 1/j\omega C}$$

$$= \left(R + \frac{r}{\omega^2 C^2 r^2 + 1}\right) + j\left(\omega L - \frac{\omega C r^2}{\omega^2 C^2 r^2 + 1}\right) \qquad (2.207)$$

As r is negative, the real part of this impedance is zero or negative for frequencies up to $(1/2\pi CR)\sqrt{(r/R - 1)}$ which is the maximum frequency at which the diode exhibits negative slope resistance and hence an upper frequency limit to use as an oscillator. Tunnel diode oscillators can provide a few milliwatts of signal power at the lower end of the microwave range to a few microwatts at the higher end.

It should be noted that by choosing a load line intersecting the diode characteristic at two points, the tunnel diode can be used as a high-speed switch and formerly was of considerable interest in this respect. More detailed consideration of the tunnel diode, its fabrication, operation and application can be found in reference 32.

2.5.2 Gunn Diode

Certain semiconductors, notably group III–V compounds such a gallium arsenide, exhibit a phenomenon whereby high-frequency oscillations are superimposed on the current flow through the material when the applied electric field exceeds a critical value of about 300 kV/m (figure 2.96a). The frequency of the oscillation, which is in the microwave region, is related to the length of the semiconductor sample. This property was first observed by Gunn in 1963 and became known as the *Gunn effect*; [33] it forms the basis of a microwave signal source.

The effect is due to the existence of two conduction bands separated by a small energy gap (≈ 0.4 eV for GaAs) and the oscillations are caused by the repetitive transfer of electrons between the two bands. When the critical field is exceeded, electrons in the lower band receive sufficient energy to transfer to the upper band in which they have a lower mobility causing a drop in current for increase in field; a negative slope resistance effect. The electric field in the semiconductor is not uniform, a high field region or *domain* being established which drifts towards the positive end of the sample. This domain collapses on reaching the end of the sample generating a pulse of current. As the rate of arrival of these pulses, and hence the frequency generated, is dependent on the time that the domain takes to drift along the sample, it follows that the frequency can be controlled by the length of the sample.

Gunn diodes are formed from a thin homogenous n-type layer (≈ 10 μm thick) grown on an n^+ substrate (≈ 100 μm square), ohmic contacts being made to the surface layer and substrate (figure 2.96b). It should be noted that the Gunn diode is a *bulk* semiconductor device, no *pn* junction is involved, the term diode referring to the *two* ohmic contacts. With reference to the mechanism of electron transfer between bands, Gunn diodes are also termed *transferred-electron devices*. Heat dissipation is a major problem, limiting the

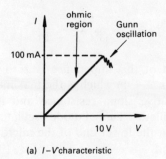

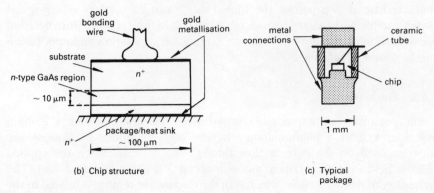

Figure 2.96 Gunn diode

lower frequency of operation of these devices. A lower frequency requires a thicker semiconductor layer which increases the problem of heat sinking. Gunn diodes are produced mainly for the 10–30 GHz operating range giving output powers of a few milliwatts for continuous wave operation. The heat problem is reduced if the device is pulsed, in which case peak powers in excess of 100 W are common. More detailed consideration of Gunn diodes can be found in references 34, 35 and 36.

2.5.3 Varactor Diode

A varactor diode is a *pn* junction diode designed to exploit the junction capacitance, the term *varactor* being derived from *var*iable re*actor*. In section 1.6.2 it is shown that the change of depletion layer space charge with applied voltage causes the junction capacitance C_T to vary with voltage (figure 2.97a). This property enables the varactor diode to be used as a control element such as in a voltage-controlled oscillator (VCO) where an input control voltage can be used to control the capacitance in a tuned circuit and hence the frequency of the generated signal. A network model representing the performance of a varactor diode is given in figure 2.97c, where $C(v)$ is the voltage-dependent capacitance of the junction (typical variation 10–100 pF), R is the resistance of the bulk semiconductor regions and L and C are the inductance and capacitance of the package.

The varactor diode is included in this section concerned with microwave signal generation since, as well as being used as a voltage-controlled capacitor at relatively low frequencies, its variable capacitance properties form the basis of the low-noise parametric amplifier used at microwave frequencies. The parametric amplifier, which is described in more detail in reference 37, operates by varying a *parameter* of the system (the capacitance) with time using an applied voltage *pump* waveform. The principle of amplification relies on the fact that the stored charge in the diode $q = Cv$ cannot change instantaneously and if the capacitance is suddenly reduced the voltage must correspondingly increase. By adjusting the frequency of the pump waveform relative to the input signal frequency, amplification can be achieved which largely avoids the introduction of noise, due to random thermal energies and the discrete nature of electrons that occurs in more conventional amplifiers. [17,18] The small amount of noise that is generated is reduced in practice by cooling the amplifier to $-268\ °C$ using liquid helium, resulting in an extremely low-noise amplifier which is used in the Earth receiving-stations of satellite communication systems.

Practical parametric amplifiers are considerably more complex than is indicated above due to the difficulty of maintaining the correct phase relationship between the pump and signal frequencies. Performance is improved if an additional tuned circuit known as the *idler* circuit is connected across the diode. Consideration of such an arrangement shows that there is an increase in frequency between the signal and idler circuits which makes the arrangement useful in high-frequency signal generation by enabling frequency *multiplication* from a relatively low-frequency source (≈ 100 MHz for example) to a higher frequency in the gigahertz region. [37] Used in this mode, the system is termed an *up-converter*.

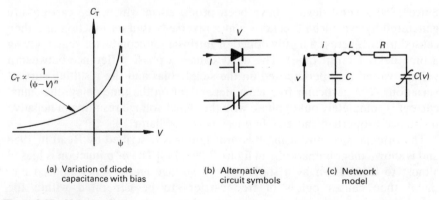

(a) Variation of diode capacitance with bias

(b) Alternative circuit symbols

(c) Network model

Figure 2.97 Varactor diode

2.5.4 Step-recovery Diode

During forward conduction, charge is stored in the vicinity of a *pn* junction. When the applied bias is reversed, a transient period exists, known as reverse-recovery, during which the stored charge is removed, before the steady-state reverse bias conditions are established (figure 2.98a). A *step-recovery diode* is designed so that the voltage transient at the end of the reverse-recovery period is very sharp (5–50 ps). Fourier analysis of such a transition shows that it has a high harmonic content and so the switching action can be used to generate high-frequency harmonics. With suitable filtering, the system is used as a frequency *multiplier* between the switching or pump waveform v_p and the output waveform, thus enabling use as a microwave signal source in the form of an *up-converter* (figure 2.98e).

The voltage transient is sharpened producing a 'snap' action by introducing a thin lightly doped *p* or *n*-type layer (typically 1–5 μm thick) between the p^+ and n^+ regions forming a p^+nn^+ or p^+pn^+ structure (figure 2.98c). Due to this fast switching action, and since the central lightly doped layer is almost intrinsic, step-recovery diodes are also called *snap* or *pin diodes*. The sharper transient at switch OFF obtained with this structure compared with a normal *pn* junction is due to the fact that the stored charge is mainly within the central '*i*' region and during switch OFF, charge removal is assisted by the electric field in this region. This does not occur in a normal *pn* junction as the stored charge under forward bias is in the essentially field-free bulk *p* and *n* regions.

The pin structure is also useful as a high-voltage rectifier. Under reverse bias, the depletion layer extends across the '*i*' region which can be made relatively thick (0.1–1 mm), reducing the electric field and giving a high reverse breakdown voltage.

2.5.5 Avalanche Transit-time (ATT) Devices

Since 1960 several devices have been produced in which free carriers are generated by avalanche breakdown at a reverse-biased *pn* junction and then caused to drift across a lightly doped or intrinsic semiconductor region giving a time delay (transit time). The delay causes a phase difference between a voltage variation superimposed on the steady bias and the resulting current variation. At a particular frequency dependent on the time delay, the signal current is completely out of phase with the signal voltage resulting in negative resistance properties and enabling use as an oscillator.

The original structure using this principle was described by Read in 1958 and is shown diagrammatically in figure 2.99a. [38] The n^+p junction is biased almost to breakdown by a static applied voltage and a superimposed a.c. signal then causes pulses of free carriers to be generated within the junction-depletion region. There is a time delay in the avalanche process due

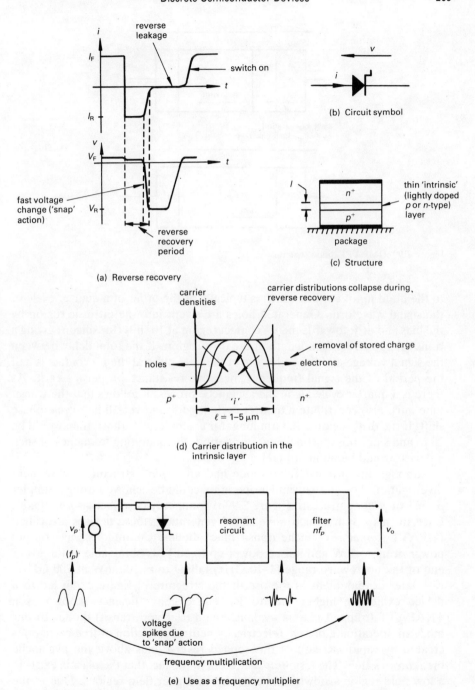

Figure 2.98 Step-recovery, snap or pin diode

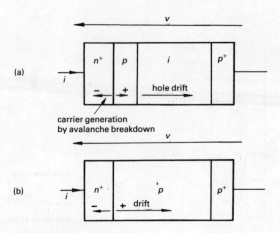

Figure 2.99 IMPATT diode structure

to the build up of current which is typically of the order of a quarter cycle of the signal waveform. Generated holes are swept into the intrinsic region by the bias and drift towards the p^+ contact region at 10^5 m/s (for silicon) giving a transit time of 10 ps/μm length of the drift region. If the total delay between the signal voltage and the collection of the hole pulse at the p^+ contact is half the period of the signal frequency, negative resistance properties exist. As there is a quarter cycle lag in the avalanche process it follows that the transit time must also contribute a quarter cycle lag for an over-all half-cycle phase shift. If the drift region is 2.5 μm long, for example, the transit time would be 25 ps and since this must be a quarter period, the operating frequency of such a device would be about 10 GHz.

Although the original Read diode had an n^+pip^+ structure, subsequent investigation showed that similar operation could be achieved using a simpler p^+pn^+ or p^+nn^+ structure (figure 2.99b) commonly described as *pin diodes*. Used in this way for microwave signal generation, these devices are called IMPATT (*imp*act *a*valanche *t*ransit *t*ime) diodes. Continuous wave output power of up to 1 W and pulsed power up to 50 W is obtainable at the lower end of the microwave range (1–10 GHz) falling to ≈ 100 mW at 100 GHz.

A later development in the use of the structure of figure 2.99b led to a device exhibiting higher d.c. to RF conversion efficiency, known as a TRAPATT (*tra*pped *p*lasma *a*valanche and *t*riggered *t*ransit) diode. In this mode of operation a *plasma* (electrically neutral collection of free carriers) is created by rapid increase of the applied reverse bias above the avalanche breakdown value. The term *trapped* refers to the fact that the plasma exists in a low field region sandwiched between two higher field regions. Due to the slower collection of the plasma charge in TRAPATT operation compared with the transit time in the IMPATT process, TRAPATT diodes are most useful at the lower end of the frequency range (1–10 GHz) where a long drift region in an IMPATT diode causes major heat sinking problems.

A related device that uses a drift period to produce a time delay is the *barri*er *t*ransit-*t*ime (BARRITT) diode which is a metal–semiconductor–metal structure utilising two Schottky barrier contacts separated by a thin semiconducting layer. If the applied voltage is sufficient to deplete the semiconductor, the forward-biased junction cannot inject electrons into the metal (as the semiconductor is depleted) but holes are injected into the semiconductor. This is because contact with the metal maintains the VB concentrations in the semiconductor at their thermal equilibrium values. The injected holes drift across the semiconductor region and at a frequency dependent on the transit time, the device exhibits negative slope resistance enabling use as a microwave signal source.

In the mid 1970s a hybrid IMPATT/transistor structure was reported which increased the power capabilities of transistors in the 1–5 GHz region. The device, known as a *c*ontrolled-*a*valanche *t*ransit-*t*ime (CATT) triode, is basically a BJT in which the C-B junction has an IMPATT diode structure. The input signal applied to the B-E junction controls injection into the base which in turn controls the avalanche process at the reverse-biased IMPATT C-B junction. This control of the IMPATT operation from a low-power source enables the device to be used as a microwave amplifier.

More detailed consideration of ATT devices can be found in references 39 and 40.

2.6 OPTICAL DEVICES

In connection with electronic devices, the term *optical* refers to the frequency range 10^{12}–3×10^{16} Hz which includes infrared (IR) and ultraviolet (UV) radiation as well as the visible spectrum. In terms of the electromagnetic spectrum (appendix D), the optical range extends beyond the radio frequency (RF) range to X-rays.

Optical devices act as electro-optical or optoelectronic transducers (table 2.14) which shift the frequency of transmitted energy from the lower-frequency 'electronic' range to the higher-frequency 'optical' range or vice versa. A selection of semiconductor optical transducers is shown in figure 2.100.

The mechanisms involved in the use of semiconductors in optical devices are associated directly with the energy gap W_g between the valence and conduction energy bands (VB and CB) of the material. In optoelectronic transducers, the absorption of incident optical energy can cause electrons to break free from their parent atoms so generating electron–hole pairs. This increase in the free carrier densities (electrons in the CB and holes in the VB) causes a change in conductivity (photoconductivity) or a generated current

272   Semiconductor Device Technology

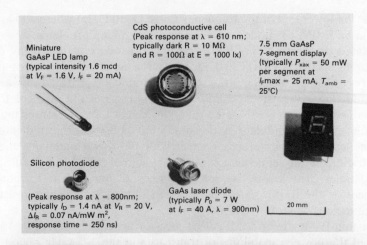

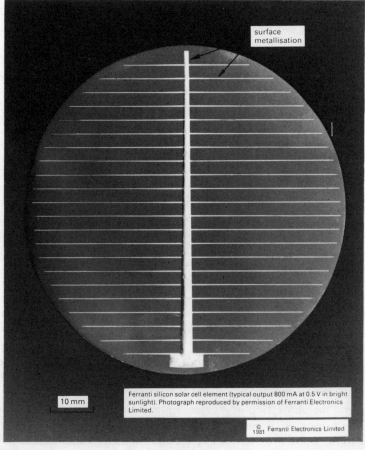

Figure 2.100   Various semiconductor optical transducers

Discrete Semiconductor Devices

Table 2.14 Semiconductor optical devices

Electro-optical transducers
    light-emitting diode (LED)
    semiconductor laser (diode)
    luminescent coatings and film devices

Optoelectronic transducers
    photodiode
    phototransistor (bipolar)
    photofet
    photothyristor
    photoconductive cell (light-dependent resistor, LDR)
    photovoltaic (solar) cell

depending on how the effect is utilised. Development of models representing the electrical properties of material and radiation (section 1.1) led Planck to suggest that radiation is emitted in discrete pulses or *quanta*. Subsequent work by Einstein on the photoelectric effect in which light (optical energy) is directed at a metal surface enclosed in a vacuum, showed that the energy of the emitted electrons is related directly to the frequency of the incident radiation. This work led to the realisation that optical energy radiates as discrete pulses of energy termed *photons*, each having an energy

$$W = hf \tag{2.208}$$

where f is the frequency of the radiation and h is the proportionality constant known as Planck's constant ($h = 6.63 \times 10^{-34}$ J s $= 4.14 \times 10^{-15}$ eV s). Radiation of optical energy can thus be considered as a stream of energy pulses, photons or 'light particles' each of energy given by equation 2.208. The effect of optical radiation on a semiconductor depends on the relation of the energy of the incident photons (hf) to the energy gap (W_g) of the material. If the frequency of the incident radiation is such that

$$hf \geq W_g \tag{2.209}$$

absorption occurs resulting in free carrier generation as shown diagrammatically in the energy diagram of figure 2.101. If $hf < W_g$ the photon energy is not sufficient to cause photoconduction and the incident photons are not absorbed. Instead they are transmitted through the material which is then said to be *transparent*. It should be noted that if the frequency of the incident radiation is progressively increased, photoconduction occurs when $hf = W_g$ providing a convenient method of measuring the energy gap of the

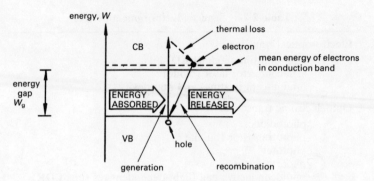

Figure 2.101 Energy absorption and emission by a semiconductor

semiconductor. In practice it is usual to refer to the wavelength λ of the optical radiation in microns (μm) or angstroms (Å), where $1 \text{ Å} = 10^{-10}$ m $= 10^{-4}$ μm, instead of its frequency

$$\lambda = \frac{c}{f} \qquad (2.210)$$

c being the velocity of propagation of electromagnetic radiation or 'speed of light' in free space, 3×10^8 m/s. Combining equations 2.209 and 2.210, the condition for photoconduction in terms of wavelength is

$$\lambda \leq \frac{hc}{W_g} \qquad (2.211)$$

Figure 2.102 lists the energy gaps of several semiconductors and indicates the corresponding radiation wavelength given by the equality in equation 2.211. Only radiation of wavelength less than or equal to this value is absorbed by the material resulting in photoconduction; higher wavelengths are transmitted. Thus silicon and gallium arsenide absorb the whole of the visible spectrum, UV radiation and a small range of IR wavelengths. Gallium phosphide, which has an energy gap corresponding to the green wavelength in the visible range, does not absorb any IR radiation and zinc sulphide absorbs only UV radiation.

If the energy of the incident radiation is considerably larger than the energy gap of the semiconductor, the generated electrons initially have energy greater than the mean energy of electrons in the conduction band. These electrons rapidly lose energy during conduction through the material until they have the same energy as the majority, that is, they are then travelling at the same velocity as other electrons. This loss, which imparts energy to the material causing its temperature to rise, may be considered in terms of the

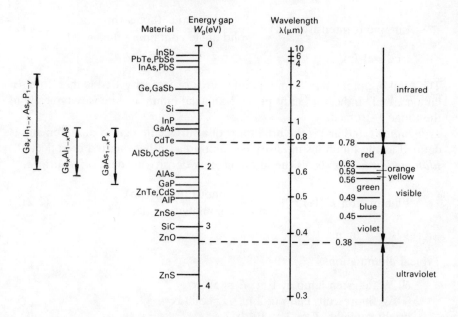

Figure 2.102 Relation of semiconductor energy gap to the optical range

particle atomic model to be due to collisions between the higher-velocity electrons and the crystal lattice.

Eventually the generated free carriers recombine, the free electron being captured by a vacancy in the binding structure of the crystal which, in terms of the band model of figure 2.101, is represented by the return of the electron from the CB to the VB. In several materials having a high proportion of direct recombination, in contrast to indirect recombination via traps (section 1.3.1), the energy released during capture is emitted from the material (electroluminescence). The wavelength of the emitted radiation is given by the equality in equation 2.211.

The variables used to quantitatively describe *visible* optical radiation are defined as follows

Luminous flux ϕ (lumen, lm) is the total visible power radiated or the rate of flow of light energy. At the peak sensitivity of the human eye ($\lambda = 555$ nm), 1 lumen $\equiv 1.47$ mW.

Luminous intensity I (candela, cd) is the luminous flux per unit solid angle in a given direction.

The unit solid angle (steradian, sr) is the solid angle subtended at the centre of a sphere of radius r by an area r^2 on the surface of the sphere. As the total surface area of a sphere is $4\pi r^2$, the solid angle around a point is 4π steradians.

Luminous intensity $I(\text{cd}) = \dfrac{\text{luminous flux } \phi \text{ (lm)}}{\text{solid angle containing flux } \omega(\text{sr})}$ (2.212)

1 cd = 1 lm/sr

The total luminous flux from a point source of intensity 1 cd is therefore 4π lumens. LED indicator lamps provide an axial luminous intensity typically in the range 1–10 mcd.

Luminance L (cd/m^2) is the brightness of a surface in a given direction and is defined as the luminous intensity per unit area, either emitted or reflected.

Illumination E (lux, lx) is the luminous flux density incident on a surface.

Illumination E (lx) = $\dfrac{\text{luminous flux } \phi \text{ (lm)}}{\text{area of illuminated surface } A \text{ (m}^2\text{)}}$ (2.213)

1 lx = 1 lm/m^2

Typical illuminations

60 W tungsten lamp at 1 m, $E \approx 50$ lx
80 W fluorescent lamp at 2 m, $E \approx 500$ lx
bright sunlight, $E \approx 3 \times 10^4$ lx

Where the radiation is not in the visible range (that is, not luminous) as in the case of infrared radiation, for example, the general term *irradiance* is used in place of illumination.

Irradiance H (W/m^2) is the radiant flux density incidence on a surface.

Irradiance $H(\text{W/m}^2)$ = $\dfrac{\text{radiant flux}}{\text{surface area irradiated}}$ (2.214)

Since, at a particular wavelength, one lumen is equivalent to a particular radiant power, the units W/m^2 and lx are directly related.

The term *colour temperature* T_C (K) is used to specify the colour of visible radiation. It is the actual or extrapolated absolute temperature at which a black-body radiator would emit radiation of the same colour. An unfiltered tungsten incandescent lamp emits radiation at a colour temperature of 2800 K.

2.6.1 Light-emitting Diode (LED)

The light-emitting diode utilises a forward-biased *pn* junction to provide continuous injection of free carriers into a semiconductor region. During subsequent recombination of the injected carriers, radiation is emitted from the surface of the semiconductor, the process being termed *injection electroluminescence*. As the intensity of the emitted radiation is dependent on the rate of injection and hence the applied bias, the LED is useful not only as a solid-state replacement of the incandescent indicator lamp but also as an optical signal source. Modulation of the applied forward bias produces a beam of radiation of modulated intensity which is useful as a low-power optical source for communication systems using optical fibres.

Figure 2.101, showing a direct transition from CB to VB, tends to indicate that a particular semiconductor having a certain energy gap has a fixed emission wavelength given by $\lambda = hc/W_g$. In practice, dopants such as Zn, Cd, Te, S or Se are added to introduce trap levels just above the VB and just below the CB which have the effect of narrowing the main or *dominant* transition. This technique is used either to slightly alter the wavelength of the emitted radiation or to cause materials, which do not emit in the pure state owing to mainly indirect recombination, to emit radiation by introducing a dominant transition (figure 2.103a). This latter effect occurs, for example, with gallium phosphide which does not emit when pure but doping with zinc and sulphur produces a dominant transition of 2.2 eV (W_g for GaP is about 2.3 eV) providing the popular green emission.

Although there are many semiconductors having energy gaps which correspond to wavelengths in the optical range (figure 2.102), many do not emit radiation. Others, such as the group II–VI compound semiconductors ZnS, CdS and CdSe, cannot be doped both p and n-type making it impossible to form a normal pn junction in these materials. This is due to the formation of crystal defects having an effective opposite type to the dopant being used, termed *self-compensation*. For example CdS can be doped n-type but if p-type doping is attempted, crystal defects having donor properties are created which compensate the p-type dopant, leaving the material effectively intrinsic. The most popular basic 'optical' semiconductors are gallium arsenide GaAs as an IR emitter and gallium phosphide GaP as an emitter in the visible range. Subsequently, the ternary compound materials GaAs$_{1-x}$P$_x$ and Ga$_x$Al$_{1-x}$ As have become extremely important. Gallium arsenide–phosphide is formed as a mixture of GaAs and GaP, its energy gap being a function of the relative proportions of the two materials. As x increases from 0 to 1, the energy gap alters from 1.4 eV for pure GaAs to 2.3 eV for pure GaP, but at $x \approx 0.45$ the high degree of direct recombination disappears and the composite material does not emit efficiently. Within this constraint, variation of the proportions of GaAs and GaP enables the energy gap and hence the wavelength of the emitted radiation to be controlled. GaAs$_{0.7}$P$_{0.3}$ is widely used as a source of red light (λ = 630–660 μm) for indicators and displays. The GaAsP red emitter is fabricated as an n-type GaAsP epitaxial layer on a GaAs substrate, the surface of the GaAsP layer being converted to p-type by the thermal diffusion of zinc. Yellow emitters ($\lambda \approx$ 580 nm) are also produced using a GaAsP on GaP structure. GaP is used as a green emitter at λ = 550–560 nm although with poor efficiency due to reabsorption. This poor efficiency is offset in practice by the increased sensitivity of the human eye at this wavelength (figure 2.104). A novel emitter

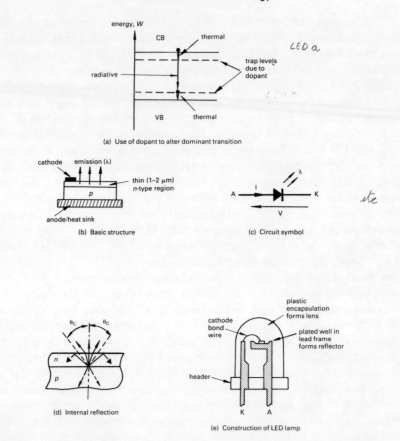

Figure 2.103 Light-emitting diodes (continued opposite)

has been developed by doping GaP with both nitrogen and zinc oxide which produces either red or green emission depending on the forward current density at the junction. One manufacturer markets a hybrid LED formed as a parallel reversed connection of red and green LED chips. The emission colour then depends on which junction is forward-biased and thus acts as a polarity indicator. Additional emission colours can be produced by *photoluminescence* (section 2.6.3) using an IR emitting GaAs diode to excite a phosphor layer, effectively creating a wavelength conversion.

The basic structure of an LED is shown in figure 2.103b. Energy is released at the junction where recombination takes place and after passing through the *n*-type region it is emitted from the surface. To minimise reabsorption of generated photons it is necessary to make the *n*-type region very thin, typically about 1 μm thick. Successful emission from the surface also depends

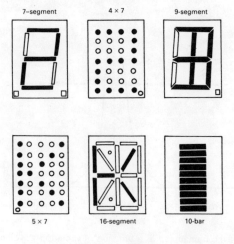

(f) Character displays

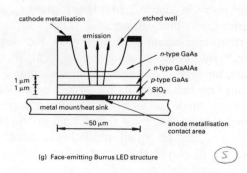

(g) Face-emitting Burrus LED structure

on the angle the photon path makes with the normal at the surface. If the angle exceeds the critical angle for the particular semiconductor-encapsulation interface, the radiation is internally reflected and not emitted (figure 2.103d). In order to improve the emission efficiency and/or viewing angle of LEDs intended as indicators or displays, the surface of the semiconductor can be domed or, more commonly, the plastic encapsulation shaped to form a lens. The plastic encapsulation/lens material is spectrally matched to the emission properties of the diode to further improve the efficiency. Typical constructional details of an LED indicator are shown in figure 2.103e. The chip is mounted in a plated well in the lead frame which forms a reflector and heat sink. Adequate light output for most indicator applications is obtained for a forward current of 5–50 mA which produces an axial luminous intensity of 1–10 mcd with a viewing angle of

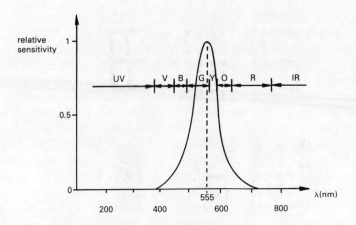

Figure 2.104 Sensitivity of the human eye as a function of radiation wavelength (colour)

20–140° depending on the type of lens incorporated. Lower current results in lower light intensity but as the human eye is most sensitive in the green region of the visible spectrum (figure 2.104) lower intensity green indicators can be tolerated better than other colours.

An important application of LEDs is in the display of alphanumeric characters and symbols either using line segments or as a dot matrix (figure 2.103f). Character heights are available in the range 3–20 mm. At the lower end of this range the line segments are formed by separate narrow rectangular diffusions into a single substrate producing a single character chip. By mounting several chips side by side in a dual-in-line (DIL) package or on a printed circuit with transparent encapsulation forming a viewing window/lens, a multicharacter display (typically 2–16 characters) is produced. Large characters are normally only available in single-character DIL packages (figure 2.100), individual LED chips being assembled on a lead frame and the light 'stretched' to the surface segment or dot size by translucent *light pipes* enclosed in the encapsulation. The seven segment configuration is widely used as a numeric (0–9) display. Nine-segment and 4×7 dot matrix configurations enable hexadecimal displays (0–9 and A–F) to be produced, while 16-segment and 5×7 dot matrix arrangements provide greater flexibility in the form of diagonal segments and greater clarity allowing full alphanumeric character sets and symbols such as the 64-character ASCII code to be produced. Individual segments are illuminated by applying a forward bias to the appropriate diode. Commerical units are available complete with a decoder which enables direct drive by a binary or binary-coded decimal (BCD) word. An additional configuration displaying a number of parallel segments or bars is also available which are end-stackable allowing production of bar–graph displays. Such displays, together with a calibrated scale, form a solid-state meter enabling convenient display of process variables such as temperature or liquid level.

In all these types of displays, clarity is improved by colouring the package to blend with off-segments and/or by the use of filters. Although a divergent beam giving a good angle of view is desirable for indication purposes, in communication systems where it is important to launch the maximum possible radiation into the fibre, a narrow angle of divergence is required. A method of reducing the width of the beam from a face-emitting structure is the use of an etched well (figure 2.103g) known as a *Burrus* structure. The etched well enables the *n*-type region above the junction to be sufficiently thin so that absorption is low, while allowing a thin *p*-type region so that the emission of radiation can be restricted to a small central portion of the junction by use of a small anode contact area. If the *p*-type region were thicker, the emission region would be larger and a simple slab geometry with both regions only a micron or so thick would not be feasible for structural reasons.

Figure 2.103g shows that the *pn* junction is formed between materials having different energy gaps, namely GaAs and GaAlAs, which is known as a *heterojunction*. The use of the GaAlAs layer reduces absorption of generated photons as they flow towards the surface. The energy gap of $Ga_xAl_{1-x}As$ increases with x and by growing a $Ga_xAl_{1-x}As$ layer with gradually changing x, a structure is created whereby the energy gap increases with distance from the junction. Photons generated near the junction have slightly less energy than that required for absorption in GaAlAs further from the junction, thus reducing over-all absorption and increasing emission efficiency. Over-all efficiency of the Burrus LED is relatively low as photons are radiated from the junction in all directions, only a small proportion being emitted through the well. The structure does lend itself to coupling to a fibre, however, either directly by mounting the end of the fibre into the well or via a lens which compensates for a certain degree of divergence allowing a greater proportion of the radiation emitted through the well to be launched into the narrow 'aperture' of the fibre. Typically, of the 5–10 mW of optical power emitted through the well only 10–100 μW is launched into the fibre. The switching speed of an LED is dependent on the recombination lifetime of injected carriers and is typically 5 ns or less. In communication applications this allows a bandwidth of 100 MHz or more. As indicators, they are bright, consume considerable power and have a restricted angle of view although this can be improved by use of an integral lens. They currently have a life expectancy for continuous operation exceeding 10^5 h (> 11 years).

2.6.2 Semiconductor Laser (Diode)

The semiconductor laser, which is also known as the *injection laser* or *laser diode*, is a *pn* junction structure that provides a higher power output than the LED with a much narrower spectral width (figure 2.105). Higher power

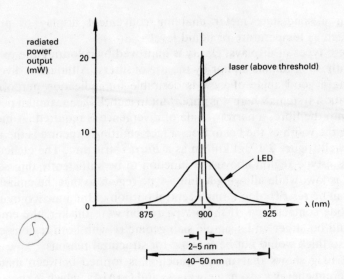

Figure 2.105 Comparative responses of a GaAs laser diode and a GaAs LED

results from the more complex mechanism of operation which utilises the chip as a resonant optical cavity.

In the case of an LED, emission is due to the release of energy from injected free carriers during recombination which can be described as the capturing of free electrons or, in terms of the band model, the return of electrons to the VB from the CB, as discussed in section 2.6.1. This return to the VB occurs randomly and the resulting emission is termed *spontaneous*. The number of free electrons at any time, that is, electrons having CB energies, is referred to as the *population* of the CB. If a mechanism could be found to substantially increase the CB population, the output power would be increased.

A laser diode is fundamentally a *pn* junction formed in degenerate (very heavily doped) semiconductor. Being degenerate, the Fermi level is inside the CB for the *n*-type material so that there is a high population of electrons in the CB (figure 2.106a). If the junction is forward-biased, the energy of these electrons is increased relative to the *p*-type side of the junction creating a population *inversion* such that there are more electrons at the upper than the lower level in the junction region. Subsequent recombination causes release of optical energy of wavelength defined by the energy gap of the semiconductor. By utilising the cleaved ends of the chip as semireflecting surfaces, some of the emitted photons are reflected back into the bulk of the chip where, by reabsorption, the rate of photon emission is increased, resulting in photon (light) amplification. If the junction bias and hence the forward current is increased, a *threshold* is reached at which the photon density is too great to be contained by the optical cavity and the radiation is emitted from the chip through the semireflecting ends. The term *laser* is derived from '*l*ight *a*mplification by the *s*timulated *e*mission of *r*adiation', the stimulation being provided by the photons reflected back into the chip, and the sudden increase of emitted power at the threshold is referred to as *lasing action*. The stimulation aspect is a major difference between the laser diode and the LED. In the laser case, recombination and hence emission is stimulated to occur rather than relying on the random spontaneous effect which is utilised in the LED.

In practice, the performance of laser diodes is considerably improved by the use of heterojunction structures in place of the basic *pn* junction in a single material. The different energy gap on either side of a heterojunction, due to different materials, causes the barrier to carrier-injection in one direction to be higher than the barrier to the other carrier in the opposite direction, resulting in efficient injection in one direction. By forming a structure of two back-to-back heterojunctions (figure 2.106b) injected carriers tend to be confined to the central region so that subsequent emission also tends to be confined to a narrow region. In addition, the difference in refractive indices between the material of the thin central region and the layers of material on either side tends to guide the photons along the plane of the junction, reducing loss due to diffraction. A typical *double-heterojunction* chip structure has a thin (≈ 0.5 μm) *p*-type GaAs layer sandwiched between p^+ and n^+ GaAlAs layers which are of the order of 1 μm thick. The structure is grown epitaxially on an n^+ GaAs substrate typically 50 μm thick which is metallised to form the cathode contact, the surface having a p^+ GaAs anode contact region (figure 2.106c). The emission region is narrow in the vertical plane due to the thinness of the central *p*-type GaAs region and is restricted in the horizontal plane by using proton bombardment on all but a narrow central strip typically 20 μm wide resulting in a *strip-geometry*. Confinement of emission to the narrow 0.5×20 μm rectangle results in an emitted beam of elliptical cross-section in contrast to the circular cross-section obtained from a face-emitting LED structure. The laser beam has a divergence of up to about 60° in the vertical plane due to the very narrow emission 'window' in this direction. In coupling the laser to a fibre, the ellipticity and divergence of the beam are compensated by the use of a rod-shaped lens which conveniently can be formed by a section of fibre mounted with its axis perpendicular to the transmission direction.

Restriction of the recombination process to a small volume within the structure reduces the spectral width of the emission to a few nanometres (typically 2–5 nm) compared with 40–50 nm for commercially available LEDs (figure 2.105). Due to the narrow spread of wavelengths, the laser output is described as being *monochromatic* (single 'colour' or frequency). This is important in fibre-optic communications as a narrow spectral width means that dispersive effects are considerably reduced. Another property of laser

284 Semiconductor Device Technology

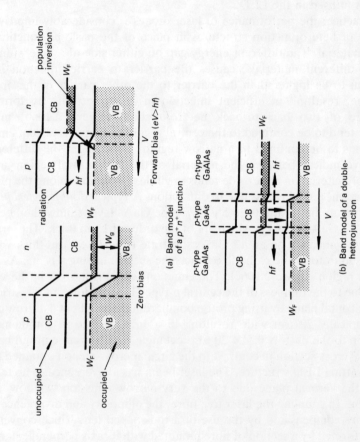

Figure 2.106 Semiconductor laser (continued opposite)

Discrete Semiconductor Devices

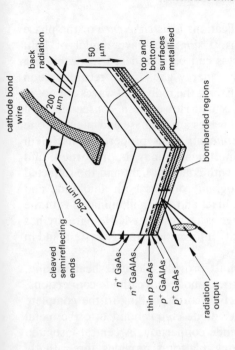

(c) Chip structure

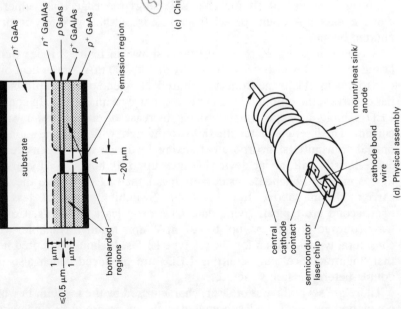

(e) Characteristic

(d) Physical assembly

emission that is important in other applications is its *coherence* or the phase uniformity between individual photons. Using the wave model of light transmission, each recombination process can be considered to generate a wave of frequency corresponding to the energy gap. If the recombination process is random as in spontaneous emission, the individual waves are not in phase and a degree of cancellation occurs. However, a consequence of stimulated emission is that the individual waves are in phase and augment one another, the emission being termed *coherent*. This property is useful in holography where a three-dimensional image may be recorded and reconstructed. An object is illuminated with coherent light and the reflected pattern (hologram) recorded. The hologram contains both amplitude and phase information of the light from the object and, when illuminated with the same coherent source, the original image is reconstructed. Holography has been of interest recently as a high-capacity information storage system and in security applications involving signatures.

The physical assembly of the laser must provide adequate heat sinking, alignment of the emitted beam and electrical anode and cathode connections. Figure 2.106d shows a popular construction corresponding to the complete package shown in figure 2.100. The laser diode chip is bonded to the metal platform which acts as heat sink and anode connection. A central insulated wire which protrudes into the mounting platform provides the cathode contact, wire bonding being used to connect to the chip surface. The mount is normally threaded to allow rigid fixing to maintain alignment in application. Alignment of the emitting face of the chip with the edge of the mounting platform is important. If the chip overhangs the edge it is vulnerable to damage and if it is mounted back from the edge a 'shadow' effect occurs in the emitted beam.

A typical laser diode characteristic is shown in figure 2.106e. Below the lasing threshold, which occurs typically at 100–500 mA, the optical output is due to low level leakage from the semireflecting end of the chip. Once lasing has occurred, the output power increases sharply with current. In contrast, an LED exhibits an approximately linear increase of output power with drive current. The laser characteristic shifts to the right with time and sophisticated optical transmitters incorporate tracking feedback to compensate for the characteristic drift, giving device lifetimes up to 10^5 hours (≥ 11 years). Laser diodes may be switched considerably faster than LEDs due to the very low carrier lifetime above the threshold. Switching times of less than a nanosecond are typical, giving data rates of at least 50 Mbit/s. Continuous wave output powers of up to 20 mW are readily available which is considerably higher than for Burrus-type LEDs. It should be noted however that 'high-radiance' edge-emitting LEDs are produced which also use the double-heterojunction structure.

There are several types of laser, characterised by the medium in which the population inversion and subsequent emission is produced, namely solid, liquid or gas. The important feature of semiconductor lasers (not to be confused with solid-state lasers such as the ruby laser) for use in communication systems is the ease with which the output can be modulated together with the convenient small size of the device.

2.6.3 Luminescent Coatings and Film Devices

Semiconductive luminescent coatings on a transparent substrate are widely used in such applications as cathode-ray tubes (CRTs) for oscilloscopes and television receivers and in fluorescent lamps. The energy gap of the material together with the degree of trapping and thermal re-excitation during the recombination of free carriers defines the wavelength (colour) and *persistence* of the emission. Two distinct processes can be identified: *flourescence* in which the recombination is mainly direct causing the emission to cease very quickly (< 10 ns) after the excitation is removed and *phosphorescence* where recombination is slow due to trapping and thermal re-excitation causing the emission to persist for seconds or even minutes after the excitation has stopped. Materials exhibiting this property, for example, zinc sulphide doped with copper and aluminium, are termed *phosphors*.

Many materials do not luminesce in the pure state but can be caused to do so by addition of a dopant (activator) which introduces trap levels into the energy gap similar to the situation shown in figure 2.103a. In addition, further dopants (coactivators) are added to control the energy released during the dominant transition to provide the required emission wavelength. For example, the blue, green and red emission phosphors used in CRTs for colour televisions are basically copper-activated zinc sulphide with different concentrations and types of coactivator dopant. The 'white' light emission required for monochrome television CRTs and fluorescent lamps is obtained by providing a trap structure allowing multiple transitions which give a range of emission wavelengths. To the human eye the net effect is a white light. A so-called 'warmer' white light, which many people find aesthetically more pleasant, is obtained by increasing the proportion of the coactivator providing red emission. Fluorescent lamps have a higher efficiency (output light power/input electrical power) than incandescent filament lamps since the emitted IR and UV radiation in the latter is wasted as far as illumination is concerned because it is not visible to the human eye. In the fluorescent lamp, the UV radiation particularly causes re-excitation of electrons which subsequently provides visible emission.

The various systems utilising luminescence employ different excitation mechanisms. In CRTs, the generation of free carriers in the phosphor is by the absorption of high-energy electrons, generated by a heated cathode, which are accelerated and focused on to the phosphor-coated screen, the mechanism being termed *cathodoluminescence*. In the fluorescent lamp, excitation is provided by photons generated by a gas discharge (*photoluminescence*) while in solid-state lighting panels, using a phosphor sandwiched between large-area electrodes, excitation is by current-flow through the material (*electroluminescence*).

Several manufacturers are currently investigating thin and thick-film electroluminescent devices for displays as well as for indicator and illumination applications. D.C. operation is not favoured due to the relatively short lifetime arising from dopant-diffusion within the film. A display panel is produced by forming the electrodes as narrowly spaced strips, the surface and rear arrays being orthogonal. Such an arrangement allows a small area of the film to be 'addressed' causing it to emit light, enabling a dot-matrix display to be produced. Display brightness increases with drive current and frequency although higher frequency leads to a shorter life. Thin-film displays currently have an operating life exceeding 10^4 hours at 2–5 kHz. The life of thick-film devices is at its maximum ($\approx$ 8000 h) at low frequency ($\approx$ 50 Hz). At higher frequency, power consumption increases and the operating life is shortened dramatically ($\approx$ 1000 h at 400 Hz).

2.6.4 Photodiode

A photodiode is essentially a *pn* junction diode with a window or lens in its encapsulation such that incident radiation is allowed to penetrate to the junction (figure 2.100). If the wavelength of the radiation is $\leq hc/W_g$ (equation 2.211) electron-hole pairs are generated in the junction region resulting in a 'photo' current which is superimposed on the thermally generated leakage current. The device can thus be used as a *photodetector* and as the photocurrent varies with the intensity of the incident radiation, it can be used as a optoelectronic transducer in fibre-optic communication systems.

There are two types of silicon photodiode in common use: the pin diode and the avalanche photodiode (APD). The basic structure of these devices is shown in figure 2.107a and b. Efficient conversion of incident energy into current relies on generated electrons and holes being swept away from each other before they recombine. Most efficient conversion therefore occurs in the depletion region where the electric field causes the generated carriers to drift apart, the bulk *p* and *n*-type regions being essentially field-free. By introducing an intrinsic layer between the *p* and *n* regions (creating a pin structure) the depletion region and hence the active area of the device is considerably increased. Photodiodes are operated under reverse bias so that a wide depletion region is created and a substantial electric field exists to sweep the generated carriers out into the neutral regions. In addition, the reverse characteristic (figure 2.107d) shows that in the third quadrant the generated current is largely independent of reverse voltage over a wide range and the increase of current with incident flux density is approximately linear. The typical sensitivity of a pin diode is 0.05–0.5 nA/mW/m^2. Greater current output from a *pn* photodiode can be obtained if it is biased to avalanche

PHOTO 1

Discrete Semiconductor Devices

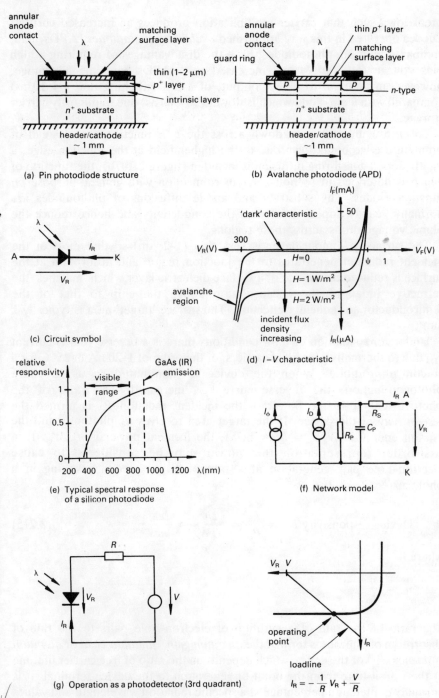

Figure 2.107 Photodiodes

breakdown such that carrier multiplication produces an increased current. Diodes operated in this way are termed *avalanche photodiodes* (APDs) and, compared with pin photodiodes, have the disadvantages of requiring a high bias voltage (50–300 V) and increased cost. APDs have the advantage, however, in addition to higher output, of a fast switching time (< 10 ns) compared with a pin diode which is slower (50–300 ns) due to minority carrier storage.

To ensure uniform breakdown across the p^+n junction area instead of premature edge breakdown due to the higher field at the junction edge, a lightly doped guard-ring diffusion is included (figure 2.107b), the function of which is described in section 2.1.1 in connection with general-purpose pn junction diodes. The substrate and anode diffusions of photodiodes are normally heavily doped to increase the conductivity and hence reduce the ohmic voltage drops across these regions.

The surface p^+ anode layer is very thin (1–2 μm) so that most of the incident radiation penetrates to the junction region and absorption at the surface is reduced by the use of a surface dielectric layer which 'matches' the refractive index of the encapsulation above the chip to that of the semiconductor so reducing reflection. The surface 'target' area is typically 1 mm^2.

Under zero radiation or *dark* conditions there is a reverse (dark) current I_D, due to thermally generated carriers, of the order of 1–20 nA at 25 °C for a silicon photodiode. When the device is illuminated so as to cause photoconduction, the reverse current is increased. The ratio of the photocurrent I_ϕ to the power of the incident radiaton P_ϕ is termed the *responsivity* of the device. If the target area receives n_ϕ photons/s and the photon energy is W_ϕ ($= hf = hc/\lambda$), the incident power is $n_\phi W_\phi$. In a first-order representation this power may be considered to cause electron–hole pair generation at a rate of n per second resulting in a photocurrent I_ϕ of $2ne$.

$$\text{Device responsivity } R_\phi = \frac{I_\phi}{P_\phi} = \frac{2ne}{n_\phi W_\phi} = \frac{2e\lambda}{hc} \eta_q \qquad (2.215)$$

where

$$\eta_q = \frac{n}{n_\phi}$$

The ratio of the rate of generation of electron–hole pairs to the rate of absorption of photons is termed the *quantum gain, quantum yield* or *quantum efficiency* (η_q) of the device which depends on the ratio of free carrier lifetime to the transit time from the point of generation to the device terminal. It is normally quoted as a percentage at a specific radiation wavelength and values in the range 20–50 per cent are common. The peak responsivity of a pin photodiode is typically 500 mA/W. As the incident power P_ϕ in equation 2.215 is equal to HA, where H is the irradiance (equation 2.214) and A is the effective detection area, the photocurrent I_ϕ is

$$I_\phi = R_\phi HA \tag{2.216}$$

and to a first-order approximation, the photodiode current well below breakdown can be written

$$\begin{aligned}I_R &= I_\phi + I_D \\ &= R_\phi HA + I_0[1 - \exp(-eV_R/kT)]\end{aligned} \tag{2.217}$$

where I_D is the 'electrical' reverse leakage current of the junction (the 'dark' current), I_0 is the theoretical reverse saturation current and V_R is the magnitude of the applied reverse bias. Photodiodes are also produced with an integral amplifier which provides a voltage output. In such cases responsivity is quoted in V/W, peak values of up to 2×10^5 V/W being typical.

The variation of spectral response of a typical silicon photodiode is shown in figure 2.107e. The response can be optimised to peak at a particular wavelength by control of the doping during manufacture. In this way efficient photodetectors can be produced to be compatible with available optical sources, for example, the response shown in figure 2.107e is particularly suited to operation in conjunction with a GaAs infrared source which has a wavelength of approximately 900 nm.

A first-order network model for a photodiode is given in figure 2.107f in which I_ϕ and I_D represent the photo and dark current components. R_P and C_P describe the properties of the reverse-biased junction and package, typical values being 10^{11} Ω and 1–10 pF for a pin diode. R_S is the resistance of the bulk semiconductor and ohmic contacts in series with the junction and is normally < 50 Ω.

Normal operation of a photodiode as a detector is shown in figure 2.107g where the operating point is determined by the intersection of the load line, which is dependent on the bias voltage V and the series resistance R, with the diode characteristic in the third quadrant. Variation of the intensity of the incident radiation causes the operating point to move along the load line.

pn photodetectors are important in applications other than as optoelectronic transducers in fibre-optic communication systems. A large array of *pn* photodiodes on a single silicon wafer is used as the target in one type of solid-state video camera tube known as a silicon vidicon. Up to 10^6 diodes are fabricated as an array in an area of about 400 mm^2 on a silicon wafer. The optical radiation is focused on the rear surface of the array and is converted to an electrical video signal by scanning the diode array with an electron beam. This tube is more sensitive, lighter and more reliable than other types of vidicon and is particularly useful in portable cameras.

By making the intrinsic region of a pin diode very thick (several millimetres), the device is useful as a detector for high energy X or γ-rays. Such devices are termed *particle detectors*, the thick *i*-region being necessary to enable absorption of particles of such high energy.

2.6.5 Phototransistor

The low output of photodiodes can be improved considerably if the photojunction forms the collector–base junction of a bipolar transistor structure forming a *phototransistor*. In the common-emitter configuration with the base open circuit, the collector current is I_{CEO} which is approximately given by $h_{FE}I_{CBO}$ for $h_{FE} \gg 1$ (equation 2.53). Illumination of the C-B junction through a window or lens in the encapsulation causes carrier generation such that I_{CBO} is increased to $I_{CBO} + I_\phi$, where I_ϕ is the

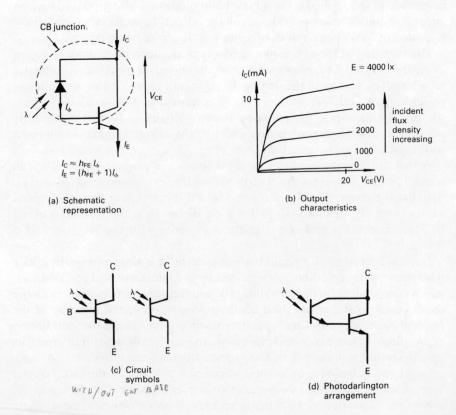

(a) Schematic representation

(b) Output characteristics

(c) Circuit symbols

(d) Photodarlington arrangement

Figure 2.108 Bipolar phototransistors

photocurrent. The collector current is thus increased to $h_{FE} (I_{CBO} + I_\phi)$ and since for practical levels of illumination $I_\phi \gg I_{CBO}$, the collector current is approximately $h_{FE} I_\phi$. Compared with a photodiode, it can be seen that transistor action effectively amplifies the photogenerated current by the CE current gain h_{FE} which is typically 100–300. Operation can be considered with reference to the schematic representation of figure 2.108a. By separating the photogeneration and collection properties of the C-B junction, it can be seen that the photogenerated carriers effectively create the base current for a normal *npn* transistor. Normal CE output characteristics result (figure 2.108b) although the controlling parameter is the illumination E instead of base current. External contact to the base region is not required and two terminal phototransistors are produced sometimes having symmetrical construction, in which case they are described as *photo-duo-diodes*. In some applications, however, connection to the base is useful as connection of an external resistor between base and emitter reduces the dark current. Circuit symbols for phototransistors with and without external base connections are shown in figure 2.108c.

Due to increased charge storage, phototransistors are considerably slower than photodiodes, having rise and fall times of the order of 1–10 μs compared to less than 50 ns for a photodiode. The generated current, however, is typically 100 times that for a pin photodiode, for example, 5 mA compared with 50 μA at 2000 lx. A phototransistor with an external base connection can be used as a photodiode simply by using the collector and base terminals and leaving the emitter open-circuit. This improves the switching time by at least an order of magnitude but the generated current is reduced.

Even higher sensitivity (generated current/incident illumination) can be achieved at the expense of switching speed by use of a *photodarlington* structure (figure 2.108d) where a phototransistor provides the base drive to a normal high gain BJT. The resultant collector current is approximately $h_{FE}^2 I_\phi$.

2.6.6 Optocoupler

The inclusion of an LED and a photodiode, phototransistor or photodarlington in a single package provides a convenient method of electrically isolating one part of an electrical system from another. The composite device is described as an *optocoupler*, *photocoupler* or *optical isolator*. A typical application is the protection of electronic equipment such as microcomputers from damage by accidentally applied high input voltages by providing electrical isolation between the sensitive circuitry and its external input connections. Another important application is the simplification of voltage level shifting, as for example in the change from negative to positive logic or vice versa between digital systems. In addition, an electrical 'shield' in the optical path between LED and photodiode reduces noise transmission

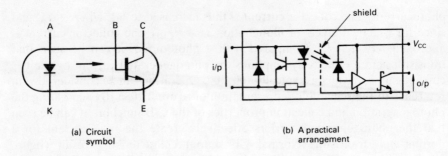

Figure 2.109 Optocoupler

making optocouplers useful in signal transmission in high induced-noise environments. Figure 2.109a shows the circuit symbol for a basic optocoupler while a practical arrangement incorporating input protection, LED current regulation and a high-gain output circuit providing TTL compatibility is shown in figure 2.109b. Various output arrangements are used providing various high speed and/or high gain specifications. Single and multichannel optocouplers are produced in dual-in-line encapsulation with switching times of the order of 20 ns which provide data transmission rates up to 10 Mbit/s and input–output isolation up to 5 kV d.c.

Other LED–photosensor arrangements either mounted separately or in a single conveniently shaped package are used as presence detectors utilising the object to be detected to break or make (by reflection) the emitter–sensor beam. *Optoelectronic pairs* are used in this way as level detectors, in limit switching, end-of-material detection, object or event counting and as rotational speed transducers.

2.6.7 Photofet

The photofet is a JFET with a window or lens in the encapsulation to allow the gate-channel junction to be irradiated. If the gate–channel junction is reverse-biased as in normal depletion-mode operation, irradiation at an appropriate wavelength causes a photo-current to be generated which increases the gate leakage current. By including a resistor R_G in the gate circuit (figure 2.110b), the generated current modifies the gate–source voltage altering the degree of channel pinch-off and hence the drain current. The voltage developed across the source resistor R_S provides a voltage output which is dependent on the incident illumination E. Assuming that the photofet is biased to pinch-off under dark conditions by suitable choice of R_S, increase of E causes reduction in V_{GS} giving output characteristics of the form shown in figure 2.110c. The sensitivity of the device can be varied over a wide range by alteration of R_G. In contrast to the phototransistor (BJT), the

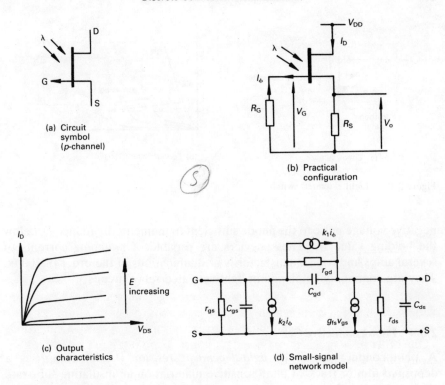

Figure 2.110 Photofet

sensitivity of the photofet decreases with increase in temperature giving poor high temperature performance. A network model of the small-signal performance of a photofet can be developed directly from the general high-frequency FET model of figure 2.86c by addition of signal photocurrents $k_1 i_\phi$ and $k_2 i_\phi$ (figure 2.110d). A detailed consideration of the photofet and its applications can be found in reference 41.

2.6.8 Photothyristor

A photothyristor or *light-activated switch* (figure 2.111) is a semiconductor-controlled switch (SCS) that may be triggered into conduction by irradiation of the blocking junction (section 2.4.3) through a window in the encapsulation. In terms of the two-transistor model of the structure (figure 2.90) for cathode gate operation, irradiation of the central junction generates a photocurrent in the C-B junction of T_1 causing conduction. If a resistor is connected between KG and K, triggering does not occur until the voltage drop across the resistor due to the photocurrent reaches about 400 mV. Switch-off can be achieved either by reducing V_{AK} to zero or by applying a

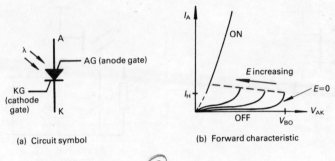

Figure 2.111 Light-activated switch

negative voltage pulse to the anode sufficient to momentarily reduce I_A below the holding value I_H. These devices are capable of switching currents of several amps and trigger satisfactorily at illuminations of the order of 100 lx. They are commonly employed in light-activated relay circuits.

2.6.9 Photoconductive Cell

A photoconductive cell or *light-dependent resistor* (LDR) comprises a deposited film or track of photosensitive material on an insulating substrate (figure 2.100). The track can be irradiated with radiation of appropriate wavelength via a window in the encapsulation causing photo-generation and the resistance of the track is reduced. Possibly the most popular track material is cadmium sulphide which has a spectral response similar to that of the human eye, making a CdS cell useful for visible light sensing. Lead sulphide and indium antimonide are sensitive in the infrared region while selenium is particularly sensitive in the blue region of the visible spectrum.

The resistance of a CdS cell varies by up to six decades from dark to maximum illumination and follows an inverse law (figure 2.112b) of the form

$$\log R \approx K - \log E \qquad (2.218)$$

where R is the cell resistance, E is the incident illumination and K is a device constant. Under equilibrium conditions, the dark resistance of a CdS cell can be as high as 10 MΩ reducing typically to 10 Ω at maximum illumination. The time taken for the cell resistance to change following a change of illumination is typically in the range 50 ms–2 s, although the value stabilises over a period of minutes or even hours causing a drift in resistance. For this reason manufacturers quote initial and equilibrium resistance values for a certain degree of illumination, specifying the elapsed time since the change in

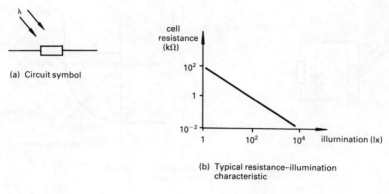

Figure 2.112 Photoconductive cell

illumination. The resistance range for PbS and InSb cells is considerably less than for CdS cells. Photoconductive cells are widely used in detection and radiation measurement such as smoke or flame-failure detection, automatic lighting or brightness control, burglar alarm systems, low-speed batch counting and radiation intensity measurement.

2.6.10 Photovoltaic (Solar) Cell

Consideration of the *pn* junction as a photodetector in section 2.6.4 shows that photogeneration at an illuminated junction creates free carriers and by operating the junction at reverse bias, these carriers are efficiently converted into a 'generated' reverse current which is approximately proportional to the illumination and largely independent of the reverse bias (figure 2.107d and g). As a photodetector, therefore, the junction is operated in the third quadrant of the *I–V* characteristic, a reverse bias being applied and a reverse current being generated. If no bias is applied, the illuminated junction being simply shunted by a load resistor, a forward voltage appears across the junction and a generated reverse current flows in the circuit dependent on the value of the load. The junction is then operating in the fourth quadrant of the characteristic (figure 2.107d) and by examining the polarity of the junction voltage and the direction of the generated current (figure 2.113a) it is evident that the junction then acts as an electrical power source, termed a *photovoltaic cell*. Under open-circuit conditions, the generation of electron–hole pairs at the illuminated junction tends to alter the balance of diffusion and drift currents compared with the equilibrium dark condition. Equilibrium is restored by a change in the junction field which causes the junction voltage to change from the contact potential value ψ and an open-circuit voltage V_{OC} appears across the junction. Increase of the intensity of illumination causes V_{OC} to increase and the effective barrier height reduces

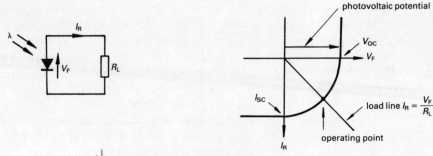

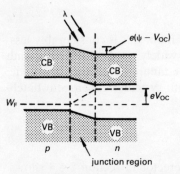

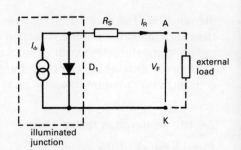

(a) Operation of an illuminated *pn* junction as an electrical power source (4th quadrant)

(b) Energy model of open-circuit junction under illumination conditions

(c) Network model

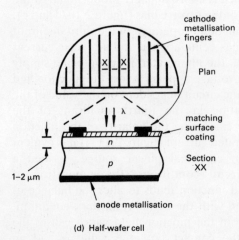

(d) Half-wafer cell

(e) Circuit symbol

Figure 2.113 Photovoltaic (solar) cell

as shown in the energy model of figure 2.113b. The generated open-circuit voltage V_{OC} is termed the *photovoltaic potential* and appears on the *I–V* characteristic as the intercept on the V_F axis (figure 2.113a). The operating point of the cell depends on the load R_L and is given by the intersection of the load line with the junction characteristic. As R_L is reduced from open-circuit to short-circuit, $V_F = V_{OC} \rightarrow 0$ and $I_R = 0 \rightarrow I_{SC}$. At a particular value of R_L the power delivered by the cell ($V_F I_R$) is a maximum.

The network model of figure 2.113c represents the performance of the cell where I_ϕ is the generated photocurrent, D_1 describes the *I–V* characteristic of the junction and R_S is the series ohmic resistance of the contacts and bulk semiconductor regions. To a first-order approximation therefore

$$I_R = I_\phi - I_0 [\exp(eV_F/kT) - 1] \tag{2.219}$$

where I_0 is the theoretical reverse saturation current of the junction due to thermal generation and the photocurrent I_ϕ is given by equation 2.216. The open-circuit voltage V_{OC} is given by V_F when $I_R = 0$, thus from equation 2.219

$$V_{OC} = \frac{kT}{e} \ln\left[\frac{I_\phi}{I_0} + 1\right] \tag{2.220}$$

Equation 2.219 enables the power delivered by the cell to be obtained in terms of V_F and differentiation with respect to V_F enables the maximum power condition to be determined.

Photovoltaic cells are of considerable interest in the direct conversion of solar radiation into electrical power, hence the more common name *solar cell*. A practical cell requires a large surface area to collect as much radiation as possible and the junction must be near to the surface of the cell so that as much of the absorbed energy as possible penetrates to the junction region where generation occurs most efficiently, the junction field separating generated electrons and holes before recombination can occur. Cells are commonly produced using either a complete circular silicon wafer (figure 2.100) or a semicircular half-wafer (figure 2.113d), up to 80 mm diameter. Using planar technology, the junction is formed 1–2 μm below the surface which is oxidised to improve absorption by reducing reflection. If the semiconductor is heavily doped, the contact potential and hence the generated voltage is increased but heavy doping reduces carrier lifetime reducing the generated current. To reduce the series resistance of the thin *n*-type surface region, the cathode metallisation is formed as an array of fingers to reduce lateral current flow. The fingers have to be narrow, however, so that the silicon target area is not appreciably reduced. Practical silicon cells of this type can provide $V_{OC} \approx 0.5$ V and I_{SC} up to 800 mA giving an operating maximum output power of about 200 mW, depending on the lighting conditions.

Solar power sources, as used to power satellites, use arrays of cells connected in series to provide the required terminal voltage and modules are parallelled to increase the current capability. Parallel-connected cells require series diodes to prevent shaded cells, which have a lower terminal voltage, form absorbing power from cells receiving higher illumination. To avoid wide fluctuations in output power as the illumination varies, solar cells are normally used to trickle-charge batteries. Silicon and gallium arsenide solar cells currently provide a power conversion efficiency of up to 15 per cent.

Recent interest in renewable energy sources for power generation on a large scale has concentrated work on electrical energy sources including solar cells. A figure of merit for an energy source is its *energy ratio* which is the ratio of the useful electrical energy produced by the fuel or device during its life to the energy expended in obtaining the fuel or manufacturing the device. Assuming a cell life of 15 years at ground level where bombardment from high energy particles has a less damaging effect than in space applications, silicon solar cells have an energy ratio of less than two compared with three for wave power, ten for a nuclear system and twelve for coal. The low energy ratio of silicon cells indicates that they are not viable for large-scale economic power generation and interest has centred on thin-film heterojunction structures such as Cu_2S–CdS which are relatively cheap to produce and give promise of energy ratios greater than ten. Reference 42 provides detailed consideration of recent developments in solar cell research and provides a detailed bibliography.

REFERENCES AND FURTHER READING

1. R.J. Smith, *Circuits, Devices and Systems*, 3rd edition, chapter 12 (Wiley, 1976)
2. M. Cirovic, *Basic Electronics: Devices, Circuits and Systems*, chapters 8 and 17 (Prentice-Hall, 1974)
3. J. Millman and C.C. Halkias, *Integrated Electronics: Analog and Digital Circuits and Systems*, chapter 4 (McGraw-Hill, 1972)
4. J. Millman, and C.C. Halkias, *Electronic Fundamentals and Applications: for Engineers and Scientists*, chapter 3 (McGraw-Hill, 1976)
5. J. Bardeen and W. H. Brattain, The Transistor, a Semiconductor Triode, *Phys. Rev.*, **74** (1948) 230–2
6. W. Shockley, The Theory of pn Junctions in Semiconductors and pn Junction Transistors, *Bell Syst. Tech. J.*, **28** (1949) 435–89
7. J. Millman and C.C. Halkias, *Integrated Electronics: Analog and Digital Circuits and Systems*, chapter 18, 695–8 (McGraw-Hill, 1972)
8. C.A. Holt, *Electronic Circuits: Digital and Analog*, chapter 16, 524–35 (Wiley, 1978)
9. J.M. Early, Effects of Space-charge layer Widening in Junction Transistors, *Proc. IRE*, **40** (1952) 1401–6

10. J. Millman and C.C. Halkias, *Integrated Electronics: Analog and Digital Circuits and Systems*, chapter 8 (McGraw-Hill, 1972)
11. C.A. Holt, *Electronic Circuits: Digital and Analog*, chapter 13 (Wiley, 1978)
12. H. Ahmed and P.J. Spreadbury, *Electronics for Engineers: An introduction*, chapter 3 (Cambridge University Press, 1973)
13. W.H. Hayt and G.W. Neudeck, *Electronic Circuit Analysis and Design*, chapter 6 (Houghton Mifflin, 1976)
14. Mullard Technical Handbook, Book 1, Part 1A, *General section: rating systems*, (Mullard Limited, 1980)
15. J. Millman and C.C. Halkias, *Integrated Electronics: Analog and Digital Circuits and Systems*, chapter 9 (McGraw-Hill, 1972)
16. SOAR: the basis for reliable power circuit design, *Mullard Technical Information No. 15*, Mullard Limited (1975) 31–2
17. F.R. Connor, *Noise*, chapter 4 (Edward Arnold, 1973)
18. A. Van der Ziel, *Noise: Sources, Characterization, Measurement*, chapters 5 and 6 (Prentice-Hall, 1970)
19. P.M. Chirlian, *Electronic Circuits: Physical Principles, Analysis, and Design*, chapter 6 (McGraw-Hill, 1971)
20. A.G. Martin and F.W. Stephenson, *Linear Microelectronic Systems*, chapter 1, (Macmillan, 1973)
21. J.J. Ebers and J.L. Moll, Large-signal Behaviour of Junction Transistors, *Proc. IRE*, **42**, (1954) 1761–72
22. J. Millman and H. Taub, *Pulse, Digital, and Switching Waveforms*, chapter 20, 765–92 (McGraw-Hill, 1965)
23. J. Seymour, *Physical Electronics*, chapter 4, 143–54 (Pitman, 1972)
24. J. Millman and C.C. Halkias, *Integrated Electronics: Analog and Digital Circuits and Systems*, chapter 11, 355 (McGraw-Hill, 1972)
25. W.H. Hayt and G.W. Neudeck, *Electronic Circuit Analysis and Design*, chapter 4 (Houghton Mifflin, 1976)
26. W. Gosling, W.G. Townsend and J. Watson, *Field Effect Electronics*, chapter 3 (Butterworths, 1971)
27. R. Regan, VMOS improves efficiency through switching performance, *Electronics and Power* (Sept 1979) 629–33
28. W. Gosling, W.G. Townsend and J. Watson, *Field Effect Electronics*, chapter 4 (Butterworths, 1971)
29. P.M. Chirlian, *Electronic Circuits: Physical Principles, Analysis, and Design*, chapter 6, 337–40 (McGraw-Hill, 1971)
30. J.E. Carroll, *Physical Models of Semiconductor Devices*, chapter 7, 150–6 (Edward Arnold, 1974)
31. L. Esaki, New phenomenon in narrow germanium pn junctions, *Phys. Rev.*, **109** (1958) 603
32. B.G. Streetman, *Solid State Electronic Devices*, 1st edition, chapter 6, 218–32 (Prentice-Hall, 1972)
33. J.B. Gunn, Microwave Oscillations of Current in III–V Semiconductors, *Solid-State Communications*, **1** (1963) 88
34. J.E. Carroll, *Physical Models of Semiconductor Devices*, chapter 8, 192–201 (Edward Arnold, 1974)
35. J. Seymour, *Physical Electronics*, chapter 8, 352–9 (Pitman, 1972)
36. B.G. Streetman, *Solid State Electronic Devices*, 1st and 2nd editions, chapter 12, 423–35 (Prentice-Hall, 1972 and 1980)
37. J. Seymour, *Physical Electronics*, chapter 8, 360–9 (Pitman, 1972)
38. W.T. Read, A proposed high-frequency negative resistance diode, *Bell Syst. Tech. J.*, **37** (1958) 401–46

39. B.G. Streetman, *Solid State Electronic Devices*, 1st and 2nd editions, chapter 12, 418–22 (Prentice-Hall, 1972 and 1980)
40. J.E. Carroll, *Physical Models of Semiconductor Devices*, chapter 8, 185–92 (Edward Arnold, 1974)
41. W. Gosling, W.G. Townsend and J. Watson, *Field Effect Electronics*, chapter 15 (Butterworths, 1971)
42. Solar Cells, Special publication, *I.E.E. Journal on Solid-State and Electron Devices (SSED)* (June 1978)
43. *Optoelectronics Designer's Catalog* (Hewlett-Packard, 1980)
44. A. Bar-Lev, *Semiconductors and Electronic Devices*, chapters 9–15, 17 and 18 (Prentice-Hall, 1979)
45. D.H. Navon, *Electronic Materials and Devices*, chapters 6–11 (Houghton Mifflin, 1975)
46. N.M. Morris, *Semiconductor Devices*, chapters 2–5, 9 and 10 (Macmillan, 1978)
47. F.R. Connor, *Electronic Devices*, chapter 4 (Edward Arnold, 1980)
48. D. Roddy, Introduction to Microelectronics, 2nd edition, chapters 3, 6, and 11, (Pergamon, 1978)
49. A. Chappell (Ed.), *Optoelectronics: Theory and Practice*, (McGraw-Hill/Texas Instruments, 1978)
50. P.G. Noble, Understanding rectifier diode data, *Mullard Technical Information No.74* (Mullard Limited, 1978)
51. P.G. Noble, Understanding thyristor and triac data, *Mullard Technical Information No.85* (Mullard Limited, 1979)

TUTORIAL QUESTIONS

Values of physical constants and semiconductor parameters are given in appendix C.

2.1 Plot, to scale, the theoretical static I–V characteristic of a *pn* junction diode at 300 K over the range from a reverse bias of -1 V to a forward current of 100 mA; the theoretical reverse saturation current is 10 nA at this temperature. If the resistance of the bulk semiconductor regions of the diode is 5 Ω, superimpose the practical I–V characteristic on the theoretical plot over the same range. What voltage must be applied to the diode in practice for a forward current of 100 mA at 300 K? Mention any effects that the above practical characteristic does not take into account.
(Answer: 919 mV)

2.2 Show that the temperature rise necessary to cause the reverse leakage current of a silicon diode to double is approximately 8 °C in the region of 25 °C and approximately 20 °C at 150 °C. Show also that for constant forward current, the forward voltage across a silicon diode decreases by about 2 mV/°C rise in temperature of $V_F = 600$ mV and $T = 25$ °C.

2.3 A diode has a thermal resistance from junction to ambient of 0.5 °C/mW. If the junction temperature must not exceed 200 °C, calculate the maximum power dissipation of the device at ambient temperatures of 25 °C and 75 °C.
(Answers: 350 mW, 250 mW)

2.4 Using piecewise-linear techniques and the data given in appendix H.1, produce a network model representing the static performance of a *typical* BAX13 diode at 25 °C.
(Specimen answer based on the model of figure 2.8a: $V_1 = 770$ mV, $r_f = 5.7$ Ω, $r_r \approx$

$3 \times 10^9 \, \Omega$, $V_B = 50$ V, $R_B \approx 10 \, \Omega$. Insufficient data is available to enable values for V_C and r_b to be determined.)

2.5 A typical BZY88 C9V1 voltage regulator diode is to be used in a voltage stabiliser circuit. Using the data given in appendix H.2, produce a PWL network model of the diode that would enable the performance of the stabiliser circuit to be analysed.
(Specimen answer based on figure 2.8c: $V_B = 9$ V, $r_z = 20 \, \Omega$)

2.6 A low-power silicon *npn* BJT has an emitter injection efficiency of 0.996 and a base transport factor of 0.999. Calculate the static collector current when the static base current is 20 µA for values of collector–base voltage well below breakdown. State any assumptions made.
(Answers: 3.98 mA; C-B leakage assumed negligible in comparison with this value of collector current)

2.7 The following data relates to a silicon planar epitaxial *npn* BJT:
Dopant densities of emitter, base and collector regions are 10^{25}, 10^{23} and 10^{21} atoms/m^3 respectively.
Distance from the ohmic emitter contact to the E-B junction = 1 µm.
Distance between the E-B and B-C metallurgical junctions = 0.8 µm.
Recombination lifetime of electrons in the base = 300 ns.
Calculate the theoretical values of emitter injection efficiency, base transport factor and the static common-base and common-emitter current gains of the transistor at 25 °C. Mention any assumptions made.
(Answers: 0.9971, 0.9997, 0.9968, 314; Early effect ignored therefore solution applies to low values of V_{CB})

2.8 With the emitter open-circuit, the C-B reverse current of the BJT of question 2.7 is 10 nA. Calculate the theoretical static emitter and base currents when the static collector current is 1 mA. Compare the ratios of I_C to I_E and I_C to I_B with the corresponding static gain values calculated in question 2.7 and hence comment on the significance of the C-B reverse leakage current.
(Answers: 1.0032 mA, 3.2 µA, negligible)

2.9 Using information in section 1.6.2 concerning depletion layer width, calculate the effective base width for the BJT of question 2.7 in normal operation when the C-B reverse bias is (a) 10 V, (b) 40 V. Hence calculate the theoretical static common-base and common-emitter current gains at these two bias levels. Comment on the relative results.
If the breakdown voltage of the collector–base junction is 50 V, calculate the theoretical static common-base current gain when V_{CB} is 40 V taking account of collector multiplication.
(Answers: (a) 0.7625 µm, (b) 0.7269 µm; 0.9970, 332; 0.99715, 350; 2.0433)

2.10 The base dopant density of a low-power silicon planar *npn* BJT is 10^{23} atoms/m^3 and the base–emitter and base–collector junctions may both be considered to be square with a side dimension of 200 µm. If the effective base width is 0.6 µm, calculate the excess electron density at the emitter side of the base when the collector current is 1 mA. Estimate the stored charge in the base. Using information from section 1.6.2, calculate the theoretical value of V_{BE} to provide this level of injection at 300 K. Mention any simplifying approximations used.
(Answers: 2.76×10^{19}/m^3, −0.053 pC, assuming a linear electron distribution in the base, that is, ignoring recombination; 604 mV, assuming an abrupt B-E junction and uniform doping)

2.11 The values of h_{FB} and I_{CBO} for a BJT are -0.995 and 20 nA respectively. What is the leakage current between collector and emitter at cut-off? If alternations to processing parameters cause $|h_{FB}|$ to increase by 0.1 per cent, what are the corresponding percentage changes in the static common-emitter current gain and the C-E leakage, assuming that I_{CBO} remains unaltered?
(Answers: 4 µA, +25.1 per cent, +25 per cent)

2.12 A BJT has $V_{BE(sat)} = 800$ mV, $V_{CE(sat)} = 200$ mV and $I_{CBO} = 10$ nA. The transistor is used in a basic switching circuit with the collector connected to a 5 V power supply via a 1 kΩ load, the base connected to a voltage source via a 100 kΩ resistor and the emitter is grounded. Calculate the minimum static common-emitter current gain of the BJT that will allow it to saturate when the base source is 5 V. What is the output voltage of the circuit in this condition if no *external* load is connected? If the base source were switched to 0 V, what would be the minimum current flow through the BJT and the corresponding maximum output voltage?
(Answers: 114.3, 0.2 V, 1.153 µA, 4.9988 V)

2.13 Under static operating conditions, the output of a BJT circuit may be considered as a constant-voltage source V_{CC}, a resistor R and the C-E port of the BJT in series. Show that the operating point of the BJT (V_{CE}, I_C) is given by the intersection of the common-emitter output characteristic of the BJT with the *load line* $I_C = -(1/R)V_{CE} + V_{CC}/R$.

If R is 400 Ω and the maximum power dissipation of the BJT is 300 mW, determine graphically the maximum value for V_{CC} for safe static operation. If the use of a heat sink increases P_{tot}max to 400 mW, what is the minimum value of R for the value of V_{CC} determined above?
(Answers: 22 V, 305 Ω)

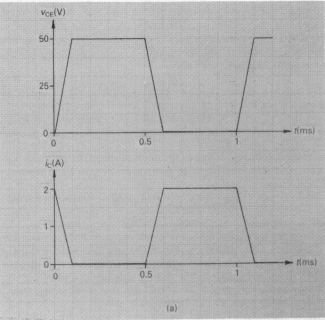

Figure 2.114 Data for question 2.16, (a) v_{CE} and i_C waveforms, (b) SOAR for $d = 0.2$ (continued opposite)

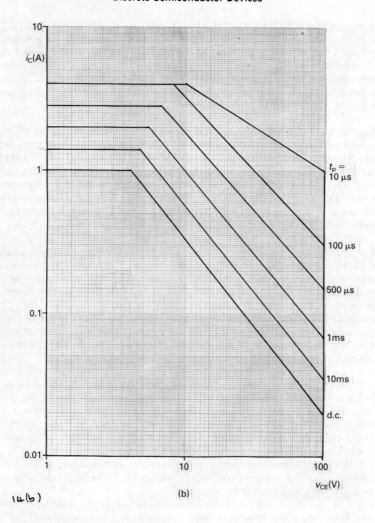

(b)

2.14 A silicon BJT having a maximum permitted junction temperature of 175 °C has a maximum power rating of 500 mW at an ambient temperature not exceeding 25 °C. If the thermal resistance between the junction and case of the device is 0.2 °C/mW,

(a) Calculate the power rating of the BJT in an environment having an ambient temperature of 75 °C if no heat sink is used, assuming linear derating above 25 °C.

(b) Repeat part (a) using a heat sink of thermal resistance 20 °C/W.

(c) When using the heat sink of part (b), what would be the C-B junction temperature when the device is dissipating 800 mW in an environment with an ambient temperature of 30 °C? Is this permissible?

(d) For the same BJT, is it possible to design a heat sink to permit the device to dissipate 1 W in an environment that may rise to 55 °C? If it is, what is the required thermal resistance of the heat sink? If not, why not?

(Answers: (a) 333 mW, (b) 462 mW, (c) 204 °C; no, as T_j would exceed $T_j$max, (d) no, as $R_{th(j-amb)}$ cannot be less than $R_{th(j-case)}$)

2.15 A silicon power BJT has the following specification: $T_j\text{max} = 150\,°\text{C}$, $R_{\text{th}(j-\text{amb})} = 100\,°\text{C/W}$ and $R_{\text{th}(j-\text{mb})} = 5\,°\text{C/W}$. The transistor is mounted on a heat sink of thermal resistance $5\,°\text{C/W}$. Electrical insulation between the transistor and the heat sink is provided by a mica washer having a thermal resistance of $3\,°\text{C/W}$. Calculate the maximum ambient temperature for safe operation when the transistor is dissipating a steady power of 10 W. What is the mounting base temperature for this condition?
(Answers: 26 °C, 100 °C)

2.16 Figure 2.114 (on pages 304, 305) shows hypothetical variations of collector current and collector–emitter voltage for a BJT during a repetitive switching sequence at 1 kHz. The changes of voltage and current during each switching interval are linear. Use the SOAR information provided in figure 2.114b to check that this repetitive switching operation is safe as far as power dissipation and second breakdown is concerned. Assuming that the collector–emitter voltage waveform remains as shown, estimate the maximum value of collector current in the ON state for safety.
(Answers: Safe, 3.2 A)

2.17 From the data given in appendix H.3 determine the transition frequency for a typical BC107 at a quiescent collector current of (a) 1 mA and (b) 10 mA. Hence determine the common-base and common-emitter cut-off frequencies for typical devices in the three gain-range types BC107 A, B and C at these current levels, when the quiescent collector–emitter voltage is 5 V.
(Answers: (a) 130 MHz (b) 300 MHz. At $I_C = 1$ mA, f_α is 130 MHz for all three gain types, f_β is 619 kHz, 394 kHz and 220 kHz for types A, B and C respectively. The corresponding values at $I_C = 10$ mA are 300 MHz, 1.4 MHz, 833 kHz and 484 kHz respectively)

2.18 Using manufacturers' data given in appendix H.3, produce the following network models for a typical BC107 B *npn* BJT for the operating conditions stated.
    (a) The *h*-parameter model representing small-signal performance in the common-emitter configuration at 1 kHz for $I_C = 2$ mA and $V_{CE} = 5$ V.
    (b) Repeat part (a) for a quiescent collector current of 5 mA, other conditions remaining unchanged.
    (c) Redraw the model derived in part (a) to represent small-signal performance of the BJT in the common-base configuration for the same conditions, in terms of the common-emitter *h*-parameters.
    (d) Convert the parameter values obtained in part (a) to common-base values and hence draw the corresponding common-base *h*-parameter model.
    (e) By conversion of the common-emitter *h*-parameter values obtained in part (a), derive the low-frequency common-emitter *y*-parameter model for a BC107 B for the same conditions.
    (f) Produce the high-frequency hybrid-π model of a BC107 B for the quiescent conditions $I_C = 2$ mA, $V_{CE} = 5$ V.
    (g) If the low-frequency hybrid-π model can be used up to frequencies at which the capacitive reactances are 10 per cent of the resistances across which they are connected, what is the highest frequency that the low-frequency hybrid-π model of a BC107 B may be used?
(Answers: (a) model of figure 2.55b with $h_{ie} = 4.5$ kΩ, $h_{re} = 2 \times 10^{-4}$, $h_{fe} = 330$, $h_{oe} = 30\,\mu\text{S}$
(b) model of figure 2.55b with $h_{ie} = 2.2$ kΩ, $h_{re} = 1.3 \times 10^{-4}$, $h_{fe} \approx 350$, $h_{oe} \approx 58\,\mu\text{S}$
(c) model of figure 2.56b with the same parameter values as part (a)
(d) model of figure 2.56a with $h_{ib} = 13.6\,\Omega$, $h_{rb} = 2.1 \times 10^{-4}$, $h_{fb} = -0.997$, $h_{ob} = 0.09\,\mu\text{S}$

(e) model of figure 2.59b with $y_{ie} = g_{ie} = 0.22$ mS, $y_{re} = g_{re} = 0.044$ μS, $y_{fe} = g_{fe} = 73.3$ mS, $y_{oe} = g_{oe} = 15.3$ μS
(f) model of figure 2.64e with $g_m = g_{fe} = 73.3$ mS, $r_{b'e} = 4.5$ kΩ, $r_{b'c} = 22.5$ MΩ, $r_{ce} = 33.3$ kΩ, $r_{bb'}$ indeterminate due to insufficient data and, anyway, negligible for a low-power BJT, $C_{b'c} = 3.2$ pF, $C_{b'e} = 65.4$ pF
(g) 22.1 kHz, limited by the reactance of $C_{b'c}$)

2.19 With reference to the switching waveforms and parameter symbols of figure 2.62, use the charge-control BJT model to show that the storage time t_s following the reversal of base drive, before the collector current begins to fall, is given by $\tau_{BN} \ln [2 \beta_N I_B/(I_C + \beta_N I_B)]$. Hence show that if a BJT is switched OFF by applying reverse base drive, the storage time is less than if the base drive is simply reduced to zero.

If the BJT is switched OFF by reversing the base drive, show that the fall time t_f, as shown in figure 2.62, is given by $\tau_{BN} \ln [(I_C/\beta_N I_B) + 1]$ where I_C and I_B refer to the values of collector and base current during the storage period t_s.
Note. Fall time has been defined as the time taken for I_C to fall to zero after the end of the storage time. In practice, fall time is normally defined as the time for I_C to reduce from 90 per cent to 10 per cent of the value at saturation.

2.20 The gate and substrate regions of a uniform-geometry n-channel silicon JFET (figure 2.72) are equally heavily doped and the dopant density of the channel region is 2×10^{22} atoms/m^3. If the channel is 10 μm long, 100 μm wide and 1 μm thick, calculate the theoretical value of pinch-off voltage and the drain current at pinch-off for zero gate–source voltage? What are the theoretical values of small-signal common-source forward transconductance of this JFET at values of quiescent gate–source voltage of zero and -2 V? If the channel length is increased by 10 per cent and the channel thickness is reduced by 10 per cent, what are the corresponding percentage changes in V_P, I_{DSS} and g_{fs} at $V_{GS} = 0$?
(Answers: -3.77 V, 5.63 mA, 3 mS, 1.4 mS, -19 per cent, -34 per cent, -18 per cent)

2.21 An n-channel lateral enhancement-type silicon MOST has the following specification: channel length = 5 μm; channel width = 100 μm; gate oxide (SiO$_2$) thickness = 0.2 μm; threshold voltage = 2 V.
Derive an expression showing how the drain current at pinch-off depends on the gate–source voltage and hence plot the transfer characteristic for operation in the saturation region for $0 \leq V_{GS} \leq 10$ V. Determine the capacitance of the gate–oxide–channel capacitor for this device and the value of the small-signal common-source forward transconductance when $V_{GS} = 4$ V.
(Answers: $I_{D(sat)} = 5 \times 10^{-4} [(V_{GS}/2) - 1]^2$, 0.0885 pF, 991 μS)

2.22 An n-channel enhancement-type silicon MOST is used as a resistor with its gate and drain terminals shorted together so that it operates in the saturated region of its characteristic beyond pinch-off. Show that the static drain–source resistance is non-linear having a value $2tl V_T V_{DS}/[\epsilon \mu_n w (V_{DS} - V_T)^2]$, where l and w are the length and width of the channel, ϵ and t are the permittivity and thickness of the silicon dioxide insulating layer, μ_n is the electron mobility in the channel and V_T is the threshold voltage. State any restriction on the value of V_{DS}. If V_T is 2 V and t is 0.1 μm, what is the aspect ratio l/w required to provide a 100 kΩ resistor at $V_{DS} = 5$ V?
(Answers: $V_{DS} > V_T$, 2.23)

2.23 Using the manufacturers' data given in appendix H.4, produce the following network models representing the small-signal performance of a *typical* 2N 5457

n-channel JFET at 25 °C for the stated operating conditions.
(a) The general y-parameter model for common-source operation at 50 MHz with the quiescent conditions $V_{GS} = 0$ V, $V_{DS} = 15$ V.
(b) A simplified model of common-source operation at 1 kHz for the quiescent conditions $V_{GS} = 0$ V, $V_{DS} = 15$ V.
(c) A simplified high-frequency common-source model for operation at 1 MHz ignoring r_{gs}, r_{gd} and C_{ds}, for the quiescent conditions $V_{GS} = 0$ V, $V_{DS} = 15$ V.
(d) If the JFET is used in a common-source amplifier circuit at 1 MHz and the small-signal voltage gain of the circuit is -10, determine the Miller capacitance. Hence simplify the high-frequency model of part (c) by representing C_{gs} and C_{gd} as an input capacitance.
(Answers: (a) model of figure 2.86a with $y_{is} = (0.03 + j\ 1.2)$ mS, $y_{rs} = (-0.01 - j\ 0.25)$mS, $y_{fs} = (4 - j\ 0.48)$mS, $y_{os} = (0.04 + j\ 0.35)$mS
(b) model of figure 2.86b with $g_{fs} = 4$ mS, $g_{os} = 10$ μS
(c) model of figure 2.86d with $g_{fs} = 4$ mS, $r_{ds} = 100$ kΩ, $C_{gs} = 3.5$ pF, $C_{gd} = 1$ pF
(d) Miller capacitance = 11 pF. Model of figure 2.86e with $C_{in} = 14.5$ pF, $g_{fs} = 4$ mS, $r_{ds} = 100$ kΩ, C_{ds} is indeterminate due to insufficient data)

2.24 Calculate the range of wavelengths of optical radiation that are (a) absorbed, (b) transmitted by a semiconductor of energy gap 2.3 eV. To which regions of the optical spectrum do these ranges correspond?

A photodiode fabricated using this semiconductor has an effective junction area of 1 mm^2 and a dark current of 0.1 μA. The diode is illuminated with a 500 nm source providing a surface illumination of 2000 lx. If the quantum yield is 30 per cent, calculate the responsivity of the diode and the diode current under these conditions. Take 1 lm = 1.4 mW.
(Answers: (a) $\lambda \leq 540$ nm; UV and blue; (b) $\lambda > 540$ nm, yellow, red and IR. 241 mA/W, 775 nA)

2.25 A solar cell comprises a single circular silicon wafer 75 mm in diameter. In bright sunlight the short-circuit current of the cell is 500 mA and its operating temperature is 30 °C. Calculate the open-circuit voltage provided by the cell for the same illumination and temperature, if under dark conditions the reverse saturation current is 0.3 μA. If the irradiance at the cell surface is 100 W/m^2, what is the responsivity of the cell? Mention any simplifying assumptions or approximations that must be made to obtain a solution.
(Answers: 370 mV, 1.13 A/W, surface metallisation area negligible)

3 Semiconductor Device Fabrication

The fabrication of semiconductor devices involves

(1) formation of *p* and *n*-type regions of the required conductivity within the semiconductor chip by doping;
(2) formation of reliable metal–semiconductor contacts on the surface of the chip;
(3) encapsulation and packaging of the chip to provide protection and a convenient method of making electrical connection.

Early semiconductor devices such as the copper–copper oxide rectifier and point-contact transistor used metal–semiconductor contacts. Performance was improved with the introduction of junction devices making use of the rectification properties of an abrupt change in the type of dopant (*pn* junction) within a semiconductor. Initially junction devices were produced by the *grown junction* technique in which a semiconductor ingot was formed by *crystal pulling* from a semiconductor melt, the junction being formed by changing the dopant in the melt from acceptor to donor during the growing process. This method was superseded in the early 1950s by the *alloy method* in which the *pn* junction was formed by heating a pellet of dopant material in contact with the semiconductor so creating an alloy between the two materials. The alloy method was used throughout the 1950s to produce semiconductor junction devices, mainly diodes and *pnp* BJTs, and it is still used on a limited scale in the production of some silicon junction diodes, particularly where surge rating and low noise are of major interest.

Two major developments took place about 1960

(1) a change from germanium to silicon as the main device material;
(2) adoption of the *planar process* in which junctions are formed by the introduction of dopants through the surface of the semiconductor.

The use of silicon gave improved device performance, particularly in terms of reverse junction leakage and operation at elevated temperature, as well as major advantages in fabrication. The planar process introduced a number of significant advantages; closer control of the doping process, reproducibility of results to within a narrow tolerance and automation which reduced device cost. The slow diffusion process enabled the position and doping of regions to be more closely controlled and BJTs could then be produced with base widths of only a few microns giving greatly improved frequency response and switching speed. The process was readily automated as devices could be fabricated as arrays through one surface of the silicon wafer. Reduced handling and mass production enabled device cost to be considerably reduced. It is generally recognised that the planar process has been the most important single factor in the advance of low-cost, high-performance semiconductor devices. During the past 20 years many refinements have taken place in the process and today the majority of semiconductor devices and integrated circuits are produced using this process.

3.1 PLANAR FABRICATION PROCESS

The process is termed *planar* as devices are fabricated through one surface (plane) of the wafer. Devices occupy the surface layer (typically < 10 μm thick) of the semiconductor material, the remainder of the 'chip', which is 100–250 μm thick, acts as a carrier.

The stages in the process are

(1) purification of silicon and preparation of wafers;
(2) epitaxial growth;
(3) oxidation;
(4) photomasking and etching;
(5) doping;
(6) contact deposition;
(7) chip testing;
(8) packaging;
(9) final testing.

In order to give some significance to the dimensions involved it is useful to note the following

1 μm (micron) = 10^{-6} m = 10^{-3} mm
1 Å (angstrom) = 10^{-10} m = 10^{-4} μm

Human hair is typically 50–150 μm thick. Silicon atoms are spaced 2.35 Å apart in the silicon crystal lattice and so in one space dimension there are approximately 4250 silicon atoms per micron.

3.1.1 Purification of Silicon and Preparation of Wafers

Silicon occurs naturally in the form of *sand*, known chemically as *silica or silicon dioxide* (SiO_2). It is estimated that silicon accounts for about 26 per cent of the Earth's crust. The first stage in the process is chemical reduction which produces impure silicon. Because the electrical properties of semiconductor devices are extremely sensitive to impurities, the silicon has to be purified to a considerable degree, typically better than one part in 10^{10}. This is achieved by *zone-refining* in which impure silicon from the reduction process, in the form of a rod, is heated to produce a small molten zone (figure 3.1a). Impurities in the zone tend to float to the zone boundaries and, if the molten region is slowly moved from one end of the rod to the other, the impurities collect at the ends and can thus be removed. Several passes are necessary to provide the degree of purity required.

For device applications, the silicon must be as flaw-free as possible, that is, it must have a regular atomic structure. The most popular method of obtaining such a *single crystal* structure is by *crystal-pulling* from a melt of the purified silicon, known as the Czochralski process (figure 3.1b) although other methods are used. [1] In this process a previously produced piece of

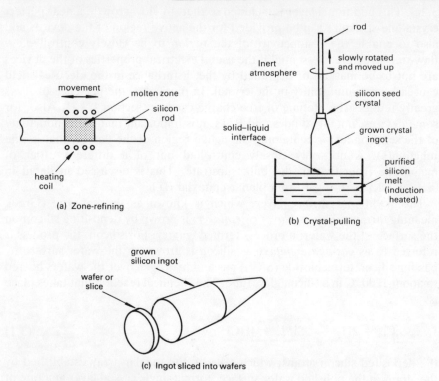

Figure 3.1 Wafer preparation

single-crystal silicon is used as a *seed crystal* which is touched on the surface of the melt and then slowly rotated and withdrawn. Silicon solidifies on the seed crystal as it is withdrawn, the atomic ordering being established by the regular array of the seed crystal. If a controlled amount of dopant is added to the melt, the grown ingot can be made p or n-type with a desired conductivity. Depending on the speed of withdrawal, which is typically less than 50 mm/hour, the grown ingot is 25–150 mm in diameter and up to 0.4 m long. The ingot is sliced into wafers (0.3–0.5 mm thick) using a diamond impregnated cutting edge (figure 3.1c) and the wafers are then ground and lapped. One face of each wafer is polished to a mirror finish and inspected under a microscope. Current production wafers for low-power discrete devices and integrated circuits are typically 75 or 100 mm in diameter and after polishing they are 0.10–0.25 mm (100–250 μm) thick.

3.1.2 Epitaxial Growth

For most devices and integrated circuits, the wafer described in section 3.1.1 forms merely the carrier or *substrate*, the devices being formed within a thin silicon layer (5–20 μm thick) grown on the polished face of the wafer (figure 3.2a). This additional layer is included so that, by slow growth, a near perfect crystalline structure can be provided for the active region of the devices and also to enable the conductivity of the region to be closely controlled. A flaw-free structure is essential if the useful electrical properties of the devices are not to be masked or degraded by the disturbance in the electrical field caused by non-uniformity in the crystal. In particular, the presence of flaws greatly reduces the lifetime of free charge carriers (section 1.3.1). Also, for some devices, notably diodes and BJTs, it is important that the conductivity of the semiconductor containing the active region of the device, that is, the junction(s), is not only closely controlled but of a different order of magnitude from that of the bulk substrate. This is discussed in detail in section 2.2.2 with regard to a planar epitaxial BJT.

This additional surface layer, which is known as the *epitaxial* (Greek meaning 'arranged upon') *layer* or *epilayer*, is grown by depositing silicon on the surface of the wafer; a process termed *epitaxy*. For silicon, the process is referred to as *gas-phase epitaxy* as silicon is grown on the wafer surface by passing silicon tetrachloride ($SiCl_4$) gas and hydrogen over the wafers heated to about 1200 °C in a silicon glass tube. The chemical reaction that takes place is

$$SiCl_4 + 2H_2 \xrightarrow{1200°C} Si \downarrow + 4HCl \tag{3.1}$$

the deposited silicon atoms, which take up the regular array established by the atoms at the polished wafer surface, form a single crystal layer at a rate of typically 1–5 μm/h. By introducing dopant gases, such as phosphine (PH_3) as

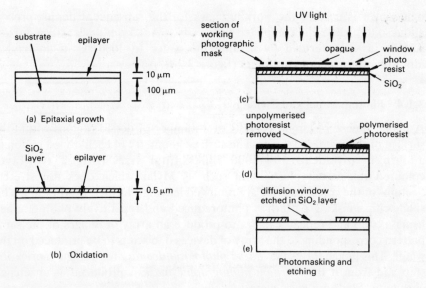

Figure 3.2 Epitaxial growth and formation of diffusion window

a source of phosphorus (donor) dopant or biborane (B_2H_6) as a source of boron (acceptor) dopant (section 1.2.1), in addition to the $SiCl_4$ and H_2, the grown layer can be formed *p* or *n*-type as required. As silicon atoms are deposited on the wafer surface, dopant atoms are trapped within the layer producing the required doping. The dopant concentration and hence the conductivity of the layer is controlled by the flow rate of the dopant gas.

Although gas-phase epitaxy is the most widely used process, crystalline epitaxial layers can also be grown by *liquid-phase epitaxy*. [1] The liquid-phase method uses the phenomenon that mixtures of a semiconductor and another element melt at a lower temperature than the semiconductor on its own, thus a semiconductor crystal can grow in a melt of a mixture containing the semiconductor. The main advantage of liquid-phase epitaxy is that it takes place at a lower temperature than the gas-phase method and avoids some of the problems of unwanted impurities which can occur in the latter method. The liquid-phase method is particularly important in the formation of devices using compound semiconductors such as GaAs and GaAlAs.

3.1.3 Oxidation

A thin layer of silicon dioxide (0.5–1.0 μm thick) protects the surface of the epilayer from contaminants including acceptor and donor dopants. Thus an oxide layer is used as the working mask for the subsequent doping process and also to protect the completed devices/ICs. A 0.5 μm thick silicon dioxide surface layer is formed by heating the wafer to 1000 °C in an oxygen atmosphere for about two hours (figure 3.2b).

3.1.4 Photomasking and Etching

A planar device or circuit consists of a number of doped regions within the surface of the prepared wafer. The surface geometry of each doped region of a chip is produced 100–1000 times final size using a two-layer coloured/transparent plastic film such as Mylar which does not stretch. Sections of the coloured layer are removed leaving transparent windows. This single-chip masking pattern is photoreduced and repetitively photographed using a 'step and repeat' camera to produce an array of images of the same pattern corresponding to the array of devices or circuits to be produced on the wafer. This image forms the *master photographic mask* for a single fabrication step and from it the *working photographic mask* is produced as an etched chromium film on a glass plate.

The working mask is used to remove areas of the silicon dioxide surface layer on the wafer so as to allow dopant to enter the epilayer. The silicon dioxide layer is coated with a thin layer of photosensitive emulsion (< 1 μm thick) known as *photoresist* which is subsequently baked. The emulsion is exposed to ultraviolet light through the working photographic mask (figure 3.2c). Windows in the mask allow the UV light to polymerise the emulsion beneath, rendering it insoluble in trichlorethylene with which the wafer is washed. Regions of the photoresist that have been obscured from the UV light by the opaque areas of the mask remain unpolymerised and are dissolved during the washing process resulting in windows in the resist layer. These windows (figure 3.2d) allow an etching solution such as hydrogen fluoride to be used to remove a region of silicon dioxide forming a window through which dopant can be diffused or implanted into the epilayer beneath (figure 3.2e). Polymerised photoemulsion is removed by washing in hot sulphuric acid. The silicon dioxide layer with windows etched through it at positions where dopant is to be introduced into the epilayer then forms the *working mask* for the doping process.

The fact that silicon dioxide forms a convenient masking medium and also is easily produced is a major advantage in the use of silicon. Other semiconductors such as germanium and gallium arsenide do not easily form compounds which can be used as a doping mask.

3.1.5 Doping

If acceptor-dopant atoms are introduced into the surface of an *n*-type semiconductor the effective donor density in the surface layer ($N_d - N_a$) is

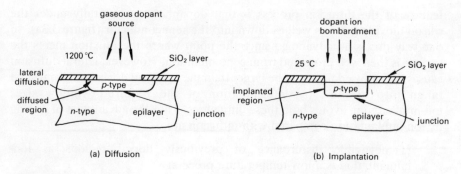

Figure 3.3 Doping methods

reduced. If the density of acceptor atoms N_a is increased to equal the density of donor atoms N_d already present, the semiconductor is said to be *compensated* and becomes effectively intrinsic. Further increase of acceptor-dopant density converts the surface layer to *p*-type and a *pn* junction is formed within the semiconductor (figure 3.3).

The most popular method of doping a semiconductor is by high-temperature diffusion. The wafer, with an appropriately etched silicon dioxide surface-masking layer, is heated to about 1200 °C and dopant in gaseous form is passed over the surface. At such a temperature some dopant atoms sink into the silicon epilayer; a process termed *diffusion*. The depth of penetration of dopant atoms into the epilayer depends on the concentration of dopant atoms in the gas, the temperature and the diffusion time. The above process is termed *constant-source* diffusion as the density of dopant atoms in the gaseous atmosphere above the wafer is maintained constant. An alternative process, termed *limited-source* diffusion, involves the deposition of a fixed quantity of dopant material on the surface of the masked wafer and subsequent heating causes diffusion into the epilayer; termed *drive-in*.

Since the mid 1970s there has been considerable development of an alternative doping method that does not require such a high temperature as diffusion. The process is *ion-implantation* in which the dopant is provided in the form of ions by a gas discharge. [2] Using an electrical accelerating, focusing and scanning system similar to that used to produce the raster on a TV tube, the masked slice is bombarded with accelerated ions which become implanted into its surface. The depth of penetration of the dopant ions is dependent on the accelerating voltage and can be closely controlled.

A major disadvantage of the diffusion method is that in a structure such as a double-diffused BJT (figure 2.11b) where it is necessary to diffuse dopant into a previously diffused region, the position of the first junction drifts during the second diffusion making it difficult to accurately control the distance between the two junctions (the base width in a BJT). Another feature of the diffusion process is that dopant moves laterally under the silicon dioxide layer as well as down into the semiconductor (figure 3.3a). In one way this is an advantage since the point where the junction meets the surface is then well protected from contamination. However, lateral diffusion causes the diffused region to be larger than the mask window and this must be taken into account in determination of the required window size; ion-implantation avoids both these difficulties. Thus the advantages of doping by ion-implantation compared with diffusion are

(1) negligible disturbance of previously doped regions as ion-implantation is a low-temperature process;
(2) due to the directional nature of the beam of ions, lateral effects are negligible;
(3) it is possible to implant a doped *buried layer* within a semiconductor as is required, for example, in integrated BJTs (figure 4.5);
(4) dopants that cannot be diffused can be used;
(5) the dopant density of implanted regions can exceed the solid solubility limit of the dopant in the semiconductor which is a limit to the dopant density obtainable by diffusion.

The main disadvantages of the ion-implantation process are

(1) the bombardment causes crystal damage which must be repaired by annealing at 400–600 °C;
(2) *channelling* can occur in certain directions through a crystal causing ions to penetrate deeper in some directions than others.

The oxidation, photomasking, etching and doping processes are repeated several times until the various doped regions within the wafer are all complete (figure 3.5). Mask alignment at each stage is of fundamental importance. Each mask has alignment marks that are used to check alignment to within about one micron using a microscope.

3.1.6 Contact Deposition

Electrical contact to the doped regions is made by forming ohmic metal–semiconductor contacts (section 1.6.1) on the surface of the chip. Using the oxidation, photomasking and etching processes, windows are opened in the silicon dioxide layer where contact is to be made to the semiconductor beneath. The surface of the entire wafer is then coated with a thin layer of aluminium or gold (≈ 1 μm thick) by vapour deposition; a process termed *metallisation*. Unwanted metal is then removed using a further photomasking and etching stage leaving the device contact pads. In the case of integrated circuits this metallisation stage also forms the interconnections between the circuit elements.

3.1.7 Chip Testing

After formation of contacts/interconnections, the individual chips on the wafer are tested for correct operation. The checks test basic operation and performance parameters and are performed automatically using a probe which makes contact with the contact pads of the device or circuit operating on a 'step-and-repeat' sequence over the complete wafer. Test results are recorded together with the type and geometric pattern of failures to enable rapid identification and correction of errors in the case of mass failure. Wafers of integrated circuit chips usually contain several *diagnostic keys* (figure 3.4a) which are test patterns used to check the correctness of mask alignment and doping levels during fabrication enabling early detection of errors. Yields can be as high as 90 per cent for discrete devices but it is not uncommon for complex circuits to have a yield of less than five per cent when first introduced. The test equipment marks defective chips with dye so that they can be discarded after separation.

3.1.8 Packaging

The wafer is broken into chips (typically 0.5–7.0 mm square) by scribing along the chip boundaries with a diamond stylus (figure 3.4a) and applying slight bending using a roller. Up to this point the chips have been processed together as wafers and production costs are low. Costs rise sharply once the wafer has been divided into individual chips due to the handling required and the cost of packaging is therefore important. The chips are sorted using a vacuum probe and the defective chips discarded. Chips are mounted on a package header or lead frame (figure 3.4b) either by forming an alloy bond between the chip and the header or by use of a heat-cured epoxy resin. Dual-in-line lead frames are formed as a ribbon stamping and individual ICs are not separated until after encapsulation. Orientation of the chip on the header or frame is important so that the contact pads can be bonded to the package leads in the correct sequence. Bonding is performed with 20–25 μm diameter gold wire using either a thermocompression joint (figure 3.4c) or ultrasonic techniques (figure 3.4d).

Device packaging must provide

(1) reliable electrical connection between the chip and the package leads;
(2) protection of the chip against chemical contamination (hermeticity);
(3) mechanical strength;
(4) adequate heat dissipation;
(5) low cost and small size.

Final encapsulation is provided by either a metal cover welded or stuck to the header or by a plastic material which is injected under pressure into a

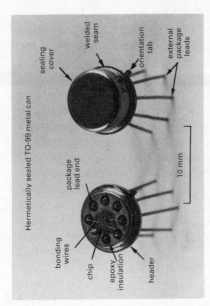

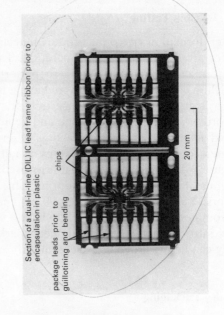

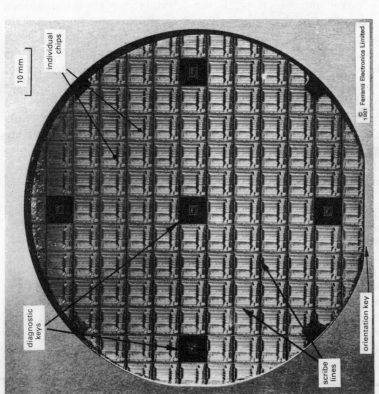

(a) Processed wafer of microprocessor chips (photograph reproduced by permission of Ferranti Electronics Limited)

(b) IC package detail

Figure 3.4 Packaging (continued opposite)

Semiconductor Device Fabrication

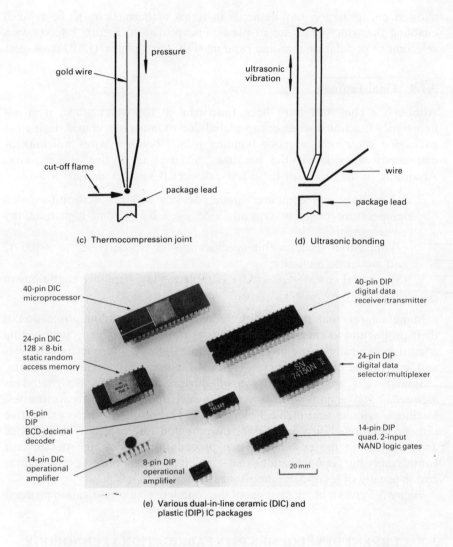

(c) Thermocompression joint

(d) Ultrasonic bonding

(e) Various dual-in-line ceramic (DIC) and plastic (DIP) IC packages

mould containing the lead frame. The welded metal cover (for example, TO-3, TO-18 and TO-39 style discrete device packages, figure 2.10, and the TO-99 IC package, figure 3.4b) provides an hermetic seal although this type of encapsulation is now little used for ICs due to the limited number of package pins and the relatively high cost. Plastic encapsulation has become the most popular type due to lower cost, although it does not provide the same degree of hermetic protection as the metal can. It is mainly due to improvements in sealing the chip surface during fabrication (termed *passivation*) using silicon dioxide and silicon nitride (section 4.2.6) that has allowed encapsulation requirements in terms of hermeticity to be relaxed enabling the widespread use of plastic encapsulation. Figure 3.4e shows a selection of popular dual-in-line ceramic (DIC) and plastic (DIP) packages.

3.1.9 Final Testing

Although a chip may have been functional at the wafer stage, it is not necessarily functional when encapsulated due to faults introduced during the packaging stage such as poor bonding joints, bonding wires touching or extraneous wires inside the package and so final testing is necessary. According to the results of these tests, devices/ICs are graded as

(1) military and computer grade: devices within specification for a temperature range of typically $-55 \rightarrow +125$ °C and high humidity environments;
(2) industrial grade: within specification for typically $-20 \rightarrow +100$ °C and moderate humidity;
(3) general commercial grade: suitable for low humidity conditions in the range $0 \rightarrow +70$ °C.

Manufacturers undertake quality control testing of a random proportion of their production to check that standards are being maintained. Such testing comprises typically X-ray inspection, temperature cycling, mechanical tests and long period electrical operation at high temperature (termed *burn-in*).

Similar tests are performed on *all* devices and circuits designated as *high-reliability* versions which are used in such applications as space probes, satellites, avionics and medical electronics. The results of these exhaustive tests on high-reliability devices are documented for each individual device and supplied to the customer. Such procedures increase the device cost considerably but, in applications that warrant such guaranteed performance, cost is usually of secondary importance.

Figure 3.5 gives a block diagram of the complete planar fabrication process.

3.2 CURRENT DEVELOPMENTS IN FABRICATION TECHNOLOGY

The most important developments are

(1) the increasing use of implantation doping techniques in place of diffusion;
(2) changes in the photolithographic process to improve resolution;
(3) a developing interest in compound semiconductors such as gallium arsenide as alternatives to silicon for high-performance devices.

Although more costly than diffusion, ion-implantation enables doped regions to be more precisely positioned. This is particularly important in

Semiconductor Device Fabrication

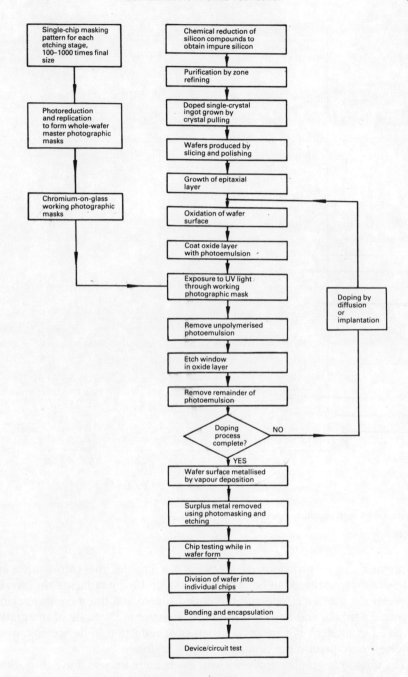

Figure 3.5 Planar fabrication process

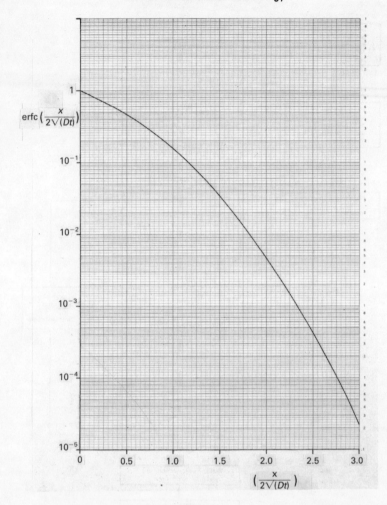

Figure 3.6 Complementary error function

improving the performance of MOSTs as it enables the gate–source and gate–drain capacitances to be reduced considerably by reducing the overlap between the gate electrode and the doped source and drain contact regions (figure 2.84b). In addition, the continual increase in the scale of integration involved in modern ICs means a progressive reduction in device size which requires more precise positioning of doped regions.

The photolithographic process which has been in universal use throughout the 1960s and 1970s has major limitations as IC complexity increases. The process involves the use of a mechanical pattern generator to produce the actual-size chromium-on-glass working photographic mask and exposure of the photoresist-coated wafer through the mask using contact lithography and UV light. The time taken to produce the mask increases as the complexity of the image increases and electron-beam pattern generators are gradually superseding the mechanical types. A major problem with contact lithography in which the mask is held in contact with the wafer during exposure, is the damage caused to both mask and wafer which becomes more serious as the circuit complexity increases. This problem is being overcome by the use of projection techniques in place of the contact method. Another difficulty is the alignment of subsequent images during the processing of a wafer which again becomes more critical as circuit complexity increases and individual device size is reduced. The basic contact process is capable of producing 3 μm line widths while optical projection methods allow 2 μm. The fabrication of current VLSICs now requires line widths of less than 1 μm and such a degree of resolution has required a change from optical exposure to electron-beam and X-ray methods which has also involved the development of suitable resist solutions. Another technique that may be used to produce line widths down to 0.1 μm does not use a conventional mask. Instead, a programmed scanning electron-beam system is used to 'write' the image directly on to the wafer. Associated with these developments in lithography are changes in the etching process. Basic wet-etching is not adequate for very narrow line widths and more sophisticated plasma-etching and ion-milling techniques are being developed.

The change of basic semiconductor from silicon to compound materials such as gallium arsenide, to take advantage of improved properties such as increased carrier mobility to improve device performance, is a major step that has many associated difficulties. The use of silicon has many fabrication advantages such as the relative simplicity of the preparation of high-quality wafers and epitaxial growth as well as the existence of an easily formed masking layer in the form of silicon dioxide. These advantages do not exist for other semiconductors which therefore have processing difficulties and increased cost imposed on them.

3.3 DIFFUSION PROCESS

The penetration of dopant atoms into a material by diffusion is described theoretically by the equation

$$\frac{\partial N}{\partial t} = D \frac{\partial^2 N}{\partial x^2} \tag{3.2}$$

where $N(x,t)$ is the dopant concentration (atoms/m^3) at a point in the material distance x below the surface at a time t. D is the diffusion coefficient (m^2/s) for the particular dopant atoms in the particular host material at the temperature concerned.

The solution of equation 3.2 gives the variation of dopant concentration with depth $N(x)$, known as the *dopant profile*, at a time t. The solution depends on the boundary conditions, namely, the variation of dopant concentration at the surface during the process $N(0,t)$ and the dopant profile at the start of the process $N(x,0)$.

If the dopant concentration at the surface of the wafer $N(0,t)$ is held constant at a value N_0 throughout the process, known as constant-source diffusion, and $N(x,0)$ is zero, the solution of equation 3.2 shows that the dopant profile within the wafer theoretically has the form of a complementray error function

$$N(x,t) = N_0 \, \text{erfc} \left[\frac{x}{2\sqrt{(Dt)}} \right] \qquad (3.3)$$

Figure 3.6 shows how the complementary error function varies for values of $[x/2\sqrt{(Dt)}]$ in the range 0–3. Alternatively if a fixed amount of dopant Q is deposited on the wafer surface and then *driven in* by heating, that is, a limited-source diffusion, the dopant profile has a Gaussian distribution

$$N(x,t) = \frac{Q}{\sqrt{(\pi Dt)}} \exp\left(-\frac{x^2}{4Dt} \right) \qquad (3.4)$$

The profile equations 3.3 and 3.4 can be used to calculate the theoretical diffusion times necessary at a certain temperature to form a device within a wafer. Figure 3.7a shows a section of a planar *npn* BJT produced by double diffusion into an *n*-type epilayer and the corresponding dopant profiles are given in figure 3.7b. The insert in figure 3.7a shows how the section relates to the complete BJT structure of figure 2.11b. The E-B junction of the device is formed typically 2 μm below the surface and the base width is typically 0.5 μm. Before diffusion of the base and emitter regions commences, the region has a uniform donor concentration of 10^{22} atoms/m^3, shown as N_C in figure 3.7b, being that of the epilayer. The *p*-type base region is formed by diffusing acceptor-dopant (for example, boron) to a density greater than the donor density already present in the epilayer, the diffusion process being continued until compensation of the donor-dopant by the acceptor occurs at the required junction depth. The C-B junction (J_1) is required to be formed at a depth of 2.5 μm and so the base diffusion continues for a time t_B until the profile is as given by N_B (figure 3.7b) such that $N_B = N_C$ at $x = 2.5$ μm. For $x < 2.5$ μm, that is, nearer the surface, the acceptor-dopant density N_B is greater than the donor density N_C and the region is thus *p*-type. For $x > 2.5$ μm, $N_B < N_C$ and that region remains *n*-type.

For purposes of this example, it is assumed that a constant-source diffusion is used having a surface concentration N_0 of 10^{25} atoms/m^3. The base profile

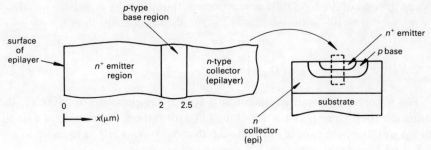

(a) Section of a planar *npn* BJT

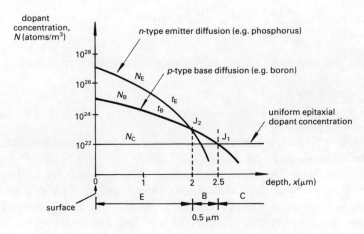

(b) Dopant profiles

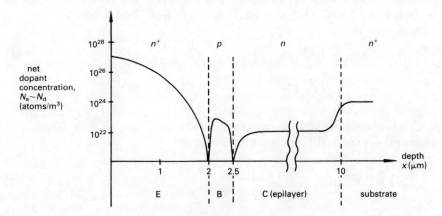

(c) Typical doping profile of a modern BJT

Figure 3.7 Formation of a double-diffused *npn* BJT

N_B at the end of the base diffusion process is thus given by equation 3.3 where $N = N_B$, $N_0 = 10^{25}$ atoms/m³ and t_B is the diffusion time, thus

$$N_B = 10^{25} \text{ erfc} \left[\frac{x}{2\sqrt{(Dt_B)}} \right] \tag{3.5}$$

For a boron diffusion in silicon at a typical temperature of 1100 °C, the diffusion coefficient D is 4×10^{-17} m²/s. To form the C-B junction at a depth of 2.5 μm, the base profile must be such that $N_B = N_C = 10^{22}$ atoms/m³ at $x = 2.5$ μm. Thus from equation 3.5

$$\text{erfc} \left[\frac{2.5 \times 10^{-6}}{2\sqrt{(4 \times 10^{-17} t_B)}} \right] = \frac{10^{22}}{10^{25}} = 10^{-3}$$

Using the erfc curve of figure 3.6, this value of erfc corresponds to

$$\left[\frac{2.5 \times 10^{-6}}{2\sqrt{(4 \times 10^{-17} t_B)}} \right] = 2.3$$

from which the required base diffusion time t_B at 1100 °C can be evaluated as 7384 s or approximately 2 h 3 min.

Having completed the base diffusion, it is then necessary to form the emitter region by diffusing donor-dopant (for example, phosphorus) such that compensation occurs at a depth of 2 μm so forming the E-B junction (J_2) at this depth, resulting in a base width of 0.5 μm. To a first-order approximation this requires compensation of the base diffusion N_B at $x = 2$ μm and it is thus necessary to determine the value of N_B at $x = 2$ μm. The base profile is given by equation 3.5 with $D = 4 \times 10^{-17}$ m²/s and $t_B = 7384$ s, thus

$$N_B (2 \text{ μm}) = 10^{25} \text{ erfc} \left[\frac{2 \times 10^{-6}}{2\sqrt{(4 \times 10^{-17} \times 7384)}} \right]$$

$$= 10^{25} \text{ erfc } 1.84 = 9 \times 10^{22} \text{ atoms/m}^3 \tag{3.6}$$

Using a constant-source emitter diffusion at 1100 °C with $N_0 = 10^{27}$ atoms/m³ (figure 3.7b), the emitter profile is

$$N_E = 10^{27} \text{ erfc} \left[\frac{x}{2\sqrt{(Dt_E)}} \right] \tag{3.7}$$

where t_E is the emitter diffusion time. To form the E-B junction at a depth of 2 μm the emitter profile must be such that $N_E = N_B = 9 \times 10^{22}$ atoms/m³ at $x = 2$ μm. At 1100 °C the diffusion coefficient for phosphorus in silicon is 4×10^{-17} m²/s thus

$$\text{erfc}\left[\frac{2 \times 10^{-6}}{2\sqrt{(4 \times 10^{-17} t_E)}}\right] = \frac{9 \times 10^{22}}{10^{27}} = 9 \times 10^{-5}$$

from which

$$\left[\frac{2 \times 10^{-6}}{2\sqrt{(4 \times 10^{-17} t_E)}}\right] = 2.8$$

giving the required emitter diffusion time t_E at 1100 °C as 3189 seconds or approximately 53 minutes.

Each region of the BJT thus contains both acceptor and donor atoms. The effective dopant concentration as far as the free carrier density is concerned is the net concentration $N_a \sim N_d$. The resulting dopant profile for a modern BJT produced by the planar process is shown in figure 3.7c. It should be noted that in practice the base diffusion is often of the limited-source type giving a Gaussian profile.

This numerical example is useful to illustrate features of the fabrication process and typical values of the parameters involved. Factors not taken into account in this theoretical analysis include

(1) movement of the C-B junction during the emitter diffusion due to continued change of the base profile;
(2) continued diffusion of dopant during cooling;
(3) the epilayer doping has not been taken into account in calculation of the position of the E-B junction;
(4) it is assumed that the erfc solution of the diffusion equation for the constant-source process accurately describes the distribution of dopant atoms.

In practice the determination of diffusion times and temperatures required to form junctions at specific depths relies more on results obtained from trial diffusions than on theory. A commonly used method of determining junction depth in a trial diffusion is the *angle-lap* technique (figure 3.8) in which the edge of a test wafer is lapped to produce a very shallow bevel (0.5–1°). The bevel is stained and microscopic examination enables the junction position to be found, from which, knowing the bevel angle, the junction depth can be determined.

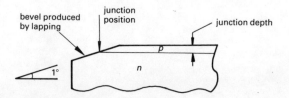

Figure 3.8 Practical determination of junction depth by the angle–lap technique

REFERENCES AND FURTHER READING

1. B.G. Streetman, *Solid State Electronic Devices*, chapter 1 (Prentice-Hall, 1980)
2. G. Carter and W.A. Grant, *Ion Implantation of Semiconductors* (Edward Arnold, 1976)
3. A.B. Glaser and G.E. Subak-Sharpe, *Integrated Circuit Engineering*, chapters 5, 10 and 11 (Addison-Wesley, 1977)
4. S.K. Ghandhi, *Semiconductor Power Devices*, chapter 6 (Wiley, 1977)
5. R.S. Muller and T.I. Kamins, *Device Electronics for Integrated Circuits*, chapter 1 (Wiley, 1977)
6. D.J. Hamilton and W.G. Howard, *Basic Integrated Circuit Engineering*, chapters 1 and 2 (McGraw-Hill, 1975)
7. J.Allison, *Electronic Integrated Circuits*, chapters 2, 3, 4 and 6 (McGraw-Hill, 1975)
8. B.G. Streetman, *Solid State Electronic Devices*, chapter 5 (Prentice-Hall, 1980)

TUTORIAL QUESTIONS

3.1 It is required to form an *npn* BJT in an *n*-type silicon epitaxial layer having a uniform dopant concentration of 5×10^{21} atoms/m^3. Constant-source base and emitter diffusions are to be used at 1150 °C with surface dopant concentrations of 10^{24} and 2×10^{25} atoms/m^3 respectively. If the C-B junction depth is to be 3 μm and a base width of 0.5 μm is required, calculate the diffusion times for the two processes mentioning any assumptions or approximations that are made. The diffusion coefficient for the two diffusions at 1150 °C is 10^{-16} m^2/s.

The erfc dopant profile given in equation 3.3 can be assumed and the erfc information of figure 3.6 will be required.
(Answers: 1.56 h, 47 min)

3.2 An n^+ diffusion is performed into a *p*-type silicon region having a uniform dopant concentration of 5×10^{23} atoms/m^3. If the dopant concentration in the gas above the water surface is maintained constant at 5×10^{26} atoms/m^3 and the process time is 30 min, calculate the depth of the *n*-type diffusion. The diffusion coefficient is 5×10^{-17} m^2/s. If the doping window in the silicon dioxide layer is a rectangle of size 20 μm × 10 μm estimate the size of the n^+ diffused region at the chip surface explaining your reasoning.
(Answers: 1.38 μm, 22.8 μm × 12.8 μm)

4 Integrated Circuit Technology

An integrated circuit (IC) is an electronic circuit formed in or on a carrier (substrate) and packaged as a single unit. The advantages of circuit integration compared with discrete component technology are

(1) greater reliability;
(2) improved performance;
(3) cost reduction;
(4) size reduction.

The electrical connections between the elements in an IC, which are formed during fabrication, are far more reliable than soldered connections on a printed circuit board (pcb) and the encapsulated IC is mechanically more robust than the discrete component circuit. Production of an electronic system using ICs involves fewer soldered connections and, since the number of pcbs is reduced, fewer connectors are required leading to further improvements in reliability. Developments in circuit design have enabled power dissipation to be reduced considerably which also provides greater reliability since, although the circuit elements are enclosed and tightly packed, the working temperature is reduced.

Miniaturisation provides improvements in electrical performance in the form of switching speed and frequency response by reducing capacitance and further improvements have been possible taking advantage of IC features such as the cheapness of active devices, the close matching of devices fabricated simultaneously and the tracking of closely positioned circuit elements.

Although processing costs are high, the degree of automation and the vast number of circuits which can be fabricated simultaneously enables the unit cost to be kept low. In addition, compared with production in discrete component form, the cost of the IC system is lower as fewer pcbs are required and the associated manual assembly such as interboard wiring is reduced.

Miniaturisation thus features in reliability improvement, performance improvement and cost reduction but in addition miniaturisation has enabled the implementation of systems which would not be feasible in discrete component form such as space technology, large computers and modern communication systems.

There are two main types of IC

(1) *monolithic* (Greek: 'single stone') ICs in which the complete electrical circuit is fabricated within the surface layer of a piece of semiconductor material typically 1–7 mm square and 0.1–0.2 mm thick, popularly known as a 'chip', using the planar process;
(2) *film* ICs in which circuit elements and interconnections are deposited on an insulating substrate such as a ceramic. Some film ICs also incorporate monolithic devices and/or ICs in which case they are termed *hybrid* ICs.

The major proportion of ICs are monolithic due mainly to economic rather than to technological factors in that more research effort was applied to the planar process than to film processes in the early days enabling monolithic technology to develop more rapidly. Film ICs, and particularly hybrid ICs, can offer certain advantages, however, such as improved performance at very high frequency, higher power dissipation and generally lower cost for small batch production.

Integrated circuits are classified as *bipolar* or *MOS* according to the type of transistor incorporated and as *digital* or *linear* according to application. Digital ICs are logic systems in which the active devices which may be BJTs or MOSTs, act as switches. Linear ICs such as amplifiers are mainly bipolar, the BJTs operating in the linear or active mode.

The first ICs were bipolar logic gates and operational amplifiers containing up to about ten circuit elements. They were developed in 1959–60 in the USA and were integrated versions of discrete component circuits. The first MOS IC was a multipurpose logic block of 16 MOSTs produced in 1962. Developments in the fabrication process during the 1960s and a new approach to circuit design enabled the complexity of ICs to increase rapidly. By 1971 reliable large-scale ICs such as calculator chips and memories containing over 1000 circuit elements were being produced. Throughout the 1970s the packing density of IC elements has progressively increased allowing microprocessor and memory chips using over 10^5 elements to be produced by the early 1980s.

4.1 ISOLATION

As the circuit elements in a monolithic IC are formed within a common conducting substrate it is usually necessary to provide isolation so that the elements are only interconnected by the surface metallisation pattern. There are four basic isolation techniques for monolithic ICs

(1) diode isolation;
(2) dielectric isolation;
(3) beam–lead technology;
(4) silicon-on-sapphire technology.

Film ICs do not require an isolation stage as the elements are deposited on an insulating substrate.

4.1.1 Diode Isolation

The circuit elements are formed within an epilayer (usually n-type) which is grown on a substrate of opposite type (figure 4.1a). The layer is divided into isolated 'islands' by forming heavily doped channels of the same type as the substrate. Using a standard photolithographic stage (section 3.1), a silicon dioxide diffusion mask is formed on the surface. For an n-type epilayer a p^+ diffusion is then performed for a time sufficient to break through the epilayer (figure 4.1b). The layer is thus divided into an array of n-type islands which are interconnected via the substrate by two back-to-back diodes. Individual circuit elements are then formed within each island and interconnected as required by surface metallisation. Figure 4.1c gives a circuit representation of two adjacent islands each containing a BJT. The n-type island forms the collector in each case which are effectively interconnected by the back-to-back isolation diodes. The diodes are maintained at reverse bias by connecting the p-type substrate to the most negative potential in the circuit via surface metallisation.

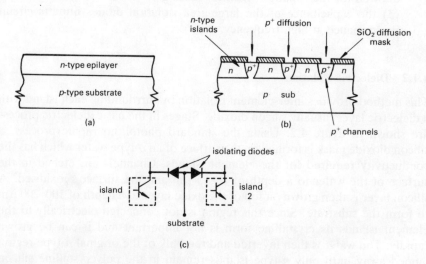

Figure 4.1 Diode element isolation

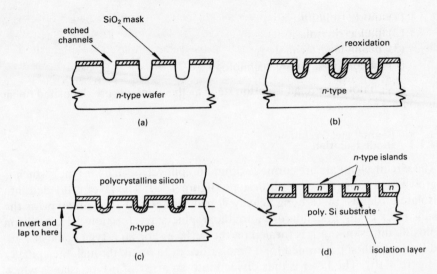

Figure 4.2 Dielectric element isolation

The basic diode isolation method simply involves additional photolithographic and diffusion stages in the fabrication process and is therefore a relatively cheap method but it has the following disadvantages

(1) as the isolation diffusion is relatively deep, the diffusion time is long;
(2) due to lateral diffusion (section 3.1.5) adequate clearance must be allowed for the isolation channels causing inefficient use of chip area;
(3) the capacitance of the large-area isolation diodes impairs circuit performance at high frequency.

4.1.2 Dielectric Isolation

This method provides inter-element isolation by surrounding each island with a dielectric layer, usually silicon dioxide. Stages in the basic dielectric process are shown in figure 4.2. Using the standard photolithographic process, a silicon dioxide mask is formed on the surface of an n-type wafer which has the conductivity required for the element islands. Channels are etched in the surface of the wafer to a depth of 10–15 μm and the surface reoxidised. A silicon layer is then grown on top of the oxide layer to a depth of 100–200 μm to form the substrate. Since this region is not connected electrically to the element islands its crystalline form is not important and it can be grown rapidly. The wafer is then inverted and the bulk of the original n-type region lapped away until only n-type islands remain in the polycrystalline silicon substrate.

Compared with diode isolation, the dielectric method provides improved inter-element isolation as leakage across the thin (1–2 μm) silicon dioxide layer is considerably less than that across reverse-biased isolation diodes. Also, the lower capacitance of the dielectric structure provides improved high-frequency performance, but the process is more costly than the diode method due to the additional stages involved.

4.1.3 Beam-Lead Technology

Excellent isolation between circuit elements is achieved if the silicon between the elements is completely etched away although the structure is consequently less robust as the elements are then held together only by the metallisation pattern. In this approach, the circuit elements are formed within the silicon wafer with no regard for isolation except for allowing sufficient space between elements for subsequent etching. A thick (10 μm) metallisation pattern is then deposited on the surface of the wafer to make the appropriate interconnections between the elements. Using standard photolithographic techniques, grooves (5–10 μm wide) are etched from the reverse side of the wafer (figure 4.3a) to remove completely the silicon between the elements leaving the elemental silicon blocks held together by the thick metallisation

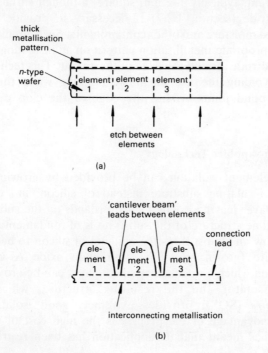

Figure 4.3 Beam-lead technology

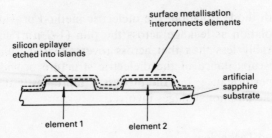

Figure 4.4 Silicon-on-sapphire technology

pattern (figure 4.3b). The structure can be encapsulated in plastic to provide mechanical strength. This technique results in high leakage resistance and low stray capacitance between elements providing excellent high-frequency performance although the additional processing steps increase cost. The name *beam-lead* derives from cantilever beam describing the connection 'beams' between elements (figure 4.3b).

Unencapsulated beam-lead ICs are produced mainly for inclusion in hybrid ICs (section 4.9.3). The absence of encapsulation enables circuit size to be kept to a minimum, typically 1–2 mm square, although surface passivation using silicon nitride (section 4.2.7) is necessary to provide the required protection against moisture and other contaminants. Such circuits are bonded directly to an appropriate metallisation pattern on an insulating substrate via the leads that protrude around the edge of the circuit. This technique has the advantage of avoiding the use of flimsy bonding wires thus improving reliability. The bond joints around the edge of the chip can be readily inspected.

4.1.4 Silicon-on-sapphire Technology

Excellent inter-element isolation can be provided by growing the silicon epilayer on an insulating substrate instead of silicon and etching away channels in the layer leaving isolated silicon islands on the substrate (figure 4.4). The crystalline structure of the substrate is of fundamental importance since it must allow an almost flaw-free epilayer of silicon to be grown on it. Artificial sapphire (grown crystalline aluminium oxide Al_2O_3) not only provides a base on which a satisfactory silicon layer can be grown but also is an excellent insulator and the resulting structure, which is termed *silicon-on-sapphire* (SOS), provides extremely good isolation between elements. A disadvantage of the process is the high cost of the substrate material which has meant that its application has been restricted to very high-frequency ICs.

4.2 MONOLITHIC IC ELEMENTS

All circuit elements in monolithic ICs (BJTs, MOSTs, diodes, resistors and capacitors) are formed within the surface of the wafer using the photolithographic and doping processes described in section 3.1. Many digital ICs are composed almost entirely of one type of element such as n-channel MOSTs in the case of an NMOS circuit or BJTs in an I^2L circuit. Other ICs, however, such as linear circuits and small-scale digital circuits, contain a variety of elements. For example, the popular 741 integrated operational amplifier contains twenty BJTs, two diodes, eleven resistors and one capacitor. To reduce the number of processing steps, elements are formed simultaneously so that diode anodes, diffused resistors and diffused capacitors are formed by the same diffusion as the bases of BJTs and similarly n-type contact regions are formed at the same time as BJT emitters. Doping profiles are usually selected to optimise the performance of the transistors in the circuit and this imposes constraints on the passive components.

The structural diagrams in the following sections assume basic diode isolation.

4.2.1 Integrated BJTs

For performance and fabrication reasons (section 2.2.1) the majority of integrated BJTs have an *npn* structure. By fabricating devices simultaneously, close matching is obtained, typically V_{BE} is matched to within 5 mV with less than 10 μV/°C drift and h_{FE} to within 10 per cent. In addition, BJTs that are closely spaced on the chip experience almost identical temperature variations and they are therefore ideal for use in the differential configuration. Compared with discrete BJTs, the integrated version has the disadvantage of requiring isolation. Also, since the collector contact of the integrated BJT has to be made on the chip surface instead of on the underside of the chip as in the discrete case (figure 2.11b), the collector current path through the relatively low conductivity n-type epiregion causes a high collector resistance and high $V_{CE(sat)}$. This effect is compensated by including a thin heavily doped (n^+) region, termed a *buried layer,* beneath the epilayer to provide a high conductivity path for the collector current. The buried layer can be formed by doping the wafer prior to epitaxial growth, by altering the doping as the epilayer is grown or it can be implanted.

Figure 4.5 shows the typical structure of an integrated low-power *npn* BJT employing diode isolation. Improvements in the fabrication process have enabled the n-type island containing the BJT to be reduced in size from typically 250 μm × 200 μm × 25 μm thick in the early 1970s to 60 μm × 50 μm × 10 μm thick towards the end of the decade, although other fabrication techniques such as the Isoplanar and collector diffused isolation processes (section 4.3) have enabled the size to be further reduced to the order of 40 μm

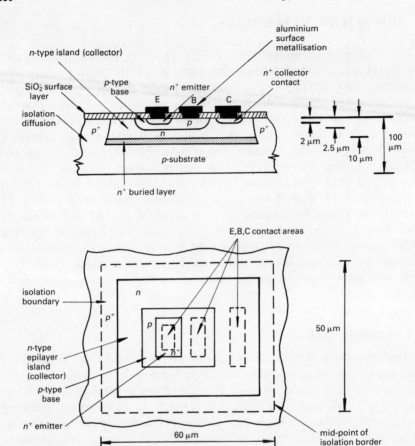

Figure 4.5 Integrated low-power *npn* BJT

× 25 μm using epilayers as thin as 5 μm. In the case of basic diode isolation, the chip area occupied by the device must include the isolation diffusion which may be up to 20 μm wide at the surface. The base region which is usually formed by a limited-source diffusion, is typically 2.5 μm deep while the constant-source diffused emitter is approximately 2 μm deep providing a base width of about 0.5 μm. This produces a *graded-base* doping profile in which the doping of the base region decreases towards the C-B junction. An electric field is thus created in the base and accelerates injected electrons towards the collector assisting carrier diffusion and thus improving performance. A heavily doped (n^+) collector contact region is necessary to enable the collector metallisation to form an ohmic contact; aluminium forms a rectifying contact with lightly doped *n*-type silicon (section 1.6.1). The n^+

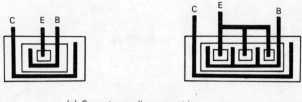

(a) Current spreading geometries

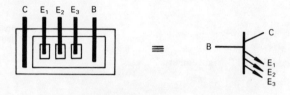

(b) Multi-emitter structure

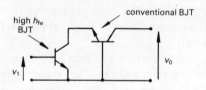

(c) High-gain cascode arrangement (a.c. model)

Figure 4.6 Integrated *npn* BJT arrangements

collector contact region is diffused at the same time as the emitter region and thus does not involve an extra processing step.

The access resistances of the base and collector regions can be reduced by 'spreading' the current flow in each region. This is achieved by employing multiple base and collector contacts (figure 4.6a) to create multiple current paths. The reduction of base current density by spreading also reduces emitter crowding (section 2.2.23) which is further reduced in BJTs intended to operate at higher current levels, such as in circuit output stages, by the use of multi-emitter arrangements (figure 4.6a).

Some applications such as integrated TTL NAND gates (figure 4.29a) require multi-emitter BJTs. These can be provided by diffusing a number of separate emitter regions into a single base region (figure 4.6b).

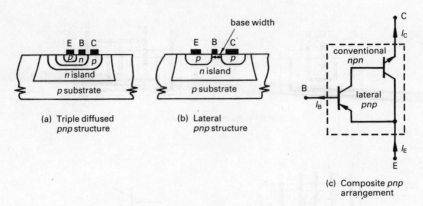

Figure 4.7 Integrated *pnp* BJT arrangements

BJTs can be produced with current gains of several thousand by diffusing the emitter region to such a depth that the base width is reduced almost to punch-through (section 2.2.19) although this results in a C-B breakdown voltage of only a few volts. The breakdown voltage can be increased to practical levels while retaining the high-gain properties by cascading the high-gain CE BJT with a conventional BJT operating in CB, termed a *cascode* arrangement (figure 4.6c).

Many IC designs require both *npn* and *pnp* BJT structures, particularly in biasing and complementary arrangements. Since an *n*-type epilayer is chosen to suit the *npn* structures required, the formation of a *pnp* structure introduces certain complications. One possibility would be a *triple-diffused* structure (figure 4.7a) although, apart from the disadvantage of an extra processing step to form the emitter, the structure has major fabrication problems. [1] One difficulty is the need for the emitter region to be *very* heavily doped since successive diffusions require progressively higher surface dopant densities to enable compensation and, apart from solid solubility setting a limit to the maximum density obtainable by diffusion, the process is difficult to control in terms of time and temperature to obtain the required result. Another problem is the low gain due to the relatively high dopant density in the base. An alternative simpler structure is the *lateral pnp* arrangement (figure 4.7b) in which the *p*-type emitter and collector regions are formed during the *npn* base diffusion. The structure is described as lateral as current flow across the base is parallel with the chip surface rather than perpendicular as in the conventional planar structure. Since the base width of the lateral device is determined by limitations in the photolithographic process instead of diffusion depth, it is relatively large, resulting in a low gain, typically less than five. The fact that the emitter is not heavily doped also contributes to the low gain. However, by using a composite arrangement of lateral *pnp* and conventional high-gain *npn* (figure 4.7c), high-gain *pnp* properties can be obtained.

4.2.2 Integrated MOSTs

The MOST is the simplest active device to fabricate, requiring fewer processing steps and less chip area than a BJT. It is consequently cheaper and offers a higher packing density than the conventional BJT. The MOST has thus become very popular for large-scale ICs. Both enhancement and depletion-type MOSTs are used in MOS ICs although the enhancement type is by far the most popular for a number of reasons. Since V_{GS} can equal V_{DS}, cascaded stages using enhancement-type MOSTs can be directly coupled, avoiding the need for level-shifting interstage circuitry and thus simplifying design. In addition, the enhancement-type MOST requires fewer processing steps and it is self-isolating.

Basic *n*-channel enhancement and depletion-type integrated MOST structures are shown in figure 4.8. The enhancement-type device is formed by

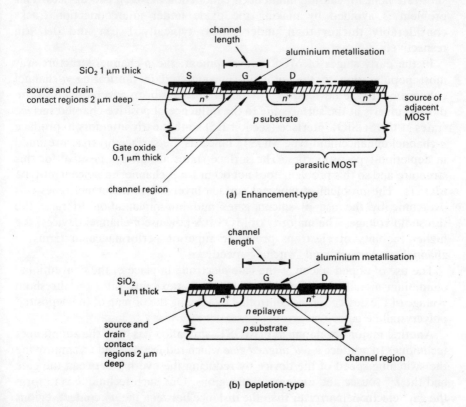

Figure 4.8 Basic integrated *n*-channel MOSTs

a single n^+ diffusion into a p-type substrate forming the source and drain contact regions. A layer of silicon dioxide forms the insulating medium while surface metallisation, usually of aluminium, forms the gate and the source and drain contacts. By applying a positive value of V_{GS}, electrons are attracted to the channel region under the gate electrode and for V_{GS} greater than the threshold voltage V_T, the surface of the substrate *inverts* to n-type forming a conducting channel between the source and drain contacts. Between the conducting n-type channel region (inversion layer) and the p-type substrate is a pn junction which, if kept reverse-biased by connecting the substrate to the most negative potential in the circuit, isolates the device from the substrate. The device is therefore *self-isolating* requiring no extra processing steps to provide isolation and a number of enhancement-type MOSTs can therefore be fabricated on the same substrate without isolation problems. However, formation of an array of MOSTs on a substrate creates a number of parasitic MOSTs (figure 4.8a) whereby the metallisation between adjacent devices can act as a gate electrode and, if the current in that metallisation is sufficiently high, an inversion layer can be created in the substrate beneath causing unintended conduction between two devices. This problem is avoided by making the oxide under interconnection tracks considerably thicker than under gates, typically 1 μm and 0.1 μm respectively.

In the early stages of MOST development, the p-channel structure was most popular due to the difficulty encountered in fabricating n-channel enhancement-type devices. Lightly doped p-type silicon covered with silicon dioxide inverts at the surface due to the influence of positive-charged surface states at the Si–SiO$_2$ interface (section 1.5) so that early attempts to produce n-channel enhancement-type MOSTs failed as the surface inversion resulted in depletion-type properties. The surface states are always positive for this structure and so this problem does not occur for p-channel enhancement-type MOSTs. The problem of inherent surface inversion in n-channel types was overcome by the use of silicon gates and ion-implantation to raise the threshold voltage. The majority of MOS ICs now use n-channel devices, the higher mobility of electrons providing superior performance in terms of channel conductivity and switching speed.

The use of doped silicon as the gate electrode in place of the conventional aluminium metallisation provides a degree of flexibility since the threshold voltage of the device can be controlled by varying the doping of the deposited polycrystalline gate.

Another major development in MOST fabrication has been the adoption of techniques to produce a *self-aligned gate* which reduces C_{gs} and C_{gd} improving the switching speed of the device by reducing the overlap between the gate and the n^+ source and drain contact regions. One such technique is to form the gate electrode narrower than the distance between the n^+ contact regions and to employ ion-implantation, using the gate electrode as a mask, to extend the n^+ regions up to the edge of the gate (figure 2.84b). In this way virtually no overlap exists and a considerable improvement in switching speed has been achieved. A similar technique can be used with a silicon gate structure, the polycrystalline gate being grown before the contact regions are formed and subsequent diffusion forms the contact regions and dopes the gate.

The depletion-type MOST (figure 4.8b) requires a conducting channel region which can be either implanted, diffused or grown as an epilayer. This additional processing step means that the depletion-type MOST is more costly than the enhancement type. Furthermore, the depletion type is not self-isolating and if an array of them were fabricated in a common epilayer some form of isolation process would be required. Arrays of depletion-type MOSTs are rarely produced but they are often used as loads for enhancement-type MOSTs in ICs, ion-implantation being used to form both types of device on the same substrate.

As with all MOS devices, the high input resistance enables static charge to build up sufficiently on input gates to cause breakdown of the thin oxide layer even with no voltage applied. Gate protection diodes can be incorporated which breakdown at a voltage lower than that necessary to cause oxide breakdown but the associated capacitance degrades performance. It is common for MOS ICs to be supplied with the package leads shorted together with either metal foil or conducting foam to provide a leakage path for induced charge.

There have been a number of structural developments in MOSTs for certain applications and to improve performance. The basic MOST structure with a gate oxide thickness of 0.1 μm has a threshold voltage of 4–5 V. Many applications require a lower V_T but reduction of the oxide thickness to below 0.1 μm is not allowable due to the presence of pin-holes which result in breakdown. However, V_T can be reduced while maintaining the thickness of the insulating layer by the use of silicon nitride (Si_3N_4) which has a permittivity approximately twice that of silicon dioxide. Direct replacement of oxide by nitride is not practical, however, because the presence of additional charged surface states at the Si–Si_3N_4 interface, compared with the Si–SiO_2 interface, cancels any improvement. This situation is avoided by the use of a very thin (≈ 0.02 μm) intervening oxide layer resulting in a *metal–nitride–oxide–semiconductor* (MNOS) structure (figure 4.9a). This structure enables V_T to be reduced to 1–3 V enabling power supply levels also to be reduced. The MNOST has been used as a non-volatile memory element using the presence or absence of a conducting channel to represent a binary 1 or 0. If a substantial positive voltage is applied between the gate and substrate of the structure shown in figure 4.9a, negative charge tunnels through the thin oxide layer and becomes trapped at the SiO_2–Si_3N_4 interface. The presence or absence of this trapped charge modifies the threshold voltage of the device which defines the state of the cell. The device has been utilised in electrically alterable read-only memories (EAROMs) in which stored information may

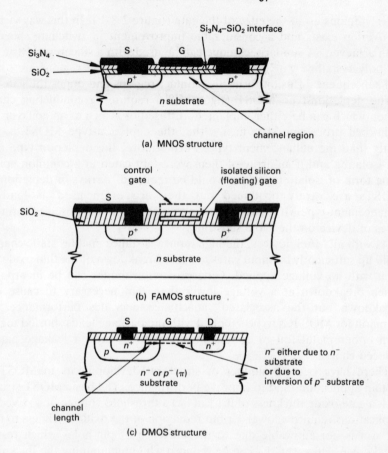
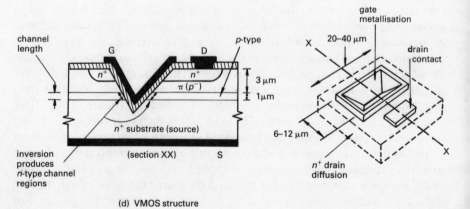

Figure 4.9 MOST structures

require occasional alteration. In such an application no degradation of the stored data can be detected after 10^6 read cycles. The structure is not suitable for random-access memory (RAM) applications which must allow continual reprogramming with fast read and write times, because the MNOS structure is inherently slow and gradual degradation of parameters occurs during repeated read–write cycles due to damage to the thin oxide layer.

Another MOS device developed particularly for erasable programmable ROM (EPROM) applications is the *floating-gate avalanche-injection* MOST structure (FAMOS) shown in figure 4.9b. As with the MNOS structure, the state of the cell, that is, the source–drain resistance, is determined by stored charge. Here the gate electrode, which is formed from polycrystalline silicon, is isolated since it is totally surrounded by oxide, no electrical connection being made to it, hence the term *floating gate*. If a short (5–10 ms) voltage pulse of about −50 V is applied between drain and source, the drain–substrate junction is driven to avalanche breakdown. Some charge from the resulting current surge tunnels through the thin oxide layer and becomes trapped on the gate. The resulting induced charge at the surface of the substrate forms a conducting channel and the device is in the ON state. Due to the isolated nature of the gate, leakage of trapped charge is very slight and it has been estimated that only 20–30 per cent of the initial charge leaks away in 10 years. Erasure of the cell is possible by exposure to UV light or X-rays and causes free carrier generation in the silicon. Subsequent recombination between the trapped and generated charge erases the stored information. The number of times the cell can be erased is limited, however, due to damage to the oxide layer. A variation of the FAMOS structure enabling electrical erasure employs an additional silicon or aluminium control gate above the floating gate (figure 4.9b); termed a *stacked-gate* structure. Both the MNOS and FAMOS structures provide non-volatile storage since, once programmed, retention of the stored information does not rely on an external power supply.

An important parameter in consideration of the performance of a MOST is channel length. In the basic lateral MOST this is limited by the tolerances in the photolithographic process to 2–3 μm. During the mid to late 1970s two structural innovations to the basic enhancement-type MOST enabled sub-micron channel lengths to be produced with corresponding improvements in switching speed. The two structures are termed *double-diffused MOS* (DMOS) and *V-groove MOS* (VMOS). The DMOST is a lateral device (figure 4.9c) in which channel length is determined by the spacing between two diffusions performed through the same window. Using a lightly doped n or p-type substrate, a p-type region is diffused through the source window followed by a shallower n^+ diffusion through the same source window and the drain window. An n-channel device is thus created with a channel length of 0.5–2 μm which depends not on tolerances in the photolithographic process but on dopant source concentration, diffusion time and process temperature.

The dopant-type of the substrate is immaterial provided that it is lightly doped as the surface will be n^- in either case. If a p^- substrate is used, the effect of the net positive charge at the Si–SiO$_2$ interface causes the surface to invert to n^-. The designations p^- and n^- are used to denote light doping corresponding to p^+ and n^+ for heavy doping. Lightly doped p-type material is also termed π-type, derived from pi (p-intrinsic).

The VMOS structure was a major development in MOS technology since it introduced a departure from lateral current flow, the current path between source and drain being along induced channel regions on the faces of an etched V-groove (figure 4.9d). The surface of the antimony-doped n^+ substrate, which forms the source, is doped with boron at a concentration of about 10^{23} atoms/m^3. As this concentration is not sufficient to compensate the donor-dopant already present, the surface of the substrate remains n-type. A π-type epilayer, typically 3 μm thick, is then grown on the substrate during which the boron 'outdiffuses' from the substrate into the epitaxial region forming a thin p-type region approximately 1 μm thick. The thickness of the p-type region, which forms the channel, is controlled by the concentration of the boron dopant. A 1 μm deep n^+ diffusion then forms the drain region and a thick SiO$_2$/Si$_3$N$_4$ passivating layer is deposited on the surface prior to formation of the groove. By arranging that the lattice structure of the silicon is suitably orientated relative to the wafer surface, an approximately symmetrical V-groove is obtained by etching. Thermal oxidisation of the groove surface forms the gate oxide and aluminium metallisation completes the structure. The short, wide channel gives the VMOS structure excellent switching speed and the chip area occupied by the device is reduced compared with lateral structures, providing an increased packing density.

Comparing the basic MOS and bipolar technologies, MOS offers lower cost and higher packing density due to the fewer processing steps and smaller chip area occupied by each device. However, MOS is slower than bipolar due to inherent capacitance and higher resistance levels necessitating higher voltage levels. The recent developments in MOS technology outlined above have, however, greatly improved its speed while retaining its high packing density although involving extra processing costs. At the same time the introduction of superintegrated bipolar technologies (section 4.6) has greatly improved bipolar packing densities and so competition between the two technologies continues.

4.2.3 Integrated JFETs

Diffused JFETs can be produced in integrated form using the standard planar process as shown in figure 4.10a. The n-type epitaxial island forms the channel with ohmic source and drain contacts at either end. A p-type diffusion forms the gate, the conductance between source and drain being controlled by the depth of penetration of the gate–channel depletion region

Integrated Circuit Technology

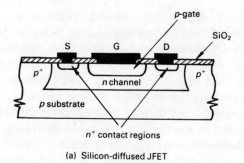

(a) Silicon-diffused JFET

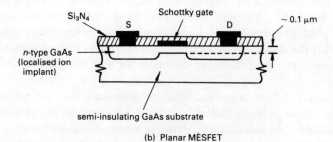

(b) Planar MESFET

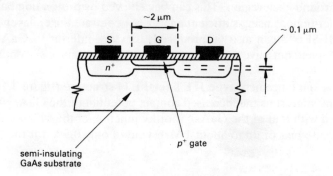

(c) GaAs enhancement-type JFET (e-JFET)

Figure 4.10 Integrated n-channel JFETs

due to V_{GS}. The absence of a buried layer makes the device simpler to fabricate than a BJT but it involves more processing steps than a MOST. Due to the relatively large capacitance of the reverse-biased gate–channel junction and the transit time of electrons in the channel, the diffused silicon JFET is a relatively slow switch and is therefore of only minor importance in comparison with BJTs and MOSTs.

A type of JFET that is currently of major interest, particularly for high-speed large-scale IC applications, uses a Schottky junction to control channel conductivity. The device is known as a Schottky-barrier FET or MESFET (*metal–semiconductor* FET) and is an integrated version of the discrete MESFET of figure 2.84a. The simple structure (figure 4.10b) promises a high packing density, while the use of GaAs and a short channel (≈ 0.5–2 μm) provides fast switching. A semi-insulating GaAs substrate ($\sigma \approx 10^{-10}$ S/m) is formed by the use of a mid-gap dopant such as chromium. The shallow *n*-type channel region (typically only about 0.1 μm thick) can then be ion-implanted into the surface of the substrate. An aluminium gate forms a Schottky junction with *n*-type GaAs while ohmic source and drain contacts are formed using multi-layer combinations such as Au, Cr, Ni or In, Ge, Au. The basic MESFET has depletion-type properties (referred to as a d-MESFET) but if the implanted channel region is sufficiently thin (that is, <0.1 μm thick), the depletion layer of the Schottky junction at zero V_{GS} can totally deplete the channel layer forming an enhancement-type device (e-MESFET).

An alternative fabrication technique in place of the use of masked ion-implantation to produce isolated *n*-type GaAs regions in the surface of the semi-insulating substrate, is to form an *n*-type GaAs epilayer over the whole chip surface and to isolate device regions by forming an isolation border around each region. This can be achieved by proton bombardment of the border regions using sufficient energy to penetrate the epilayer. Silicon d- and e-MESFETs can also be produced; they are inferior to GaAs types in terms of speed but have the advantage of using the mature silicon fabrication process.

A GaAs enhancement-type JFET (e-JFET) structure (figure 4.10c) is also of current interest as the increased contact potential of the GaAs *pn* junction compared with that of the GaAs Schottky junction of the MESFET, allows a forward gate bias of up to about 1 V instead of only 0.5 V for the MESFET.

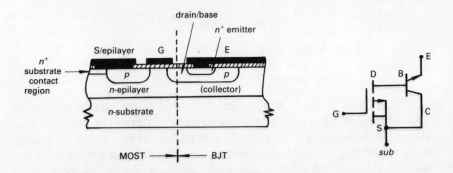

Figure 4.11 Hybrid MOS/bipolar device

Enhancement-type properties are obtained by using ion-implantation to produce a very thin channel (≈ 0.1 μm thick) which is totally depleted by the gate–channel junction depletion region for $V_{GS} = 0$. Channel lengths of 1–2 μm are attainable giving excellent switching speed.

GaAs e-MESFETs and GaAs e-JFETs are of major interest for ultra-high-speed LSI and VLSI circuits of the future. The particular attraction is the low-power dissipation attainable by virtue of the enhancement-type (normally OFF) properties coupled with fast switching speed possible using the small, simple geometry. These devices provide the lowest predicted power-delay product (1–10 fJ) for uncooled logic gates.

4.2.4 Hybrid Active Devices

Input and output impedance levels of MOS circuits are considerably higher than for bipolar circuits and this introduces problems when the two types of circuits are interconnected. A convenient method of impedance-matching between MOS and bipolar circuits is to provide the MOS circuit with a bipolar output stage using a hybrid MOS/bipolar device (figure 4.11). The device is basically a *p*-channel enhancement-type MOST with an n^+ region formed within the *p*-type drain forming a bipolar *npn* structure with the epilayer. This is an example of *superintegration* where doped regions have dual roles. In this case the *p*-type drain region of the MOST also forms the base of the BJT and the epilayer forms the BJT collector which is connected to the MOST source via the surface metallisation. MOS ICs with this type of output stage can be interfaced directly with bipolar ICs such as TTL; they are described as being 'TTL compatible'.

4.2.5 Integrated Diodes

The base–emitter and collector–base junctions of BJT structures are used as *pn* junction diodes although, due to different doping levels and junction areas, the two junctions have different properties particularly with regard to switching speed and breakdown voltage. Five configurations are possible as shown in figure 4.12a–e. Structures *a* and *b* using the C-B junction have a high breakdown voltage of about 40 V whereas those utilising the B-E junction (*c*, *d* and *e*) break down at about 5 V due to the heavy emitter-doping. The collector access resistance associated with structure *a* causes a switching time of about 100 ns which can be reduced to 80 ns or less by shorting emitter to base providing an extra charge removal path. This structure, *b*, is the most suitable if high breakdown voltage is required. Structure *c* gives a switching time of about 80 ns which can be more than halved if the collector is shorted to the base (structure *d*) providing the fastest-switching junction diode. By connecting the two junctions in parallel (structure *e*) and operating at reverse breakdown, a low-voltage reference diode is formed, the current-spreading

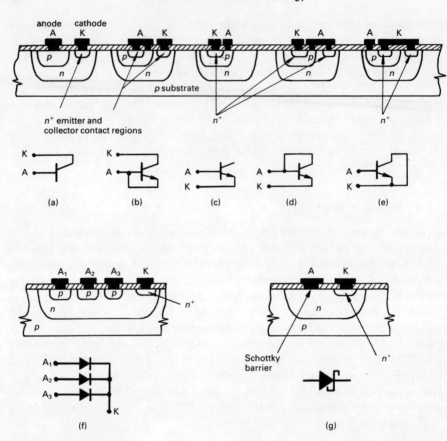

Figure 4.12 Integrated diodes

provided by the parallel connection improving heat dissipation. This arrangement has poor switching performance due to the high capacitance of the paralleled junctions.

The diode structure chosen may depend on the application rather than on electrical parameters. For example, a common-cathode arrangement can be conveniently formed by selecting the n-type epiregion as the cathode (figure 4.12f). Similarly common-anode arrays can be formed in a single 'base' region. In other cases, selection of a particular structure enables a diode to be conveniently included in the same isolation island as another circuit element.

An aluminium/n-type silicon contact forms a low-capacitance Schottky-barrier diode (figure 4.12g) which is used as a fast switch (few picoseconds) and as a voltage clamp to improve the switching speed of bipolar logic circuits (sections 4.7.5 and 4.7.7).

4.2.6 Integrated Passive Circuit Elements

Integrated resistors and capacitors are fabricated at the same time as active devices. In each case there are two basic types: a diffused element which is bipolar-compatible and an MOS element which is compatible with the MOS fabrication process. Integrated inductors are generally impractical; inductors of a few nanohenrys which are useful at very high frequency can be formed by a metallisation pattern on the chip surface [2] but they have a low Q-factor due to the resistance of the thin metallisation. Larger values are not practical using conventional planar techniques due to the flux leakage associated with a two-dimensional surface pattern and also the large chip area occupied by such a pattern. Two approaches have been adopted regarding inductors in ICs:

(1) circuit design techniques are employed to avoid the use of inductors;
(2) where it is imperative that an increase of impedance with frequency is provided, either a discrete component is used or, alternatively, inductive properties can be provided by an operational amplifier with suitable resistive and capacitive input and feedback components.

Diffused Resistor

Diffused resistors use the bulk resistance of a diffused region, the resistance value being controlled by the surface geometry of the diffusion. The p-type BJT 'base' diffusion is commonly used to provide resistances in the range 50 Ω–10 kΩ (figure 4.13a). The 'emitter' diffusion is also used but, due to its higher conductivity, it provides only low-value resistances (1–100 Ω). The resistor diffusion is formed as a narrow track of length L and width W opening out into square contact areas at either end. Assuming uniform doping with depth and ignoring the contribution of the contact areas to the total resistance, the resistance of the diffused track is

$$R = \rho \frac{L}{A} \tag{4.1}$$

where A is the cross-sectional area of track and ρ is the resistivity of the region. The contact areas effectively add about 0.5 W to the track length for resistance calculation purposes.

Figure 4.14 shows part of the resistance track assuming a rectangular cross-section and thickness t. The resistance of a square section of the track R_{sq} in the direction of current flow is

$$R_{sq} = \rho \frac{l}{A} = \rho \frac{W}{Wt} = \frac{\rho}{t} = \rho_S \tag{4.2}$$

since, being a square section, $l = W$. Thus R_{sq} depends on the resistivity of the diffused region and its thickness, not on the size of the square. This is termed

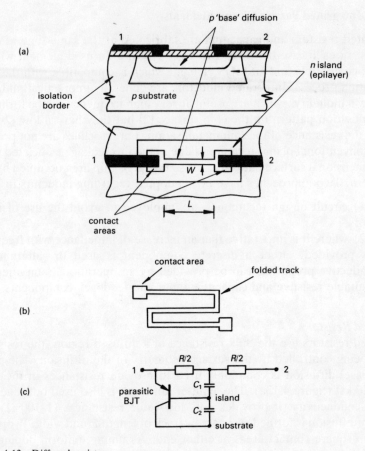

Figure 4.13 Diffused resistor

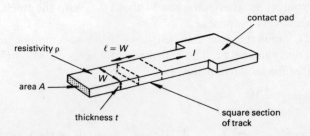

Figure 4.14 Sheet resistance

the *sheet resistance* ρ_S of the diffusion and is quoted in Ω/square. For the total track length L, the resistance R is thus

$$R = \rho \frac{L}{Wt} = \rho_S \frac{L}{W} \tag{4.3}$$

where L/W is the *aspect ratio* of the track. Sheet resistance is a useful parameter from the IC-layout designer's point of view because its value takes account of region resistivity (and hence doping level) and region depth. Thus, knowing the value of ρ_S for the diffusion, the designer need only be concerned with the surface geometry (aspect ratio) required to give the required resistance value. For example, for $\rho_S = 100$ Ω/square, a 1 kΩ resistor is provided by a track ten times as long as its width. For a 10 μm track width, the required track length would therefore be 100 μm. The sheet resistance for the BJT base diffusion is typically 100–200 Ω/square while that for the emitter diffusion is 1–10 Ω/square. The dopant density and hence conductivity of a diffused region varies with depth so that the current density within the resistive track is not uniform, the majority of the current being confined to the surface layer. This non-uniformity is included in the value of sheet resistance which is obtained by practical I–V measurements on a trial diffusion.

The track width W is typically 5–20 μm; a larger value provides improved thermal properties but a longer track length is required for a certain resistance value. The lower limit is imposed by tolerances in the photolithographic and etching processes. Due to these tolerances, the values of individual resistors can only be produced to within about ± 20 per cent although pairs of resistors produced simultaneously can be formed to within ± 1 per cent of each other. This leads to an IC design rule whereby *resistance ratios should be used in preference to relying on specific values*. High-value resistors having a high track length are usually folded (figure 4.13b) to suit chip layout. The sheet resistance of the 'base' diffusion can be increased to the order of 1 kΩ/square by including the emitter diffusion to reduce the effective cross-sectional area.

Figure 4.13c shows a network model representation of the diffused resistor. C_1 and C_2 are the capacitances of the inter-region *pn* junctions which degrade performance at high frequency. The structure includes a parasitic *pnp* BJT which must be maintained at cut-off by connecting the *n*-type island to the most positive potential in the circuit. Note that a number of resistors can be formed in a single island.

MOS Resistor

If a MOST is suitably biased it exhibits a slightly non-linear I_D–V_{DS} characteristic providing a non-linear resistor (section 2.3.2) termed a MOSR, which can be used as a load in MOS circuits. The MOSR provides substantial resistance values using only a small chip area; for example, a 20 kΩ MOSR

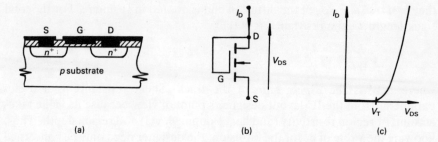

Figure 4.15 MOS resistor

occupies only about five per cent of the chip area of a 20 kΩ diffused resistor. The non-linear nature of the resistor is insignificant in the vast majority of MOS ICs which provide digital functions and do not therefore require linear operation.

If the gate is shorted to the drain (figure 4.15a and b) such that $V_{GS} = V_{DS}$, the device operates on a characteristic similar to the saturation line on the output characteristic (figure 2.76a) but shifted along the V_{DS} axis by the threshold voltage V_T since no conduction can take place until $V_{GS} \geqslant V_T$ (figure 4.15c). From the relationship between I_D and V_{DS} for the ohmic region of the output characteristic of a MOST (equation 2.178) it can be seen that I_D is dependent on the ratio of channel width to channel length (W/l); thus to obtain a substantial resistance the channel length is increased and its width decreased. The MOSR therefore has the same structure as a MOST but its aspect ratio on the chip surface is different.

Diffused Capacitor

A diffused capacitor utilises the depletion layer capacitance of the base–island or base–emitter junctions, where the terms base and emitter refer to p and n^+ regions formed at the same time as the BJT base and emitter regions. Figure 4.16 shows the structure and a network model representation using the base–island junction. To ensure low leakage, the junction must be maintained at reverse bias and so terminal 2 must be more positive than terminal 1. The useful capacitance C (figure 4.16b) depends on doping, junction area and bias, the base–island junction providing typically 100 pF/mm² at a reverse bias of 1 V. Due to the high doping of the emitter diffusion, the base–emitter junction provides a higher capacitance per unit area, typically 1000 pF/mm², although the breakdown voltage of this junction is only about 5 V compared with 40 V for the base–island junction.

The resistance R in the network model is the resistance of the island region between the junction and terminal 2; this can be reduced to about 10 Ω by including an n^+ buried layer beneath the island similar to that used in the BJT structure (figure 4.5). The parasitic capacitance C_S of the reverse-biased

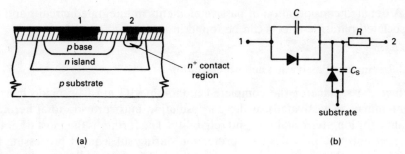

Figure 4.16 Diffused capacitor

island–substrate junction degrades performance at high frequency. Parallel connection of the base–emitter and base–island junctions increases the capacitance obtainable per unit area but the arrangement has higher leakage and a low breakdown voltage.

MOS Capacitor

The MOS capacitor (figure 4.17) is a parallel-plate device with the surface metallisation forming the upper plate, a shallow high-conductivity diffusion forming the lower plate and a thin SiO_2 or SiO_2/Si_3N_4 insulating layer forming the dielectric. The useful capacitance is given by $\epsilon_o \epsilon_r A/t$ where ϵ_r is the relative permittivity of the dielectric layer (4 for SiO_2; 7.5 for Si_3N_4), A is the effective plate area and t is the dielectric thickness. A capacitance of up to 1000 pF/mm^2 with a breakdown voltage of 100 V is obtainable using a dielectric thickness of 0.1 μm which is the minimum thickness allowable to avoid breakdown problems due to pin-holes. For the structure of figure 4.17a using basic diode isolation, terminal 2 must be biased positively relative to the substrate to maintain isolation between the island and the substrate. In the network model of figure 4.17b, resistance R is the bulk resistance of the n^+ diffusion (typically of the order of 10 Ω) and C_S is the parasitic capacitance of the island–substrate junction.

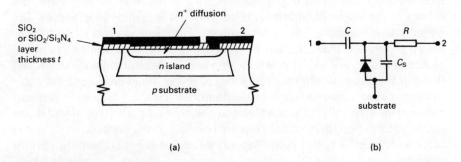

Figure 4.17 MOS capacitor

A detailed consideration of passive elements in integrated circuits and the associated parasitic effects can be found in references 2 and 3.

4.2.7 Surface Passivation and Metallisation

Either the surface of the completed monolithic IC must be made inert to contaminants (passivated) or the encapsulation must provide total hermetic sealing for long-term stability and reliability. The former is the most desirable since it not only provides chip protection during subsequent processing but also enables the encapsulation specification to be relaxed allowing the use of non-hermetic plastic encapsulation or removal of the need for encapsulation altogether as in the case of certain beam-lead ICs (section 4.1.3). Lateral diffusion (section 3.1.5) causes the point where the junction meets the chip surface to move under the oxide layer providing a degree of protection although this introduces other problems in some cases, such as parasitic capacitance, and the use of ion-implantation to avoid such problems also removes the protection advantage. Silicon dioxide provides good protection against the ingress of moisture and the majority of contaminants but certain ions, particularly sodium, can migrate through the oxide layer via oxygen vacancies and contaminate the junction. The use of an additional silicon nitride layer on top of the dioxide produces an almost totally impervious layer, termed the *sealed junction* method, providing very effective surface passivation.

As described in section 3.1.6, interconnections between the circuit elements in a monolithic IC are formed during the metallisation stage. The majority of silicon ICs use aluminium metallisation which, as well as being a good conductor, is relatively inexpensive compared with the alternative — gold. In addition, aluminium adheres well to both silicon and silicon dioxide if the temperature is raised to the melting point of the silicon–aluminium alloy after deposition; a process termed *micro-alloying*. Although structural diagrams included in this section generally show the metallisation filling the connection window and an even layer being formed on the silicon dioxide, it must be appreciated that in practice, after a number of processing steps, the surface of the silicon dioxide layer is far from flat on a microscopic scale. The metallisation therefore has to fill 'valleys' and cover corners or 'peaks' as shown in figure 4.18a. Care has to be taken to ensure that the angles of the etched areas are not so sharp as to form voids between the metallisation and the chip since they would result in poor reliability. Final etching of the 1 μm thick vacuum-deposited layer of aluminium leaves the inter-element connections (typically 10 μm wide) and the contact pads (typically 100 μm square) at the periphery of the chip for bonding to the package.

As with printed circuit-board layout, design of the metallisation pattern often involves topological problems requiring cross-overs. In simple cases, cross-overs can be introduced at circuit elements where the metallisation

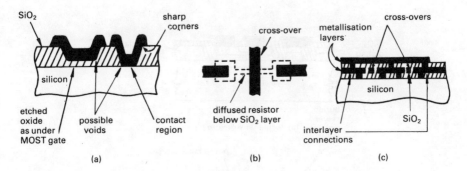

Figure 4.18 Metallisation

track terminates at a contact. For example, a metallisation track can cross over a diffused resistor (figure 4.18b) perpendicular to the resistance track between the contacts at either end of the resistor or alternatively a low-value resistor can be introduced solely to provide a cross-over point. In complex cases such as LSI memory or microprocessor chips, multi-layer metallisation is usually necessary, each layer of metallisation being separated from others by a silicon dioxide layer and interconnection between the layers being provided via windows (figure 4.18c). Multi-layering is a disadvantage of complex circuits since it not only introduces additional processing steps and hence extra cost but also reduces reliability and yield.

4.3 MONOLITHIC IC FABRICATION TECHNIQUES

The various stages of the planar fabrication process are described in chapter 3 and the basic structures of the individual circuit elements using diode isolation are considered in section 4.2. Although the diode-isolation method (section 4.1.1) offers simplicity and hence economy, it has a number of disadvantages, namely, long diffusion time for the deep isolation regions, considerable chip area devoted to isolation and relatively large parasitic capacitance. Manufacturers have developed structures that retain the basic simplicity but avoid these major disadvantages. A widely used technique for bipolar ICs is known as the *Isoplanar process*; an example of an Isoplanar BJT is shown in figure 4.19a. The process uses a thin epilayer (≈ 5 μm thick) and employs thermally grown silicon dioxide in etched channels to isolate the device in the epilayer but retains conventional diode isolation between the device and the substrate. A silicon nitride masking layer is used during etching of the isolation channels and subsequent oxide growth takes place only in the channels since the nitride layer prevents growth elsewhere. An additional feature is the oxide region separating the base from the collector contact. The

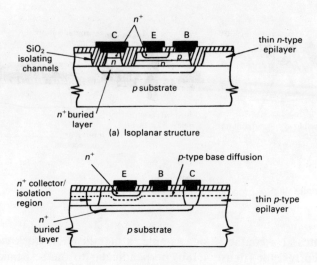

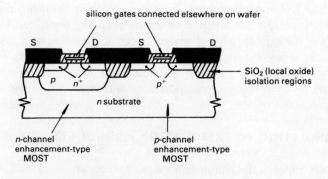

Figure 4.19 Improved device structures

base is thus surrounded by oxide which allows larger mask-alignment tolerances for the emitter diffusion and the collector and base contacts. This tolerance relaxation allows the chip area occupied by the device to be reduced to about one-third of that required for basic diode isolation with the same photolithographic constraints. Thus a low-power Isoplanar BJT can be fabricated in a chip area typically 40 μm × 25 μm compared with 60 μm × 50 μm for basic diode isolation (figure 4.5). The reduced size allows increased packing density which is of particular interest for large-scale bipolar ICs such as memories.

Another development of the basic planar structure is termed *collector-diffused isolation* (CDI); the structure of a CDI BJT is shown in figure 4.19b. In this technology a thin p-type epilayer (≈ 5 μm thick), in contrast to the normal n-type layer, is grown on a p-type substrate after formation of an n^+ buried layer. An n^+ diffusion through the epiregion down to the buried layer provides device isolation and also forms the BJT collector. The p-type epilayer forms the base region although a p-type diffusion is necessary to modify its dopant profile to that required for a BJT base. As the isolation/collector region is heavily doped, the p-type diffusion can be non-selective (unmasked) simplifying the process. A shallow n^+ emitter-diffusion and surface metallisation completes the structure. The CDI process offers a number of advantages: fewer processing steps providing reduced cost and improved yield; small chip area compatible with the Isoplanar process and improved performance resulting from lower capacitance.

Similar structural improvements have taken place in MOS technology, an example being the reduction of size using silicon dioxide isolation regions. Figure 4.19c shows a complementary MOS (CMOS) structure comprising p and n-channel enhancement-type MOSTs on the same substrate. The particular importance of this combination is explained in section 4.7.3, but in terms of structure the use of oxide isolation regions in place of more conventional guard-ring diffusions (figure 4.27) enables the chip area occupied by the combination to be reduced and hence provides an improvement in packing density. The manufacturers describe the oxide regions as 'local' oxide and hence the structure is termed *local oxide isolated CMOS* (LOCMOS).

4.4 MONOLITHIC IC DESIGN

The stages in the design of a monolithic IC are

(1) selection of the system to be integrated;
(2) choice of technology to be used;
(3) circuit design;
(4) layout design.

System Selection
The manufacturer is concerned primarily that the IC to be produced will have a substantial market and so provide an adequate return on the development outlay. Selection of the system to be produced as a single integrated 'block' is important in this respect since, if the selected system is too highly specialised in terms of the function it performs, it may well be that its usefulness to electronic equipment manufacturers is so limited that sales are poor. In such a

situation it could be advantageous to integrate part of the system which, having a more general-purpose nature, could have a variety of applications and thus be a better proposition economically. The term 'specialised' does not necessarily imply complexity, for example, a circuit for a certain specialised application could have very limited usage but it may not be complex as far as the number of circuit elements is concerned. Alternatively, a microprocessor is one of the most complex ICs currently produced although it is general-purpose in nature as it has a wide variety of applications from microcomputers and industrial process control equipment to domestic appliances and toys. The choice of electronic function to be provided in integrated form is thus important for the economic success of the product.

Technology Choice

Having selected the system to be integrated, a detailed performance specification must be produced. From this specification, the type of system (digital or linear) and the complexity (scale of integration) involved, a choice is made as to which of the available technologies is most suitable for implementation.

Some fundamental properties of the basic MOS and bipolar technologies are compared in table 4.1. The extremely small chip area occupied by a MOST enables MOS technology to offer the highest packing density making it the most suitable choice for the majority of large-scale digital ICs. Linear ICs and small-scale digital circuits do not require high element packing densities since the circuits use relatively few circuit elements and it is not usually practicable to use a chip size of less than about 0.5 mm square due to the poor active/bonding pad area ratio and difficulties in packaging. In such cases, therefore, the packing-density capability of MOS is not an advantage as it cannot be exploited and some other criterion must be used.

Table 4.1 Relative properties of basic MOS and bipolar technologies

	MOS	Bipolar
Packing density	higher	lower
Switching speed/frequency response	lower	higher
Input impedance	higher	lower
Output impedance	higher	lower
Power dissipation	lower	higher

Operating speed, whether in terms of switching speed in digital cases or frequency response in linear systems, is a fundamentally important property. Due to the inherent capacitance in the MOS structure it is generally slower than bipolar. There is thus a trade-off between system speed and complexity; a faster operating IC can be produced using bipolar technology but the integrated system must be less complex due to the lower packing density and the maximum chip size of about 7 mm square that can be conveniently encapsulated in the popular package styles.

Other factors that are significant are the current loading (input impedance), current drive (output impedance) and power dissipation properties of the two technologies. MOS circuits have a very high input impedance and thus draw very little current from a preceding (driving) circuit; the MOS circuit thus does not 'load' the driving circuit. Bipolar circuits generally have a much lower input impedance and loading can be a problem that requires the use of buffer stages. The lower output impedance of bipolar circuits provides good output current drive capability, however, so that substantial current can be drawn by a following stage without impairing the performance of the driving stage significantly. Alternatively, a number of separate circuits can be driven by a single bipolar circuit. This requirement occurs frequently in digital systems where the maximum number of circuits it is possible to drive is termed the *fan-out* of the driving circuit. The higher output impedance of MOS circuits provides very limited current drive capability although this is not a significant problem provided that only high input impedance circuits (for example, other MOS circuits) are to be driven. If a substantial output current capability is required from an MOS circuit it is necessary to include a bipolar output stage resulting in a hybrid MOS–bipolar circuit. Power dissipation is important in battery-powered systems particularly miniature portable equipment relying on low-capacity batteries; high current drain means frequent recharging or replacement of batteries. Additionally in complex systems, substantial power dissipation may create heat problems within the chip. The basic MOS switching circuits (PMOS and NMOS) consume rather less power than the established bipolar circuits and CMOS circuits offer a substantial power saving making this technology popular for low-power consumption systems. There is a trade-off, however, between power consumption and speed; circuits switch faster if they are allowed to consume more power.

The choice of technology to be used in a particular application thus depends on the necessary packing density to enable the system to be implemented on a reasonable-size chip together with speed requirements and power consumption limitations. At any time there are limitations as to what is feasible in terms of complexity, speed and power; complexity can be traded for speed and speed for power, but some combinations of high complexity, high speed and low-power consumption are not feasible. The choice of technology is further complicated by continuing developments. For example, the introduction of the VMOS structure improved the speed capability of MOS tending to rival bipolar. Correspondingly, the packing-density advantage of MOS for complex ICs has been partially eroded by the development of superintegrated bipolar structures such as I^2L which can also provide low-power dissipation similar to that of CMOS.

Table 4.2 Relative chip area occupied by integrated circuit elements

Element	Relative area
MOST (lateral)	1 (400 μm^2)
BJT (Isoplanar)	2
Diode	2
Resistor (10 kΩ diffused)	10*
Resistor (10 kΩ MOS)	1
Capacitor (10 pF diffused)	25†, 250‡
Capacitor (10 pF MOS)	25§

* using the base diffusion of 200 Ω/square with a minimum track width of 5 μm
† using the E–B junction, providing a low breakdown voltage (≈5 V)
‡ using the C–B junction, providing a breakdown voltage of about 40 V
§ using a minimum dielectric thickness of 0.1 μm

Circuit Design

Early integrated circuits (1960–5) were mainly integrated versions of discrete component circuits and the same circuit design techniques could be used. In the mid-1960s consideration of the advantages, limitations and economics of circuit elements in integrated form led to a set of general circuit-design rules and to a new approach to circuit design. In discrete component form, active devices are in general more expensive than passive components. In integrated form the cost of a circuit element is related directly to the chip area it occupies and with reference to table 4.2 it can be seen that capacitors and diffused resistors are typically an order of magnitude more expensive than transistors. This led integrated-circuit designers to maximise the proportion of active elements and to avoid the use of capacitors and high-value resistors where possible. Furthermore, since integrated transistors are relatively cheap, designers were able to improve circuit performance using additional transistors with very little increase in cost. The close matching of element characteristics provided by simultaneous fabrication and good temperature tracking due to the proximity of elements on the chip are additional factors that the designer can use to advantage. Where it is necessary to use diffused resistors, their values should be kept small to reduce the chip area required and due to the tolerance limitations explained in section 4.2.6, circuit designs should depend on resistance ratios where possible rather than on absolute values.

The fundamental circuit-design rules may therefore be summarised as

(1) maximise the proportion of active devices in the circuit, avoiding the use of capacitors and large-value diffused resistors where possible;

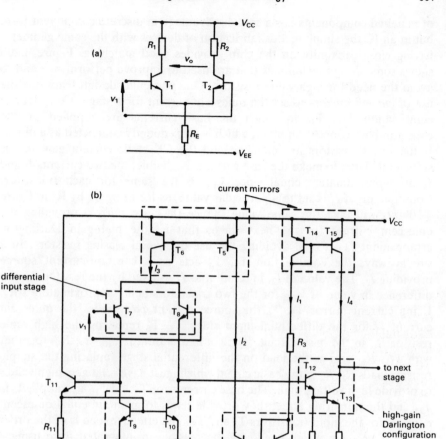

Figure 4.20 Influence of integrated circuit-design rules

(2) take advantage of component matching and temperature tracking to improve performance;
(3) use resistance ratios rather than absolute values.

The influence of these rules on IC design is shown in figure 4.20. The basic differential amplifier arrangement of figure 4.20a gives improved performance compared with a single transistor stage particularly in terms of stability with temperature variation although ideally transistors T_1 and T_2 should have identical properties and R_E should have a high value. [4] The use of matched components has a cost disadvantage in discrete component form but in an IC the simultaneous fabrication of devices with the same geometry having close proximity on the chip provides good matching. Figure 4.20b shows some design techniques that are used to improve performance and to avoid the need for high-value resistors. One particular design feature is the use of *current mirrors* to set the quiescent current for a stage. Consider the combination T_{14}, T_{15} in which the base currents are supplied by the collector–base connection of T_{14} which is consequently connected as a diode. If the two transistors are matched and the CE static current gain h_{FE} is sufficiently large to make the base currents negligible, the two currents I_1 and I_4 are approximately equal since V_{BE} is the same for each transistor. Consequently if I_1 is set to a particular value, as for example by R_3 in figure 4.20b, I_4 will take the same value (hence the term *mirror*) providing the quiescent current for the next stage, that is, the high-gain Darlington arrangement (T_{12}, T_{13}), avoiding the use of several biasing resistors. In a similar way, the combination T_3, T_4 acts as a constant-current source providing I_2. The value of I_2, in relation to I_1, is fixed by the resistor R_4, the difference in value of V_{BE} for the two transistors being approximately I_2R_4. Using current mirror T_5, T_6 the source current I_2 provides the quiescent current I_3 for the differential input stage T_7, T_8 replacing the high-value resistor R_E in the basic circuit. The configuration T_9, T_{10}, R_9, R_{10} together with T_{11}, R_{11} acts as the load on the differential stage replacing the simple resistor loads R_1, R_2 in the basic circuit which must have relatively high values to provide high voltage gain. The base currents for T_9 and T_{10} are supplied via T_{11} and the mirroring effect they provide tends to maintain equal quiescent currents for the input transistors T_7 and T_8. The effective load is of the order of 1 MΩ, a value which would not be feasible in integrated form using a diffused resistor. The use of a Darlington stage (T_{12}, T_{13}) not only provides high gain, for which T_{15} acts as a high resistance load, but due to its high input resistance it does not load the driving stage enabling the high gain of the latter to be maintained. Interstage coupling is direct, avoiding the use of coupling capacitors and in addition provides appreciable gain at low frequency. The stability problems encountered in direct-coupled circuits are avoided using the differential arrangement. In other circuits additional transistors/diodes are included to compensate temperature variations in the main circuit elements.

The objective of the circuit design is to meet the performance specification taking account of the design rules. For both digital and linear circuits this inevitably means attempting to optimise speed, power consumption, impedance levels and noise performance.

A first-generation IC design of the early 1960s could be tested to a reasonable degree of accuracy by implementation in discrete form. As the technology developed, discrete component evaluation became less accurate as parasitic effects within the integrated structure were difficult to represent.

Eventually the technique became impractical due to the size of the integrated system and also due to the fact that integrated devices such as multi-emitter BJTs and superintegrated elements could not be represented in discrete form. At this stage computer simulation was introduced to provide design evaluation before continuing to the layout stage (figure 4.21). Considerable work has taken place in device-modelling to enable simulation routines to provide sufficiently accurate predictions of the response of a circuit design when implemented in integrated form and this procedure is widely used.

Layout Design

When the computer simulation shows that the circuit designer can be reasonably confident that the design will meet the required performance specification when fabricated in integrated form, the design sequence passes to the layout stage which involves the geometrical layout of the circuit elements on the chip.

Each fabrication process (basic diode-isolated bipolar, Isoplanar bipolar, lateral MOS, VMOS, etc.) has a set of layout constraints which specify the minimum widths, separations and overlaps that may be used with safety for that particular process taking account of processing tolerances in mask alignment, etching and doping. The objective is to devise a layout of the circuit elements which minimises chip area, produces a suitable aspect ratio for the chip and minimises the number of cross-overs required. The layout constraints specify minimum doped region separation at the chip surface, minimum isolation region width, minimum element contact region size, minimum resistor track width, minimum metallisation line width and spacing and chip contact pad size. The absolute minimum active chip area (that is, not including bonding pads) required for a particular circuit may be determined initially by evaluating the total area occupied by transistors, which usually have a standard geometry for a particular power range and fabrication process, together with the area occupied by the total resistance and capacitance in the circuit taking account of any isolation required. In practice, the final active chip area will be larger than this calculated minimum value since it is not usually possible to avoid wasted area in the form of additional low-value resistors providing cross-unders (figure 4.18b) or increased spacing to accommodate metallisation tracks. However, such a calculation does give an indication of the order of size to be used as an objective.

Stages of the layout process are as follows:

(1) The electrical circuit is redrawn to suit the required pin configuration for the package. In some cases part of the pin configuration, such as power supply connections, may be predetermined particularly where the new circuit is an addition to a range of ICs already in production. During this redrawing stage, elements required to be matched are

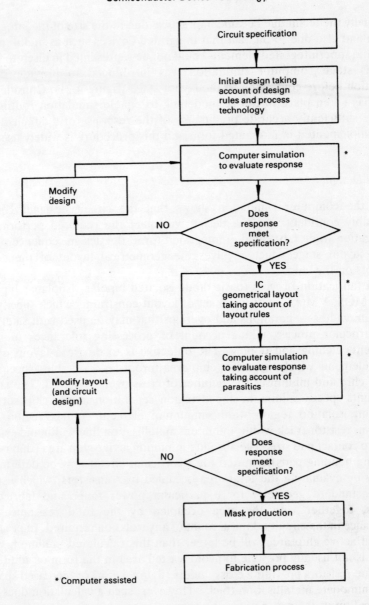

Figure 4.21 Integrated circuit-design sequence

positioned adjacent to each other and the number of cross-overs is reduced as much as possible.

(2) The minimum number of isolation islands required is then determined. Islands for individual active devices generally have a specific geometry for a particular process. For resistors and capacitors

the geometry is flexible; only the area will be known from the parameter values required.

(3) The islands are positioned relative to one another according to the redrawn circuit diagram, modification being made to obtain an approximately square chip (1:1 aspect ratio). During rearrangement of islands, care must be taken not to destroy the pin configuration nor to greatly increase the number of cross-overs required. In some cases a suitable aspect ratio can be obtained by using irregular-shaped resistor or capacitor islands to complete the shape. Ratios of up to 3:2 are usually acceptable, higher ratios can introduce mechanical and packaging problems. Most manufacturers use computer-aided graphics to assist in layout, particularly for large-scale ICs.

(4) The geometry of doped regions and contact windows within the islands and the surface metallisation pattern, including the bonding pads at the periphery of the chip, can then be included.

(5) Once the layout has been formalised, the values of parasitic elements can be calculated and computer simulation techniques used to assess if the circuit/layout combination will meet the required performance specification. If the results are unfavourable, the layout and, possibly, the circuit must be modified and the simulation repeated (figure 4.21).

(6) When a satisfactory arrangement has been devised, the layout is translated into master masks for each of the etching stages in the fabrication process.

Reference 5 gives a detailed consideration of the layout constraints and includes an illustrative layout example using the above procedure.

4.5 SCALE OF INTEGRATION

Scale of integration refers to chip complexity, that is, the number of circuit elements per chip. The terms 'small', 'medium', 'large' and 'very large-scale' integration (SSI, MSI, LSI, VLSI) are used to describe the order of complexity. The advantages of integration compared with discrete component implementation are discussed in the introduction to this chapter. Similar advantages apply to an increase in the scale of integration. Considering size reduction as an example, an SSIC such as a 7400 series logic IC in a 14 pin DIP package (dual-in-line style with plastic encapsulation) occupies an area of approximately 200 mm^2 on a printed circuit-board although the chip area is only about 4 mm^2 so that 98 per cent of the area is wasted as regards electronic processing. The lower limit to package size is set by minimum pin spacing and handling requirements. Clearly an increase in chip complexity using a similar package size makes more efficient use of the

Table 4.3 Scale of integration

Chip complexity (introduced)	Circuit elements/chip	Gates/chip (equivalent)
SSI (1960)	<50	1–10
MSI (1965)	50–1000	10–300
LSI (1971)	10^3–3×10^4	300–10^4
VLSI (1976)	>3×10^4	>10^4

space occupied by the package. Complex ICs such as logic arrays, processors and memories usually require a larger package to accommodate the number of pins required, 24 and 40 pin DIP packages being popular (figure 3.4e). In such cases the increase in space occupied by the package is offset by the vast increase in system complexity provided on the chip; if such a system were to be implemented in ICs using a smaller scale of integration the space occupied would be orders of magnitude greater.

Table 4.3 gives an indication of the chip complexities corresponding to the various scales of integration together with the year of introduction. Although the scale of integration advances, the less complex ICs continue in volume production since applications require all levels of complexity, for example, SS logic–gate and operational amplifier ICs, MS digital adders and decoders, LS low-capacity memories and calculator ICs and VLS high-capacity memories, logic arrays and microprocessors. Only digital systems have been able to benefit from the larger scales of integration as linear (analog) systems, comprising, for example, amplifiers, comparators, converters, regulators, timers, drivers and analog switches, do not require a large number of circuit elements. Due to the predominance of digital ICs there is a tendency to refer to complexities in terms of gates/chip or equivalent gates/chip (table 4.3)

A problem related to an increased scale of integration is that of *yield*, that is, the percentage of functional chips per wafer. Although extreme precautions are taken during fabrication regarding crystal growth, surface preparation and cleanliness of the environment, flaws do occur in the crystal structure and surface dirt during processing causes faults in the pattern. As the scale of integration is increased, element size is reduced and to increase the processing ability of each IC, the chip size has been increased. Both these factors tend to reduce the yield as illustrated in figure 4.22. If the size of the chip is increased, a certain-size flaw or density of flaws is *likely* to cause a smaller proportion of functional chips (figure 4.22a). Similarly if the feature size of individual elements is reduced as in the case of the BJT in figure 4.22b, a certain-size fault is more likely to cause an element failure.

There are two basic approaches to the problem of yield in large-scale digital ICs:

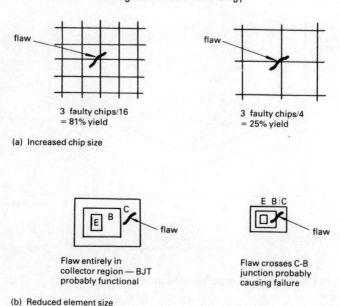

Figure 4.22 Reduction of yield with increase in scale of integration

(1) *(Assumed) 100 per cent yield approach.* The array of gates is fabricated on the chip, individual gates are not tested and a fixed metallisation pattern is used to interconnect gates as required. Completed ICs are tested and faulty circuits discarded.

(2) *Discretionary wiring.* The array of gates is fabricated on the chip as in (1) but each has test pads allowing individual gates to be tested using step-and-repeat equipment. Computing facilities can be used to store the grid positions of faulty gates and individual metallisation patterns created for each chip avoiding the faulty gates, hence producing a high yield.

The disadvantages of the discretionary wiring technique are that the incorporation of test pads for individual gates reduces the packing density and the use of computing facilities and individual metallisation patterns greatly increases costs. This technique was first used in the late 1960s when complex LSICs were produced using whole wafers. More recently the vast reduction of computing costs and the development of direct-write-on-wafer exposure systems using electron-beam methods avoiding the use of masks make this technique more attractive.

Figure 4.23 shows the relationship between function cost and IC complexity. At a particular stage in technology development, for example, at the present day, the cost/electronic function in the IC first reduces as

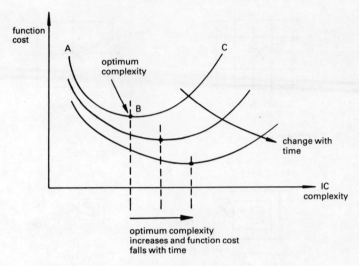

Figure 4.23 Function cost—IC complexity relationship

complexity increases due to cost savings such as in packaging (curve AB, figure 4.23). If circuit complexity is progressively increased, however, function cost eventually rises due to the fall in yield (curve BC), resulting in an optimum complexity at that stage of development providing a minimum cost/function. At some future time, developments in the technology enable higher yields to be obtained and the optimum complexity increases. At the same time, the increased IC complexity causes the individual function cost to fall so the curve moves downward and shifts to the right with time.

4.6 SUPERINTEGRATION

The term *superintegration* refers to the merging of circuit elements on the chip as a means of increasing packing densities and improving performance. Superintegrated circuit elements overlap on the chip surface with doped regions performing a dual role; that is, a single region forms part of one circuit element and part of another at the same time.

Figure 4.24a shows a *pnp-npn* BJT combination that forms the basis of an integrated-injection logic (I^2L) gate (section 4.7.7). The chip structure of this arrangement in superintegrated form (figure 4.24b) shows that the *n*-type epilayer forms both the base region of the lateral *pnp* BJT and the emitter of the conventional *npn* device. Also the *p*-type diffusion doubles as the collector of the *pnp* device and the base of the *npn* BJT. The two devices are thus *merged* and as no isolation is required between them, a substantial saving

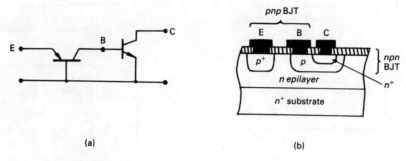

Figure 4.24 Superintegration

in chip area is obtained. The two-transistor combination can be accommodated in the area occupied by a conventional *npn* BJT together with its surrounding isolation area. Another example of superintegration is the hybrid MOS/bipolar device of figure 4.11.

4.7 MONOLITHIC IC TECHNOLOGIES IN PRACTICE

A variety of IC technologies have developed based on MOSTs and BJTs with particular emphasis on digital applications. The fundamental performance parameters in these applications are as follows:

(1) *Gate propagation delay*. The time delay between the change in logic level at the gate input and the corresponding change in logic level at the gate output.

(2) *Gate power dissipation*. The power dissipation per gate when operating on a specified ON–OFF duty cycle, usually unity, at a specific frequency.

(3) *Power–delay product*. The product of power dissipation and propagation delay for a gate. A logic gate should have a low propagation delay to permit high-speed operation and low-power dissipation so that power requirements and chip temperature can be kept low. The power–delay product, often termed the *speed–power product*, is used as a figure-of-merit to compare the performance of the various technologies. The product of power and time is energy and the power–delay product is quoted typically in *picojoules*.

(4) *Maximum operating frequency*. This is limited by the propagation delay. It is often quoted as the maximum clock frequency for a flip-flop circuit.

(5) *Gate-packing density*. The maximum number of gates that can be fabricated per unit surface area of the chip.

370 Semiconductor Device Technology

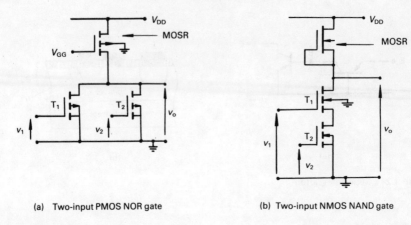

(a) Two-input PMOS NOR gate   (b) Two-input NMOS NAND gate

Figure 4.25 PMOS and NMOS technologies

4.7.1 PMOS

PMOS digital ICs consist entirely of p-channel enhancement-type MOSTs; no other type of circuit element is involved. PMOS was the first MOS technology to be adopted for ICs and provided a moderate element-packing density and cheap processing with no major difficulties. The process involves the formation of an array of p-channel MOSTs by diffusing p-type source and drain regions into an n-type wafer. Individual devices are self-isolating (section 4.2.2) so that isolation methods are not required and the process thus requires only four masking stages for the p-type diffusion, gate oxide growth, contact windows and the surface metallisation pattern. The simplicity of the basic process produced a moderate packing density of about 50 gates/mm^2, a good yield and low cost, making the process popular for first generation LSICs. PMOS is still used in applications where low cost is the overriding consideration and where high performance, in terms of speed, and low-power consumption are secondary. Figure 4.25a gives an example of a two-input PMOS NOR gate using negative logic with V_{DD} typically -15 V, the logic levels being $0 \equiv -1$ V, $1 \equiv -15$ V. If T_1 and/or T_2 are turned ON by the application of v_1 or v_2 of -15 V (logic 1), v_o is reduced to about -1 V (logic 0). The gate uses an MOSR as the load with its gate voltage V_{GG} fixed. The simplest arrangement uses $V_{GG} = V_{DD}$ so that the MOSR gate can be shorted to V_{DD} (figure 4.15b) although switching speed can be improved by making V_{GG} more negative, reducing the effective load. The threshold voltage of a MOST in basic PMOS technology is of the order of -4 V requiring a relatively high value of V_{DD}. A lower threshold (≈ -1 V) is advantageous as performance is improved and V_{DD} can be reduced although the additional processing necessary to achieve this for metal-gate MOSTs detracts from the basic simplicity and low cost of the process. The replacement of the metal gate by a doped silicon gate is a simple method of controlling the threshold voltage. Compared with the basic technology, silicon-gate PMOS provides a reduction of gate-propagation delay from about 80 to 40 ns, a reduction of power–delay product from about 480 to 200 pJ and an increase of packing density from 50 to 100 gates/mm^2. A detailed appraisal of the performance of PMOS can be found in reference 6.

4.7.2 NMOS

In NMOS technology, digital ICs are formed entirely of *n*-channel MOSTs. The higher mobility of electrons compared with holes ($\mu_n \approx 2.4\,\mu_p$ for silicon at typical doping levels) gives an immediate speed improvement compared with PMOS and also power dissipation is reduced. NMOS developed later than PMOS due mainly to the difficulties in fabricating *n*-channel enhancement-type MOSTs (section 4.2.2). Although the improvements in MOS technology can be applied equally to PMOS as well as to NMOS, the trend has been that PMOS has remained virtually in its basic form to maintain its simplicity and cheapness while NMOS has become regarded as the high-performance MOS technology. Various innovations have been made to improve the performance of NMOS as much as possible, such as the introduction of silicon gates and the use of ion-implantation to produce a self-aligned structure, as described in section 4.2.2 The additional processing makes NMOS dearer than PMOS although the improvement in performance is considerable; gate-propagation delay is reduced to about 15 ns with a power–delay product of only 45 pJ and a packing density of the order of 150 gates/mm^2. The threshold voltage of *n*-channel enhancement-type MOSTs is typically about 1 V allowing a power supply voltage (V_{DD}) of 5 V to be adequate. The improvement in packing density is a consequence of the lower threshold voltage since lower voltage means a narrower depletion region around the device. Usually positive logic is used with NMOS, the logic levels being $0 \equiv +0.5$ V, $1 \equiv +5$ V. Figure 4.25b shows a two-input NMOS NAND gate. The output voltage v_o is low (+0.5 V, logic 0) only when both T_1 AND T_2 are turned ON by application of v_1 and v_2 of +5 V. This arrangement uses a depletion-type MOSR as the load, in contrast to the enhancement-type load of figure 4.25a, which avoids the need for a second voltage supply. However, the use of a depletion-type structure introduces an additional processing step to form the conducting channel region. The term *active pull-up* is often used to describe the use of a transistor structure as a load, 'pull-up' referring to the increase of the output voltage to V_{DD} when the gate is OFF.

By forming series/parallel combinations of enhancement-type MOSTs with a common load, combined NAND/NOR functions which use less chip area than separate interconnected gates can be obtained.[6]

The VMOS derivative of NMOS (section 4.2.2) showed particular promise in the late 1970s as regards improvements in performance and packing density

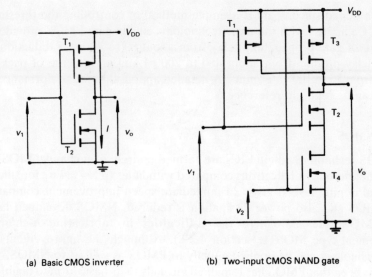

(a) Basic CMOS inverter   (b) Two-input CMOS NAND gate

Figure 4.26 CMOS technology

but, in IC form, cost factors appear to have made it non-viable in competition with the lateral NMOS structure.

4.7.3 CMOS

*C*omplementary *S*ymmetry MOS (COSMOS or CMOS) technology combines p and n-channel enhancement-type MOSTs in such a way as to provide very low static power dissipation. The basic CMOS inverter is shown in figure 4.26a. V_{DD} is positive, typically in the range 5–15 V, and the logic levels are $0 \equiv 0$ V, $1 \equiv V_{DD}$. The gate–source voltage for T_2 is the input voltage v_1 while that for T_1 is $v_1 - V_{DD}$. As the two MOSTs are of different channel type, one of the transistors will be OFF and the other ON under static conditions. When $v_1 = 0$ V (logic 0) T_1 is ON and T_2 OFF so that T_1 provides a low resistance path from V_{DD} to the output, v_o thus being approximately V_{DD} (logic 1).

The converse is true for $v_1 = V_{DD}$, T_1 is OFF and T_2 ON, so that T_2 provides a low resistance path between output and ground, v_o being approximately 0 V. The arrangement thus acts as a digital inverter which consumes very little power (a few nanowatts) in the static state since T_1 or T_2 is OFF, the static current I (figure 4.26a) being a leakage. The circuit dissipates more power during the switching operation when both transistors are partially ON, the average power dissipation increasing with switching frequency. For $V_{DD} = 5$ V, the basic CMOS gate dissipates typically 2 μW at 1 kHz, 20 μW at 10 kHz increasing to about 2 mW at 1 MHz. Increasing V_{DD}

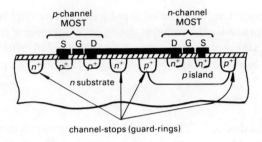

Figure 4.27 CMOS structure

improves the switching speed and noise immunity but also increases the power dissipation. For example, at $V_{DD} = 15$ V the above dissipations are increased by an order of magnitude. The maximum allowable power dissipation in a particular application using CMOS can therefore determine the maximum operating frequency. The low-power dissipation at low operating frequency, compared with other technologies, makes CMOS attractive for battery-powered equipment where high-frequency operation is not required, particularly where only low-capacity batteries can be used, as for example in digital watches. An example of a CMOS gate is the two-input NAND gate of figure 4.26b. The output is low (0 V, logic 0) only when both inputs v_1 AND v_2 are high (V_{DD}, logic 1) causing T_1 and T_3 to be OFF.

The formation of both p and n-channel MOSTs in the same substrate requires not only the creation of an island for one of the devices, commonly a p-type island in an n-type substrate to accommodate the n-channel MOST (figure 4.27), but also requires some means of preventing channels forming between adjacent devices. This was originally done by surrounding each device by a heavily doped *channel-stop* or *guard-ring* difffusion as shown in figure 4.27. The packing density has been improved however by using recessed oxide in place of the guard-ring diffusions (figure 4.19c) termed *local oxide-isolated CMOS* (LOCMOS). A reduction in chip area of about 30 per cent is claimed and in addition capacitance is reduced improving the switching speed. Silicon-gate CMOS has a propagation delay similar to that of NMOS (≈ 20 ns) but the power-delay product is typically two orders of magnitude lower than that for NMOS at typically 0.4 pJ, due to the low power dissipation. The necessity for separate p-type islands and guard-ring diffusions or oxide strips are disadvantages in CMOS technology causing a CMOS gate to occupy a greater chip area than the corresponding gate in PMOS or NMOS. Typically the packing density for CMOS is about 60 gates/mm^2 compared with 100 for PMOS and 150 for NMOS.

Implementation of CMOS using silicon-on-sapphire technology (SOS-CMOS) in which the p and n-channel MOSTs are formed in separate silicon islands formed on an artificial sapphire substrate (figure 4.4) provides considerably improved performance although at increased cost. A packing density of over 200 gates/mm^2 is possible in SOS-CMOS, with a propagation delay of 5 ns or less and a power–delay product of only 0.1 pJ.

4.7.4 CCDs

A charge-coupled device (CCD) is an MOS structure comprising an array of very closely spaced gate electrodes (figure 4.28a). If a gate (G_1) is biased positively with respect to the substrate, a region under the gate is depleted of majority carriers within a few microseconds. Provided that the gate voltage is greater than the threshold voltage of the structure, minority electrons are attracted to the surface region and an inversion layer eventually forms. However, this takes several hundred milliseconds and in the period between formation of the depletion region and inversion taking place, a non-equilibrium state exists during which minority charge beneath the gate is free. If, during this 'free' period, a greater bias voltage is applied to the adjacent electrode (G_2, figure 4.28b) a deeper depletion region is formed under G_2 and provided that the gate electrodes are sufficiently close (≤ 2 μm) for the depletion regions to merge, the minority charge transfers to the region under G_2 due to the lateral field created by the larger voltage V_{G_2}. It is evident, therefore, that minority charge can be transferred along an array of closely spaced electrodes if the applied voltages progressively increase provided that transfer takes place before the inversion layer forms. Thus there is a minimum transfer rate below which minority charge becomes fixed

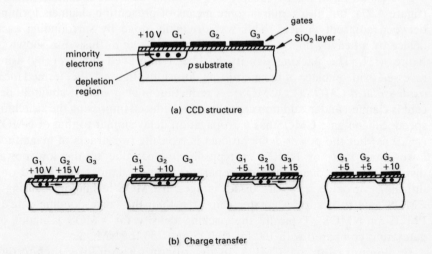

Figure 4.28 Basic CCD structure and operation

forming an inversion layer and it cannot then be transferred. As the minority charge must be kept on the move, a CCD is described as being *dynamic*. The requirement that the gate voltages must progressively increase is inconvenient not only due to the number of voltage levels required but also due to the limitation imposed by the maximum voltage that the MOS structure can withstand before breakdown. Once the charge has transferred to G_2, however, V_{G_1} and V_{G_2} can be reduced from 10 V and 15 V to 5 V and 10 V respectively without substantially altering the situation, the voltage differential being maintained. V_{G_3} can then be increased to 15 V and the charge will be transferred to the region under G_3. V_{G_2} and V_{G_3} can then be reduced to 5 V and 10 V respectively and 15 V applied to G_4 to continue the transfer. A gate voltage sequence of 15, 10, 5, 15, 10, ..., is therefore required in this system; the sequences applied to the three gates being the same but shifted in phase relative to one another. V_{G_1}, V_{G_2} and V_{G_3} are termed *clock signals* and the above system is therefore said to use a *three-phase clock*.

The minority charge under G_1 at the start of the sequence can be provided by a forward-biased diode fabricated at the input of the gate array and linked to the region under G_1 by an enhancement-type MOST. Using presence of minority charge to represent a binary 1 and absence a binary 0, a string of binary digits can be introduced successively at the input of the array by switching the MOST ON and OFF as required. After transfer along the array the 'packets' of minority charge representing binary 1s flow across a reverse-biased diode causing a voltage pulse to be generated across a series load resistor. The array of electrodes thus operates as a dynamic shift register, each group of three electrodes acting as a one bit store. Three electrodes are necessary to keep successive bits apart, once one charge packet has been transferred from G_3, the next packet can be introduced to G_1. It should be noted that where a binary 0 appears in the information string, no minority charge is transferred.

The maximum operating frequency of a CCD depends on the efficiency of charge transfer from one gate position to the next. As the charge transfer takes a finite time, operation at high frequency causes only part of the charge packet to be transferred at each step, the remainder being left behind and subsequently merged with the following packet. Incomplete charge transfer due to a short transfer time introduces noise and, if the transfer time is too short, information is lost, resulting in a cut-off frequency for the device. The CCD structure described above is termed a *surface-channel CCD* since the charge transfer takes place very near the surface. The charge transfer efficiency of this type is impaired by the presence of surface trapping states at the Si–SiO_2 interface resulting in a maximum operating frequency of about 5 MHz. To avoid the problem of surface states, an implanted buried channel is used to cause charge transfer within the bulk semiconductor rather than at the surface. Such a structure is known as a *buried-channel CCD* and allows operation up to 100 MHz although, in practice, operation is limited to about

20 MHz by drive circuit speed limitations. An absolute low frequency limit is set by the time taken for the inversion layer to form after application of the gate voltage. However, the practical low frequency limit is determined by the thermal generation of free carriers, known as the dark current effect, which superimposes additional charge on to the signal charge packet causing noise and eventual loss of information.

The concept of forming a shift register by controlling the transfer of charge 'packets' within a homogeneous semiconductor by voltages applied to surface electrodes was introduced in 1969 and since then a variety of applications has been exploited. The shift register structure is used as a serial memory, the stored information being continually circulated in a number of shift register 'loops', for example a currently available 64k device comprises 16 addressable 4094-bit loops. By forming electrode patterns that allow charge packets from different digital signals to be combined, logic functions can be performed although such devices have limited application at present. The CCD array is also used in signal processing both as a time-delay and as a filter. To delay an analog signal, it is first sampled and the individual voltage samples used to control the input gate of the CCD. Thus a sequence of charge packets is established of magnitude proportional to the original voltage samples. After transmission along the array, the signal is reconstituted after being delayed by a time dependent on the number of electrodes in the array and the clock frequency. Delay times between 100 ns and 1 s are obtainable using up to 1000 elements and clock frequencies between 1 kHz and 10 MHz. By modifying the electrode structure from the basic linear array and combining the signals from different parts of a split-electrode structure, the device provides signal filtering.

Possibly the main application area for CCDs is in *imaging* where the minority charge is produced via carrier generation by incident radiation. The image is focused on to an array of MOS elements, the charge generated in each element being proportional to the intensity of the radiation. Subsequent output of the charge packets line by line using the shift register principle, produces an electrical video signal corresponding to the image. Such techniques are being applied to solid-state television cameras and are of particular interest for low light level and infrared imagers.

Charge-coupled devices as described above are the most important group of devices relying on charge transfer although the transfer of charge has also been used in *bucket-brigade devices* (BBDs) which can be formed in either bipolar or MOS technology. BBDs consist of a cascade of transistor switches, the capacitance of each switch (C_{BC} in bipolar types and C_{GD} in MOS types) being used as a charge store. Using a two-phase clock to control alternate transistors, charge is transferred from one stage to the next. CCDs and BBDs are sometimes collectively described as *charge-transfer devices* (CTDs). Detailed consideration of the properties and applications of CTDs, and of CCDs in particular, can be found in references 7 and 8.

4.7.5 TTL

Transistor–transistor logic (TTL) is a popular integrated bipolar technology capable of providing high-speed and low-power dissipation. TTL first appeared in 1963 and was a direct development from the DTL (diode–transistor logic), and previously RTL (resistor–transistor logic) technologies, which were used in the early integrated logic circuits (1960–3). The improvement in logic IC performance provided by TTL established the technology as the most popular high-speed logic technology of the 1960s and early 1970s.

The *basic* TTL NAND gate is shown in figure 4.29a and uses a multi-emitter input transistor having up to ten emitters. The V_{CC} supply is typically +5 V and using positive logic, the logic levels are $0 \equiv 0.2$ V and $1 \equiv 4.95$ V. When any or all inputs are low, the positive base supply ensures that T_1 is saturated and so $v_{C_1} = v_1 + V_{CE(sat)} \approx 0.4$ V which is not sufficient to turn T_2 ON. Thus for any input at logic 0, T_2 is cut-off and v_o is equal to V_{CC} less the voltage drop across R_2 due to leakage through T_2, that is, $v_o \approx 4.95$ V (logic 1). When all the inputs are high, the B-E junctions of T_1 are reverse-biased and the base current supplied via R_1 flows across the forward-biased C-B junction of T_1 providing a base current for T_2 which consequently saturates causing v_o to fall to $V_{CE(sat)}$ for T_2, that is, about 0.2 V (logic 0). It should be noted that in this state T_1 operates in the inverse action mode. Although TTL is a saturated logic technology in that BJTs are driven into saturation causing increased charge storage (section 2.2.13), moderately fast switch-OFF occurs as T_1 operates in the normal active mode (E-B junction forward-biased, C-B junction reverse-biased) when an input is switched low, causing removal of base charge from T_2 and hence a fast switch to cut-off.

The practical version of a *standard* TTL NAND gate (figure 4.29b) includes an output stage which provides good output current drive capability and also improves switch-off time. The collector and emitter voltages of T_2 are used to drive the output transistors T_3 and T_4. When T_2 is saturated, T_3 is OFF but T_4 is ON due to the voltage across R_3, the base current for T_4 being provided via T_2, and v_o is thus low. Diode D_1 provides sufficient voltage drop to ensure that T_3 remains OFF. When T_2 is cut off, its high collector voltage causes T_3 to switch ON, base current being provided via R_2, and T_4 switches OFF, its base supply being removed. The output voltage thus rises to almost V_{CC}. The rate of rise of v_o depends on the load on the gate which is invariably capacitive. During the switch, however, T_3 moves through its active region and thus it provides a low impedance current path from V_{CC} enabling a substantial current to be supplied to rapidly charge the capacitance and v_o rises sharply. The use of T_3, described as *active pull-up*, in place of the simple resistive load R_2 in the basic circuit thus improves the switching speed. The topology of the series R_4, T_3, D_1, T_4 output stage gives rise to the term

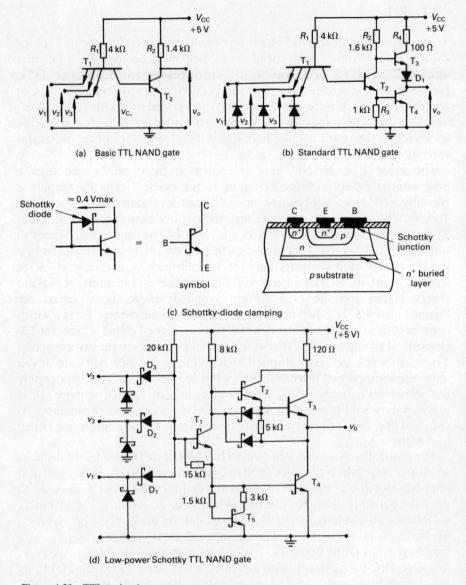

Figure 4.29 TTL technolgoy

totem-pole output. Clamping diodes are normally included between each input and ground which conduct if the input voltage goes negative. They thus limit negative transients which can occur due to inter-gate stray capacitance. The standard TTL gate is capable of operation up to about 35 MHz, having a propagation delay of approximately 10 ns and power dissipation of the order of 10 mW/gate.

Over the years, various improvements have been made in TTL technology regarding speed and power dissipation resulting in a range of logic families known as high-speed TTL (H-TTL), Schottky-clamped TTL (S-TTL), low-power TTL (L-TTL) and low-power Schottky-clamped TTL (LS-TTL). By replacing T_3 and D_1 in the standard circuit by a high-gain Darlington pair (section 2.2.18) the gate-propagation delay can be halved providing a high-speed TTL version, although power dissipation is typically doubled.

The combination of a Schottky-barrier diode (section 2.4.2) and a BJT forms a BJT that does not saturate; the resulting reduction in stored charge in the base and collector regions provides a considerable improvement in switching speed. The Schottky diode is fabricated between the collector and base of the BJT by extending the aluminium base metallisation on to the n-type collector with which it forms a rectifying contact (figure 4.29c). As the forward voltage drop across the Schottky junction is about 0.4 V, the voltage between base and collector of the BJT is limited to this value. The BJT does not enter saturation until V_{BC} reaches about 0.7 V when the B-C junction begins to conduct appreciably. The Schottky junction thus prevents the BJT entering saturation when switched ON, the additional base drive being diverted via the Schottky diode to the collector of the BJT. The low capacitance of the Schottky diode ensures that the rise time of the gate is not degraded. *Schottky clamping* has been used very successfully in TTL ICs reducing the propagation delay to about 3 ns by fabricating Schottky diodes between the base and collector of the BJTs in the high-speed TTL gate, the logic family being described as *Schottky (-clamped) TTL*. The power dissipation of this type of gate is of the order of 20 mW.

A low-power version of standard TTL (L-TTL) has been produced by increasing resistance values in the circuit by up to an order of magnitude. This enables power dissipation to be reduced to about 1 mW/gate although the propagation delay is increased to about 30 ns. By applying the same principle to Schottky TTL, a low-power version has been produced (low-power Schottky TTL, LS-TTL) which has an average propagation delay of about 10 ns while restricting power dissipation to only 2 mW/gate. Figure 4.29d shows a typical LS-TTL NAND gate. Moderately high resistance values restrict power dissipation while high speed is maintained by the use of a Darlington arrangement in the output stage (T_2, T_3), the use of Schottky-clamped BJTs and the replacement of the multi-emitter input BJT by a DTL-type diode input (D_1, D_2, D_3). T_1 is OFF only when one or more of the inputs is low such that the current supplied via the 20 kΩ resistor is diverted from the base of T_1 by the conducting input diode(s). The diodes, including the input clamping diodes, are all low capacitance Schottky junctions. Low-power Schottky TTL is currently the most popular bipolar logic technology providing high speed with fairly low power consumption. It is particularly popular at the small and medium levels of integration, but also finds application to a lesser extent for LSI processors and low-capacity memories.

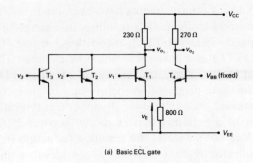

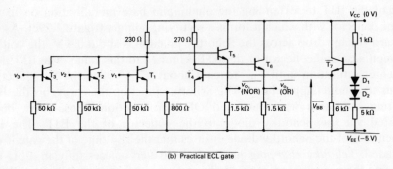

Figure 4.30 ECL technology

4.7.6 ECL

Emitter-coupled logic (ECL) is a non-saturating bipolar technology which provides the ultimate in high-speed logic circuitry although at the expense of relatively high power dissipation. The gate propagation delay is typically 2 ns with a power dissipation of 25 mW/gate, although the delay can be reduced to 1 ns or less with power levels of up to 100 mW/gate.

The difficulty of producing a bipolar switching circuit that does not saturate is due to the fact that the static current gain h_{FE} of a BJT is very sensitive to variation of manufacturing parameters and so h_{FE} cannot be guaranteed to a close tolerance. For this reason the avoidance of saturation by control of base current is not practical. In ECL, saturation is avoided by controlling the emitter current using a common-emitter resistor shared by the input BJTs and a reference BJT. Figure 4.30a shows the basic ECL gate.

Current flows through the emitter resistor either through T_4, which has a constant bias, or through one or more of the input transistors. With all the input transistors (T_1, T_2, T_3) OFF, output v_{o_1} is high ($\approx V_{CC}$) and T_4 conducts due to V_{BB}. The voltage at output v_{o_2} is thus low, its value being dependent on the ratio of load to emitter resistance. If any or all of the input voltages are made greater than V_{BB}, the corresponding input transistor(s) will conduct and current is diverted from T_4 to the input transistor(s), the total current drawn by the stage being maintained approximately constant by the common-emitter resistor. Output v_{o_1} thus falls to a value dependent on the ratio of load resistance to emitter resistance and v_{o_2} rises to approximately V_{CC}. Output v_{o_1} is high and v_{o_2} is low only when all inputs are low, thus v_{o_1} provides a NOR function and v_{o_2} an OR function.

Typically $V_{CC} = 0$ V, $V_{EE} = -5$ V, $V_{BB} = -1.3$ V and the two logic levels are -0.9 V (logic 1) and -1.7 V (logic 0). It should be noted that the logic levels are equally spaced about the fixed reference V_{BB} and that the logic swing, that is, the difference between the two logic levels, is small facilitating fast transitions. With $V_{BB} = -1.3$ V and T_4 ON, $v_E = V_{BB} - V_{BE} \approx -2$ V. If all inputs are low (-1.7 V), the B-E voltage of the input transistors ($+0.3$ V) is insufficient to cause them to conduct. However, when an input is high (-0.9 V), the larger B-E voltage causes the input transistor to turn ON. Since the input voltage (-0.9 V) is more positive than V_{BB} (-1.3 V), v_E increases from -2 V to -0.9 V $- V_{BE}$, that is, about -1.6 V, causing T_4 to turn OFF. The fact that v_E changes slightly is partly offset by the unequal load resistances. Saturation is avoided by arranging the ratio of load to emitter resistance to be such that V_{CE} is always greater than $V_{CE(sat)}$ for the BJTs so that the C-B junctions cannot become forward-biased. Figure 4.30b shows a practical ECL gate. Emitter-follower output stages (T_5 and T_6) provide low impedance outputs (increasing the fan-out capability of the gate) and also introduce a voltage shift of V_{BE} providing the logic levels given above. The V_{BB} reference is supplied by the constant-bias emitter-follower circuit incorporating T_7, diodes D_1 and D_2 compensating for changes in temperature and V_{EE}. An important feature of ECL is that since the current flow from the power supply is diverted from one path to another, varying only slightly, instead of being switched ON and OFF, noise generation is low. From a fabrication point of view, the fact that the operation of the circuit relies mainly on resistance ratios and not on the precise values of current gains of BJTs is a major advantage.

ECL provides the fastest semiconductor-based circuitry available commercially and is used in applications where high speed is essential such as for fast computer processors and memories, fast analog-to-digital converters and high-speed instrumentation although cost, power requirements and heat sinking are major disadvantages. Recently the speed supremacy of ECL has been challenged by Schottky TTL, ISL and SOS-CMOS (figure 4.33).

4.7.7 I²L

The development of integrated-injection logic (I^2L), also known as merged-transistor logic (MTL), in the early 1970s was a major advance in bipolar digital ICs. I^2L technology employs element-merging whereby a single doped semiconductor region forms part of more than one circuit element,

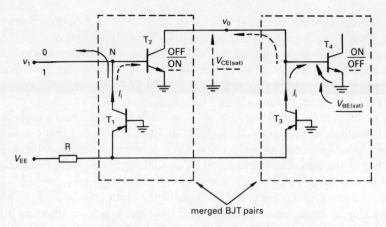

(a) Basic I²L inverter

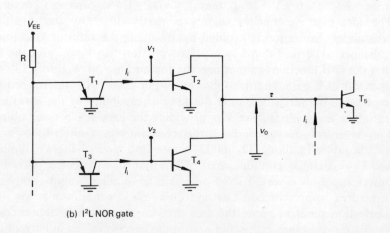

(b) I²L NOR gate

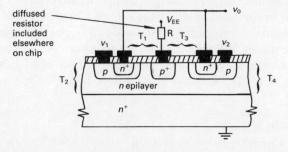

(c) Structure of I²L NOR gate

Figure 4.31 I²L technology

termed *superintegration* (section 4.6). Considerable savings in chip area are obtained not only from element-merging but also from the reduction in inter-element isolation. Packing densities of the order of 200 gates/mm^2 are possible in I^{2}L exceeding the scale of integration possible which the conventional MOS technologies and comparable to that obtained with VMOS.

The basic I^{2}L cell comprises a lateral *pnp* BJT and a conventional vertical *npn* BJT, the *n*-type epilayer forming the *pnp* base and the *npn* emitter while a *p*-type diffusion forms the *pnp* collector and the *npn* base (figure 4.24). Figure 4.31a shows the use of two of these merged BJT pairs as a logic inverter. In each stage the *pnp* BJT acts as a current injector, the value of current being determined by the fixed voltage V_{EE} and resistance R. The injected current I_i supplied by T_1 is *steered* at node N either away from T_2 or into the base of T_2 by the voltage level v_1 applied at the gate input. The two logic levels are provided by $V_{CE(sat)}$ (≈ 50 m V, logic 0) and $V_{BE(sat)}$ (≈ 0.7 V, logic 1) of the *npn* BJTs. If v_1 is low, I_i is *sunk* by the input source (solid arrows, figure 4.31a) and T_2 is OFF. Consequently the injected current supplied by T_3 is steered into the base of T_4 which saturates and v_o, the collector voltage of T_2 relative to ground, is $V_{BE(sat)}$ of T_4. Alternatively, if v_1 is high, I_i is steered into the base of T_2 (broken arrows, figure 4.31a) which being saturated is able to sink the current from T_3. T_4 therefore is OFF and v_o is $V_{CE(sat)}$ of T_2. The arrangement thus provides logical inversion.

Figure 4.31b shows how this principle is extended to provide a NOR function, the corresponding chip structure being given in figure 4.31c. The output v_o of the gate is high ($V_{BE(sat)}$) only when T_2 and T_4 are both OFF so that T_5 is saturated by the injected current. T_2 and T_4 are both OFF only when the inputs v_1 and v_2 are low, sinking the currents injected by T_1 and T_3. Thus the output is a logic 1 only when all inputs are at logic 0: a NOR function. If T_2 and T_4 are provided with dual collectors, additional outputs providing the inverse of the inputs v_1 and v_2 are also available with little increase in chip area occupied by the gate.

The fabrication process for I^{2}L requires only five masks which makes it cost-competitive with the other logic technologies. Its power–delay product can be varied over a range by control of the injector current via V_{EE} and R. A minimum value of about 1 pJ is obtainable for gate dissipation below 10 μW/gate giving, for example, gate delays of 100 ns and 1 ms at gate dissipations of 10 μW and 1 nW respectively. Fastest operation occurs at about 1 mW/gate when the propagation delay is of the order of 20 ns. The trade-off between power dissipation and speed by adjustment of injector current allows a combination to be selected to suit the application.

Although I^{2}L provides a high gate-packing density coupled with low power dissipation, its high propagation delay compared with other technologies limits its usefulness (table 4.4). There are a number of factors that contribute to the slow speed

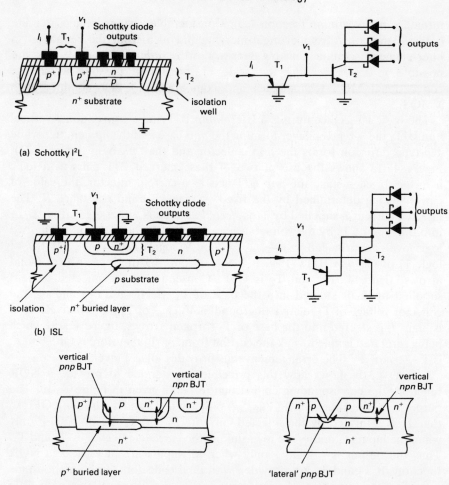

Figure 4.32 Developments in I²L technology

(1) The emitters of the *npn* BJTs are formed by the epilayer (figure 4.31c) and are more lightly doped than the base regions, resulting in low gain;
(2) lateral loss of electrons injected into the epilayer;
(3) saturation of the *npn* BJT.

A number of developments has been made to the basic I²L structure aimed at improving the switching speed such as

(1) the use of Schottky output diodes to reduce the logic swing and to improve multi-output performance;

(2) improvement of the doping profile of the *npn* BJT to increase gain;
(3) the use of isolation regions to reduce lateral loss;
(4) improvement of the current gain of the *pnp* BJT.

The first three of these improvements are incorporated in the Schottky I^{2}L gate of figure 4.32a. The formation of Schottky output diodes reduces the logic swing as the output voltage cannot fall below about 0.45 V ($V_{CE(sat)}$ for T$_2$ + Schottky diode drop). In standard I^{2}L the use of multiple collectors is unsatisfactory due to the interaction between collectors resulting from the low *npn* BJT gain. The use of Schottky output diodes not only isolates the outputs from one another but also allows the use of a lightly doped collector enabling an improved doping profile and hence an increased gain to be obtained for the *npn* BJT. The oxide isolation well in figure 4.32a reduces the lateral loss of injected electrons. Structures using isolating oxide wells have been termed I^{3}L, *Isoplanar integrated-injection logic*.

A recent commercial development allows injection of current directly into the base of the *npn* BJT and uses the merged *pnp* BJT as a clamp to restrict saturation. The structure incorporates Schottky output diodes (figure 4.32b) and is termed *integrated Schottky logic* (ISL). The performance of ISL has been very promising (table 4.4), providing a gate delay as low as 4 ns at a dissipation of 1 mW/gate. As with standard I^{2}L, the speed of ISL can be traded for power economy, for example, if a gate delay of 10 μs can be tolerated, gate dissipation can be reduced to about 50 nW; a power–delay product of only 0.5 pJ. Due to the structural change, however, ISL achieves only about half the packing density of standard I^{2}L.

Various structures have been proposed to improve the performance of the *pnp* BJT and hence reduce the gate delay of I^{2}L. A structure called *vertical injection logic* (VIL) [9] uses a p^+ buried layer to form a vertical *pnp* BJT (figure 4.32c) replacing the low-gain lateral device of standard I^{2}L. By controlling the out-diffusion of the acceptor-dopant from the buried layer into the epilayer, a narrow-base moderate-gain *pnp* BJT is produced. Another proposed structure with the same objective uses an accurately etched V-groove to produce a narrow-base *pnp* BJT (figure 4.32d). The structure has been termed *V-groove I^{2}L* or V-I^{2}L.

4.7.8 Summary

As the performance of logic circuits in terms of propagation delay and power dissipation depends on the particular operating conditions such as power supply voltage, frequency and output loading, it is not possible to state precise values. Table 4.4 lists *typical* values for the various technologies together with possible gate-packing densities providing a comparative guide. Figure 4.33 gives typical ranges of delay and dissipation showing the performance of each technology in relation to others. The lower the

Table 4.4 Performance of logic technologies

Technology	Gate propagation delay (ns)	Gate power dissipation (mW)	Power-delay product (pJ)	Gate-packing density (gates/mm^2)	Maximum frequency (MHz)
Basic PMOS	80	6	480	50	5
Silicon-gate PMOS	40	5	200	100	10
Silicon-gate NMOS	15	3	45	150	20
Silicon-gate CMOS	<20	0.02†	<0.4	60	15
SOS–CMOS	<5	0.02†	<0.1	200	>100
VMOS	5	2	10	200	100
Standard TTL	10	10	100	20	35
High-speed TTL (H-TTL)	6	20	120	20	50
Schottky TTL (S-TTL)	3	20	60	20	120
Low-power TTL (L-TTL)	30	1	30	15	3
Low-power Schottky TTL (LS-TTL)	10	2	20	20	40
ECL	<2	>25	50–100	40	>200
I^{2}L	>20	<1	1–20	200	10
ISL	>4	<1	0.5–4	100	40

†static operation

Integrated Circuit Technology

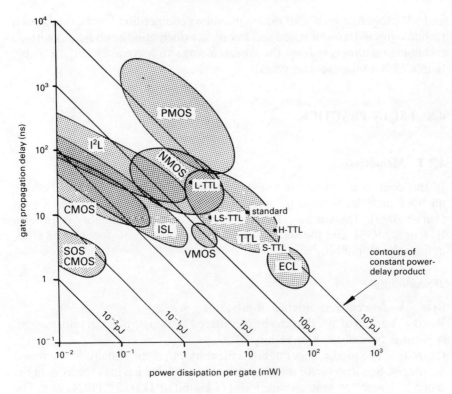

Figure 4.33 Speed-power properties for the various logic technologies

power–delay product of a technology, that is, the nearer the origin it appears in figure 4.33, the better its over-all performance. It can be seen that PMOS offers the poorest performance although, due to its basic simplicity and hence low cost, it is used in low-speed general-purpose applications such as pocket calculators. NMOS and TTL are well-established medium performance technologies offering a wide speed-dissipation range and thus find application in the majority of logic applications. ECL provides the ultimate in speed although at the expense of high power dissipation and thus it is used only where high speed is essential. At present it is the only commercial semiconductor logic technology capable of sub-nanosecond switching delays; speeds which are necessary for logic systems operating with data transfer rates in excess of 1 Gbit/s. Standard CMOS is the popular technology for applications requiring very low power dissipation although in order to take advantage of this property the operating frequency is limited to below 1 MHz. SOS-CMOS provides the most impressive over-all performance although its high cost is a restriction to its use except in a few highly specialised applications. The newest technologies, I^2L and its derivatives such as I^3L, ISL and VIL, together with VMOS, are providing competition for the established technologies in terms of speed and low power dissipation although, as with all technologies, progress from the research stage to a commercially available family of ICs takes several years.

4.8 LSI IN PRACTICE

4.8.1 Memories

In this context a *memory* is a store of digital information. Each *cell* of the memory provides storage for one unit of digital information, termed a *bit* (*bi*nary dig*it*). The storage element in a semiconductor memory is either a BJT or a MOST and the state of conduction of the transistor, ON or OFF, defines the stored bit.

Terminology

Byte A consecutive sequence of bits, often 8.
Word A group of bits that can be considered as a single piece of information. A word is typically 4, 8 or 16 bits.
Capacity The total number of bits a memory can store, usually 2^n where n is an integer. Semiconductor memory chips are available having capacities in the range 16 bits to 256 kbits, where 1 kbit (1 kilobit or 1k) is 2^{10} (1024) bits. The letters k and M are used as *binary* prefixes to represent $\times 2^{10}$ and $\times 2^{20}$ respectively (appendix C), analogous to their use as *decimal* prefixes, k ($\times 10^3$) and M ($\times 10^6$). A 1k (bit) capacity memory thus has 1024 (2^{10}) cells, a 4k has 4096 (2^{12}) cells, an 8k has 8192 (2^{13}) cells and so on. Commercially available semiconductor memories have capacities of 16, 32, 64, 128, 256, 512 bits, 1k, 2k, 4k, 8k, 16k, 32k, 64k, 128k and 256k, the 1–16k range being particularly popular at the time of writing. The cells of a memory are grouped into words; a 1k memory may be *organised* as 128 words each of 8 bits or 256 4-bit words. These memories would be described as having 128 × 8 and 256 × 4 organisations respectively. A 1024 × 1 version is a 1k memory in which individual bits rather than groups of bits can be selected. Memory capacity is sometimes quoted in *bytes* rather than in *bits*; a 16 kbit memory organised as 2k × 8 bytes can be described as a 2 kbyte memory and the shorthand description of memory capacity as 2k, for example, is therefore ambiguous.
Address Each cell or group of cells in the memory has a code known as its *address* enabling it to be selected. The cells are arranged in rows and columns, and the address comprises a row address and a column address. Using an array of logic gates known as a *decoder,* [11] the address is converted to a logic 1 on one of the row select lines and one of the column select lines. These signals close transistor switches connecting the addressed cell to the input/output (I/O)

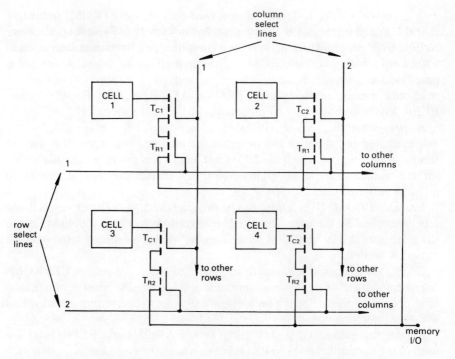

Figure 4.34 Cellular memory array with row and column switches

line. Figure 4.34 shows part of a memory array with MOST row and column select switches T_R and T_C. Only the cell having row and column address lines at logic 1 is connected to the I/O line via the conducting row and column switches.

Write The setting of addressed bits or words in the memory to particular binary values.

Read Determination of the information (binary values) stored in particular bits or words in the memory.

The most significant performance parameters of a memory are its *access time*, that is, the time between addressing a cell and the stored information appearing at the output, and its *power consumption*.

There are two basic memory applications; one in which it is necessary to frequently change the stored information, the other where the stored information is permanent or semi-permanent.

A *random-access memory* (RAM), also termed a *read–write* memory, can be repetitively rewritten as required by a 'scratch-pad' store for data during a computation. Semiconductor RAMs are described as *volatile* as the stored information is lost if the supply voltage to the memory is removed.

A permanent store is described as a *read-only memory* (ROM) indicating that the stored information can be read but not altered. Typical applications include fixed program instructions for a computer or a fixed data store used as a 'look-up table'. The basic ROM is programmed during manufacture and is purchased as a programmed unit for a particular application. Other types of read-only memories, designated PROMs, UV EPROMs, EAROMs and EEPROMs, allow the user to program/reprogram the memory.

A *programmable ROM* (PROM) is supplied unprogrammed by the manufacturer and the user can program the memory *once* only. It is used in similar applications to the basic ROM but where it is not economically viable for the memory to be programmed during fabrication due to the small quantity required.

Erasable PROMs (EPROMs) can be programmed, bulk (block) erased and reprogrammed by the user. Erasure is by exposure to ultraviolet light, hence the designation UV EPROM. The number of times the memory can be erased is limited.

EAROMs (*electrically alterable ROMs*) and the more recent EEPROMs (*electrically erasable and reprogrammable ROMs*) are the most sophisticated read-only memories allowing bulk erasure or selective rewriting of individual words by electrical means.

Reprogrammable ROMs (UV EPROMs, EAROMs and EEPROMs) are used in similar applications to PROMs but where the stored information may need *occasional* alteration. EAROMs and EEPROMs cannot be used in RAM applications requiring fast repetitive rewriting due to their relatively slow operation and limited life (*endurance*). All ROMs are *non-volatile* in that they retain the stored information when electrical power is removed.

Random-access Memories

Although the term 'random-access' can be applied to both read–write and read-only memories since cells or words in both types of memory can be addressed (accessed) at random, by popular usage the name RAM is taken to refer to the type of memory for which there is no practical limit to the number of times it can be rewritten.

The electronic operation of a RAM is described as being *static* if the memory retains stored information without further operation, or *dynamic* if the stored state has to be reconstituted to avoid information being lost.

The cells of a static RAM are latch circuits (flip-flops) fabricated in one of the IC technologies, namely PMOS, NMOS, VMOS, CMOS, TTL, ECL or I^{2}L. Early MOS RAMs used *p*-channel technology but this was soon superseded by higher performance NMOS. A latch circuit comprises two cross-coupled logic inverters (output 1 connected to input 2 and output 2 to input 1) and has two stable states enabling a binary 0 or 1 to be stored. This type of memory is termed *static* as once the latch is set to one state, that is, a binary 0 or 1 is stored, it will remain in that state, until reset, provided that the

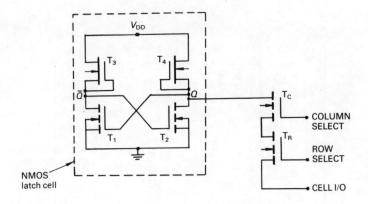

(a) NMOS static RAM cell

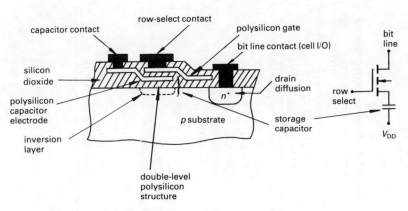

(b) Single transistor double-level polysilicon NMOS dynamic RAM cell

Figure 4.35 NMOS RAM cells

power supply to the circuit is maintained. Figure 4.35a shows a basic NMOS static RAM cell, the latch being formed by *n*-channel MOSTs T_1–T_4, T_3 and T_4 being depletion-type loads. Row and column switches T_R and T_C driven by the address decoders, connect the cell to the I/O line. After the cell has been addressed (connected to the I/O line) it is programmed or written (logic level Q is set) by applying the appropriate logic level to the input line. Reading the stored information is the reverse process, with T_C and T_R selected, logic level Q appears on the output line. An NMOS static RAM consists of an array of this type of cell connected as shown in figure 4.34.

Typical values of access time and power dissipation/bit for static RAMs

Table 4.5 **Memory performance**

Memory type	PMOS	NMOS	CMOS	TTL	ECL	I^2L
Static Ram	200–1000 (0.2–2)	35–500 (0.05–1)	100–1000 (0.01–0.1†)	25–70 (0.1–2)	5–25 (1–10)	≈50 ≈0.05
Dynamic RAM	–	70–400 (0.01–0.2)	–	–	–	≈150 ≈0.1
ROM	600–1000 (500–2000)	100–500 (300–1000)	400–600 (40–100)	25–250 (500–1000)	–	–
PROM	–	–	–	25–250 (500–1000)	10–25 (400–1000)	–
UV EPROM	400–1000 (500–1000)	200–500 (400–800)	300–500 (200–400)‡	–	–	–
EAROM	1000–2000 (1000–1500)	–	–	–	–	–
EEPROM	–	200–500 (200–500)	≈500 (≈5)	–	–	–

Access time (ns)
Operating power dissipation (mW/bit)
(Total operating power dissipation, mW)

† at 1 MHz
‡ NMOS array, CMOS peripheral circuitry.
– not available

fabricated in the various IC technologies are given in table 4.5. The most popular types are Schottky TTL and silicon-gate NMOS. TTL RAMs provide faster operation in general with access times typically in the range 25–70 ns compared with 35–500 ns for NMOS versions. Due to packing-density limitations, however, TTL RAMs have a maximum capacity of 4k whereas NMOS static RAMs are available up to 16k. The VMOS derivative of NMOS has been shown to be capable of TTL operating speeds but is not cost-competitive with lateral NMOS. Of the various technologies, ECL provides the fastest operation with access times below 25 ns but the high power dissipation limits the capacity to 4k due to temperature rise. Lowest operating power consumption of the commercially available technologies is provided by CMOS although at the expense of speed. CMOS static RAMs of up to 16k capacity are available and have the additional advantage that on standby, that is, when not being written or read, total power consumption is negligible (< 1 μW). Static RAMs fabricated in I^2L technology are still at the development stage but are expected to provide the speed of TTL at much reduced power consumption. Memory capacity is limited either by the maximum allowable power dissipation, as in the case of ECL RAMs, or more commonly by the maximum chip size allowable to give a good process yield and to enable encapsulation in a standard dual-in-line package.

Dynamic RAMs utilise a simpler cell configuration than the static type, often a single transistor, enabling higher memory capacities for the same chip size. Dynamic semiconductor RAMs are available with 1k, 4k, 8k, 16k, 32k and 64k capacities, with a 256k version under development. The 16k size is particularly popular at the time of writing for microprocessor-based systems. The storage mechanism in a dynamic RAM cell is the charge stored in a capacitor and, due to leakage, the state of charge has to be replenished or 'refreshed' typically every 2 ms. The need for refresh circuitry adds a complication to the use of this type of RAM but the increased capacity provided by the smaller cell size is an overriding advantage. At present, all RAMs of capacity greater than 16k are dynamic and, for such capacities, the saving in cost/bit, compared with the use of several lower capacity static RAMs to provide the same over-all storage capacity, far outweighs the cost of refresh circuitry. A popular one-transistor dynamic RAM cell is the double-level polysilicon NMOS structure of figure 4.35b. The cell is a modified n-channel enhancement-type MOST incorporating an additional polysilicon electrode which when appropriately biased, forms an n-type inversion layer in the substrate. The inversion layer then acts as the source of the MOST while the polysilicon electrode and the inversion layer form a storage capacitor (≈ 0.05 pF), the state of charge determining whether a logic 0 or 1 is stored. The drain diffusion acts as the input/output of the cell; all cells in the same column have a common drain diffusion and all column drain diffusions are connected to the memory I/O via column select MOST switches. Thus each cell does not have a metallised bit line contact as shown in figure 4.35b; instead, the n^+ diffusion forming the common bit line for a column of cells runs under the oxide and contact is made at the end of the column. The polysilicon gate electrodes of cells in the same row are connected together via surface metallisation which provides the row-select. Thus energisation of a particular column-select switch connects the bit line of a column of cells to the memory I/O and application of a voltage greater than the threshold voltage (1–2 V) to a row-select line forms a conducting channel connecting the storage capacitor of a particular cell to the bit line. The inversion layer is formed by the bias on the capacitor electrode. A logic 0 or 1 is then written into the cell by setting the bit line low (0 V) or high (+10 V) respectively and addressing the particular cell via the column and row-select lines enabling the free charge state of the inversion layer to be altered. The read operation is the reverse process except that the bit line is precharged to the mean of the logic 0 and logic 1 voltage levels. On addressing a cell, the bit line voltage increases or decreases by about 200 mV depending on the charge state of the cell. This fluctuation is converted to an output 0 or 1 by the sense amplifier which is a latch circuit connecting each column line to the memory I/O via the column-select switches. The read operation is *destructive* in that the storage capacitor loses its charge state and so a *rewrite* MOST switch has to be included to enable the sensed charge state to be automatically written back into the cell.

Leakage across the reverse-biased depletion layer between the inversion layer and the substrate causes the storage capacitor to discharge. In physical terms this corresponds to the inversion layer approaching equilibrium for the electrode voltage applied. Before the charge state has altered sufficiently to cause the stored bit to be lost, the cell is *refreshed* by reading the cell on a repetitive basis, the rewrite mechanism described above then restores the non-equilibrium charge state of the inversion layer. The use of the non-equilibrium state of the inversion layer as a storage cell in a dynamic RAM is similar to its use in a CCD (section 4.7.4), the refresh operation in the RAM case corresponding to the shift operation in the CCD, that is, the charge must be replenished or shifted from the region before equilibrium occurs.

The I^2L structure of figure 4.24 is also used as a dynamic RAM cell; the capacitance of the *npn* B-E junction/*pnp* C-B junction providing the storage medium. Using a *p*-type substrate, the structure is formed on an n^+ buried layer which forms the column-select line and makes contact to the *pnp* base and *npn* emitter. Adjacent columns are isolated by oxide wells penetrating to the substrate while adjacent cells in the same column are isolated by shallower oxide wells. The collector of the *npn* BJT is the row-select and the cell is written and read by biasing the *pnp* emitter and row-select appropriately to charge or sense the charge of the junction capacitor. Repetitive refreshing is necessary to replenish charge lost by leakage and is performed by a rewrite operation. The merged transistor I^2L cell occupies a similar chip area to the single transistor NMOS cell and operates from a single voltage supply, typically 5 V. Until recently, NMOS dynamic RAMs required three voltage supplies, typically a capacitor bias of +12 V, a −5 V substrate bias to provide a threshold voltage of 1–2 V and the +5 V logic level. However, processing modifications have enabled NMOS power requirements to be simplified to a single 5 V supply.

Although the need for refresh introduces a complication in the use of dynamic RAMs at present, it seems likely that dynamic memories will soon be produced with on-chip refresh circuitry so that, as far as the user is concerned, the memory will appear as a high-capacity static.

Typical values of access time and power dissipation for dynamic RAMs fabricated in the various IC technologies are given in table 4.5. The problem of volatility is often overcome by the provision of battery back-up, in which case low power dissipation, particularly on standby, is important.

Read-only Memories

The basic ROM cell stores a logic 0 or 1 by having a BJT or MOST fabricated at that location or by being left open circuit. Figure 4.36a shows one storage cell in an NMOS ROM. By fabricating a MOST at this location the point X is grounded when T_3 is turned ON by the row select signal. If this cell has been addressed via the column-select transistor T_2 the output will be approximately at ground (logic 0), the supply voltage V_{DD} being dropped across the load T_1. Alternatively, if the cell had been left blank, that is, no transistor fabricated at this location, there would not be a conducting path between point X and ground and the output would remain at V_{DD} (logic 1) when the cell was addressed. Thus a ROM is programmed during manufacture by fabricating transistors at particular cell locations and the ROM therefore provides fixed storage.

The NMOS ROM is fabricated as an array of *n*-channel enhancement-type MOSTs with strip n^+ diffusions providing the sources and drains with surface gate metallisation providing the row-select. Figure 4.36b shows two adjacent cells. For convenience MOST structures are fabricated at *all* cell positions and working MOSTs created where required by etching away the thick (1 μm) oxide and by growing thin (0.1 μm) oxide in its place. Thus in figure 4.36b, location 1 contains a working MOST but location 2 does not since, although there is a MOST structure at location 2, the thick oxide causes a high threshold voltage so that in practice it cannot switch ON. The programming step is thus contained in the etching and formation of thin gate oxide at specific locations and for this reason such memories are said to be *mask-programmed*.

Bipolar ROMs are also produced, usually employing Schottky-clamping, to provide fast access. In this technology the BJT base is connected to the row-select line and the emitter to the column-select line by the surface

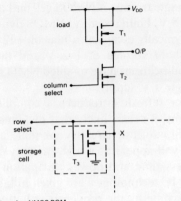

(a) Section of an NMOS ROM

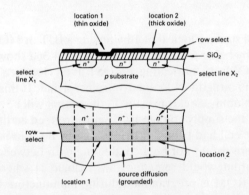

(b) Implementation of NMOS ROM cells

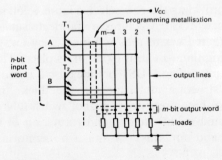

(c) Bipolar ROM as an encoder

Figure 4.36 Read-only memories

metallisation. As in the NMOS case, programming occurs during fabrication and is therefore fixed. The term *hardwired* is often used to describe the fixed metallisation that defines the storage. The collectors of the BJTs are common and they can therefore be fabricated in a common n-type island which is in fact connected to the voltage supply V_{CC}. As storage does not rely on the electrical supply, ROMs provide *non-volatile* storage.

ROMs are available commercially with capacities of 1k, 4k, 8k, 16k, 32k, 64k, 128k and 256k, mostly organised with word lengths of 4 or 8 bits. Typical performance figures for ROMs fabricated in the various technologies are given in table 4.5. Bipolar versions, listed in table 4.5 under TTL, provide the fastest access time but are limited to 32k whereas the popular lateral NMOS structure, although slower by a factor of about three, has only half the total power dissipation of bipolar types and permits capacities up to 256k. The VMOS structure can provide the high capacity of NMOS while halving the access time. CMOS ROMs are noted for their low power dissipation but are slower, while PMOS ROMs have been largely superseded by the superior-performing NMOS versions.

Although a ROM has been considered above to be a store of digital data in the form of a set of words which may be addressed to determine the value or contents, it may also be considered as a logic circuit which in response to a binary input (the address) provides a binary output. Figure 4.36c shows a bipolar ROM using multi-emitter transistors which provides an m-bit output word from an n-bit input word, the conversion depending on the connections between the emitters and the output lines. The ROM can thus be programmed to provide a complex logic function which is simpler to produce, cheaper, but slower than implementation in combinational logic. Considered in this way, a ROM acts as a code converter, often termed an *encoder*, and is used in such applications as character generation for displays and 'look-up tables' of function values.

Programmable ROMs
In some applications it is convenient for the user to program the ROM, particularly where only a small number of memories are required. This facility is provided by forming an array of BJTs each with its emitter connected to the column line via a fusible link (figure 4.37). The link is either a metal alloy (for example, nickel–chromium) or doped polysilicon and is either left intact or open-circuited by the user depending on the required storage. The fusing operation is normally performed on a *PROM programmer* which allows cells to be addressed and a short current pulse of about 20 mA to be applied to fuse the link. Although the stored information can be altered by fusing links which have been left intact, fused links cannot be remade and so a PROM should be regarded as a memory that cannot be reprogrammed. The majority of PROMs use Schottky TTL technology and are available in capacities of 256 bit, 1k, 2k, 4k, 8k and 16k with 4 or 8-bit

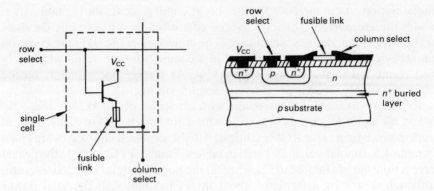

Figure 4.37 Basic PROM cell

word lengths. Access times are typically in the range 25–250 ns with total power dissipation 500–1000 mW. Low capacity ($\leqslant$ 1k) ECL PROMs are produced using fusible-link technology having low access times of 10–25 ns although power consumption/bit is high.

Erasable PROMs

By using the FAMOST device (figure 4.9b) as the storage element, a ROM is produced which can be electrically programmed and *bulk* erased by the user. The memory is termed an *erasable programmable read-only memory* (EPROM) and is particularly useful for applications where the stored information may need *occasional* modification. The FAMOST cell stores a logic 1 or 0 by the presence or absence of stored charge on the floating polysilicon gate which controls channel conduction. As with other types of memory, an EPROM has row and column-select lines driven by address decoders. A logic 1 is written into an addressed cell (section 4.2.2) by applying a short voltage pulse between drain and source causing avalanche breakdown of the drain–substrate junction. Some charge from the current surge tunnels through the thin oxide layer and becomes trapped on the gate. Since the floating gate is totally isolated, leakage is very small and manufacturers claim that the device will retain the stored information for at least 10 years under normal operating conditions. The memory is non-volatile as charge remains on the floating gate when electrical power is removed.

Bulk erasure of all the cells in the memory is achieved by exposure of the chip to UV light or X-rays via a window in the encapsulation. The resulting carrier generation and recombination with the trapped charge erases the stored information and the memory can then be reprogrammed. For convenience, erasure by UV light is most popular and this type of erasable PROM is therefore normally described as a *UV EPROM* to distinguish it from EPROMs that can be erased electrically (EAROMs and EEPROMs).

Integrated Circuit Technology

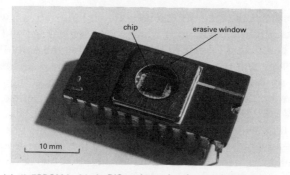

(a) 4k EPROM in 24-pin DIC package showing erasive window

(b) 4k (512 × 8 bit) UV EPROM chip

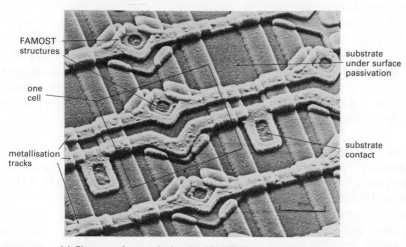

(c) Electron micrograph showing UV EPROM cell layout

Figure 4.38 UV EPROM package, chip and cell layout

The number of times the memory can be erased is limited to a few hundred by radiation damage to the oxide layer which increases leakage and reduces the storage life. The original UV EPROMs introduced in the early 1970s utilised PMOS technology and required multiple power supplies. As the technology developed there was a switch to NMOS to improve access times and power requirements were simplified by most manufacturers to a single 5 V supply. Currently, UV EPROMs are available with capacities of 1k, 2k, 4k, 8k, 16k, 32k, 64k and 128k with a 256k version under development. They are usually organised as 8-bit words, some as 4-bit, and most use double-level polysilicon NMOS technology giving access times in the range 200–500 ns and total active power dissipation of 400–800 mW. VMOS versions have been produced but have been found to be uneconomic in competition with lateral NMOS. To reduce power dissipation, some UV EPROMs are produced with an NMOS memory array but with CMOS peripheral circuitry (address decoders, sense amplifiers, and control logic), enabling power consumption to be halved.

Figure 4.38 shows details of the package and chip layout of a 4k UV EPROM. The two rectangular areas in figure 4.38b are the 4096 bit cellular storage array organised as 512 8-bit words with the address decoders, sense amplifiers and control logic included along two sides and across the centre of the chip. The 4 mm square chip is bonded to the 24-pin dual-in-line ceramic package using 25 μm thick gold wire. Detail of individual storage cells is shown in the scanning electron micrograph of figure 4.38c. Each diamond-shaped structure is a single FAMOST storage cell; surface irregularities due to etching during fabrication and surface metallisation can be clearly identified.

A disadvantage of UV EPROMs is that they usually have to be removed from circuit to be erased and that erasure is time-consuming, requiring irradiation for 10–40 minutes depending on the material thickness covering the floating gate. The programming time for a modern device is typically 1–10 ms per word (byte) enabling an erased memory to be reprogrammed in under two minutes. This type of memory is used particularly in ROM program debugging during system development. The stored program can be modified until the system is entirely functional at which time the program can be transferred to permanent ROM or PROM storage.

Electrically Erasable ROMs
In about 1973 a ROM was introduced using the MNOS structure (figure 4.9a) that permits electrical erasure: an *electrically alterable ROM* (EAROM). As with the FAMOS structure used in UV EPROMs, the state of an MNOS cell depends on the presence or absence of trapped charge. The substantial difference being that for the MNOS structure the charge is trapped mainly by the surface states at the oxide–nitride interface rather than on an electrode and the stored charge can be removed, that is, the cell erased, by applying a sufficiently high voltage of appropriate polarity to the gate.

EAROMs are mostly p-channel devices and are available up to 8k capacity. Most types must be block erased before reprogramming, the erase time and the byte program time being of the order of 20 ms. Compared with UV EPROMs, the EAROM provides the convenience of electrical erasure which is fast and can be performed without removing the memory from circuit. In addition, the number of times the memory can be rewritten, termed the number of *endurance cycles,* is typically 10^5, far in excess of the few hundred for a UV EPROM. However, from the user's point of view, EAROMs have three disadvantages: they are slow, having a typical read access time of 2 μs, they dissipate considerable power (typically > 1 W) and, in addition to a 5 V supply, they require multiple higher voltage supplies (of the order of 20 V and 30 V) for programming, reading and erasure.

Recently an electrically erasable ROM has been introduced, using a modified form of the stacked-gate FAMOS structure of figure 4.9b. The modification is that a small region of the oxide under the floating gate is etched to reduce its thickness so that charge can tunnel more easily from the channel region to the gate permitting lower operating voltages than for the MNOS structure. These *electrically erasable, reprogrammable ROMs* (EEPROMs) are fabricated in NMOS. A logic 1 is written into an addressed cell by applying about 15 V between the control (upper) gate and the source to form an n-type channel and at the same time applying a similar voltage between drain and source. Under the influence of the transverse electric field created in the oxide between the channel and the floating gate, some charge from the resulting current flow tunnels through the oxide and becomes trapped on the floating gate. If V_{DS} is maintained at zero during this write operation there is no current flow through the channel and so no charge tunnels through to the floating gate, the cell thus stores a logic 0. Stored charge is erased by arranging V_{DS} to be zero, setting the control gate potential to zero and pulsing the p-type isolation island containing the device to about 15 V for up to 10 ms. The reverse transverse field beneath the floating gate causes the stored charge to tunnel back to the channel region leaving the floating gate uncharged.

The newer EEPROMs using the stacked-gate structure retain the convenience of in-circuit electrical erasure of the earlier MNOS EAROMs but have access times similar to other memory types (200–500 ns), lower power-dissipation and simpler, lower voltage power supply requirements. Manufacturers claim that the endurance of EEPROMs is similar to that of EAROMs ($\approx 10^5$ cycles) with data retention better than 10 years under normal operation. They are available up to 16 k capacity, some allowing erasure of individual words rather than requiring block erasure.

Shift Registers
A shift register is a cascade of memory cells in which stored bits are progressively transferred along the array at a rate determined by an applied

'clock' frequency. Such an arrangement can be used as a memory, data being input and output in serial form, that is, as a string of bits one after the other. Both static and dynamic versions using bipolar and MOS technologies are produced. The static type uses a latch cell whereas the dynamic type uses a device capacitance as the storage medium in which case the rate of circulation of data within the register loops must be greater than a lower threshold value to avoid loss of information. The charge-coupled device (CCD) described in section 4.7.4 is used in the form of a shift register as a memory. High-capacity memories (up to 64k) of this type are available, organised as a number of separate addressable loops (for example, 16 4k-loops). They operate at a clock frequency in the range 1–5 MHz using a two-phase clock and have a read–write cycle time of the order of 300 ns. Total operating power dissipation is typically 250 mW which reduces to below 50 mW during standby recirculation.

4.8.2 Programmable Logic Arrays

During the discussion of read-only memories in section 4.8.1 it was stated that a ROM can be considered as a code converter providing a certain binary output word in response to a certain binary input word which is in fact the memory address. In this way the ROM provides a logic function which can be used in place of a logic circuit implemented in conventional logic gates. If this principle is extended so that one ROM array is used to drive another on the same chip, complex logic functions can be implemented using fewer cells than would be required using a single ROM array. Such a combined array is mask-programmed during fabrication to provide a given logic function and is termed a *programmable logic array* (PLA). The two arrays are often described as the (logical) product or AND array and the (logical) sum or OR array.

The use of fusible-link programming technology in place of mask-programming during manufacture, as described in connection with PROMs in section 4.8.1, enables PLAs to be produced which can be programmed by the user. This type of array is referred to as a *field-programmable logic array* (FPLA). Apart from substantial economic advantages, the implementation of complex logic functions using PLAs can provide improvements in speed.

A related, though simpler structure described as a *field-programmable gate array* (FPGA) comprises a number of AND/NAND, OR/NOR gates and similarly uses fusible-link programming. They find application typically as random logic, address decoders and code detectors. The general name *integrated fuse logic* (IFL) has been used to describe these field-programmable arrays.

4.8.3 Microprocessor

In the context of digital electronics, a *processor* is a logic circuit which operates on binary input data according to a set of instructions described as a *program* and provides an output in binary form. The physical electronic circuit that performs the operation is referred to as *hardware* while the set of instructions (the program) which describes the operation to be performed is termed *software*. The 'library' of instructions that are logic functions, are stored in a ROM, the programmed ROM often being described as *firmware*, that is, a hardware unit containing software information. A *microprocessor* (μP), which has popularly become known as 'a micro' or '*the* chip', is a single integrated programmable logic circuit capable of operating on input data according to a set of instructions selected by the user.

An electronic system capable of processing information, whether numeric or not, is termed a *computer*. A computer consists of three basic parts: a *memory* to store the software instructions, the information or data to be processed and the result of the processing, a *central processing unit* (CPU) to implement the processing and *input/output* (I/O) hardware enabling the information to be input to and obtained from the system. A microprocessor is thus a single chip CPU and a computing system using a μP as its CPU is termed a *microcomputer* (μC). It must be emphasised that a μP cannot operate alone; it requires an external memory together with the software written into that memory and the software must be considered as an integral part of the system. Further, due to the progressive decrease of hardware costs and the escalation of the cost of software development which is labour-intensive, the software/hardware cost split for a μP-based system is typically 60 per cent/40 per cent. The economics of software development are therefore of vital importance.

From an electronic engineer's viewpoint, a μP is regarded as an electronic device which performs logical/arithmetic operations and can be included in an electronic system as a logic circuit or computing facility. The substantial difference between the μP and traditional electronic devices is that it is multioperational and the user has to set the operational sequence by developing suitable software. Users, therefore, have to be skilled in both hardware and software aspects and it is this combined requirement that may be responsible for the relatively slow exploitation of the enormous processing ability of the device. The μP was introduced in 1971 but it was not until the mid-1970s that its significance became apparent and since then its intrusion into our way of life has greatly overshadowed that of the transistor almost two decades before.

Figure 4.39 shows the basic *architecture* of a μP. The internal memory is a set of registers used to store data and to program instructions fetched from main memory. Typically they comprise

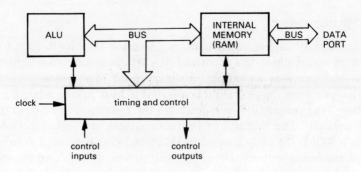

Figure 4.39 Microprocessor architecture

(1) *'scratch-pad' registers* to store data between stages in a computation;
(2) a *memory address register* to store the address of the memory location currently being accessed;
(3) a *program counter* to store the address of the next program instruction to be fetched from main memory;
(4) an *instruction register* to store the next instruction to be performed (executed).

The *arithmetic/logic unit* (ALU) performs all the arithmetic and logical processing on data words supplied from the scratch-pad internal registers, the result of the operation being returned to a scratch-pad register. The library of instructions includes such operations as arithmetic ADDITION of two words supplied from the internal memory and logical OR performed bit by bit on two data words.

Data and instruction words are conveyed between the functional blocks of the system via *buses,* often described as highways or channels. Each bus must provide one physical connection for each bit of the word, typically 4, 8 or 16, allowing parallel transfer. The flow of data and instruction words along a bus is controlled by enabling/inhibiting gates at either end of the bus.

The *timing and control unit* provides over-all control of the flow of data and synchronises operation using a clock frequency typically in the range 1–10 MHz. On command from the control unit, a program instruction is transferred from main memory (external to the μP) to the instruction register. This instruction word is decoded within the control unit to provide the necessary signals to implement the instruction. If appropriate, data words are transferred from the main memory via the internal data registers to the ALU which executes the instruction and the result is returned to a data register. The program counter is then updated and the system is then ready to deal with the next instruction. During implementation of a program, the control unit provides output signals indicating the *state* of the system and in

Integrated Circuit Technology

(a) Ferranti F100-L microprocessor prior to sealing with a metal lid

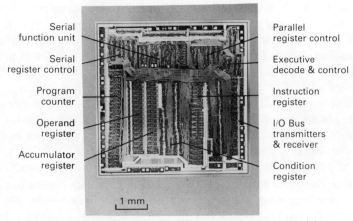

(b) Ferranti F100-L microprocessor chip (photograph reproduced by permission of Ferranti Electronics Limited)

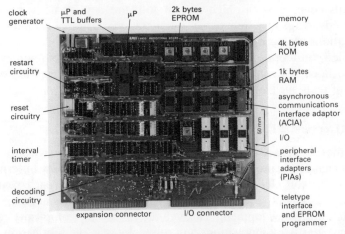

(c) Single-board microcomputer using an AMI 6800 microprocessor

Figure 4.40 Microprocessor

addition it is capable of receiving external control signals, such as an interrupt, to modify the sequence of operations.

Most μPs use NMOS technology although some are produced using CMOS and there are a few bipolar (Schottky TTL, ECL and I^2L) types. The original PMOS μPs of the early 1970s have largely been superseded by up-dated NMOS versions. The majority are available in 40-pin dual-in-line encapsulation. Figures 4.40a and b show the package and layout of the bipolar 16-bit Ferranti F100-L μP; the chip is approximately 6 mm square and contains approximately 9000 transistors. A single-board μC based on the 8-bit AMI 6800 NMOS μP is shown in figure 4.40c. First-generation types, which until the early 1980s have been used in the vast majority of applications, have an 8-bit word length. Compared with the speed and programming flexibility of mini and main-frame computers, 8-bit μPs are time-consuming to program and slow in operation although for many applications, particularly where the full processing ability of the μP is not used, this is not a significant problem. In the late 1970s two developments took place; the introduction of 16-bit μPs and bit-slice μPs.

The most important feature of the 16-bit second-generation 'super μPs' in comparison with earlier 8-bit types, apart from the greater computing power and faster operation, is the ease of programming which simplifies and hence reduces the cost of software development for the system.

Bit-slice μPs [12, 13] are basic 2, 4 or 8-bit processors each containing an ALU and scratch-pad registers and designed so that they can be parallelled via interconnecting buses to provide an n-bit system. Thus a 16-bit μP system can be formed from four 4-bit slices. The advantage of this facility is that the bit length can be chosen to suit the application and the system can be readily expanded at a later date should the need arise. Relating to previous comments concerning the high cost of software development, possibly the greatest advantage to system expansion in this way is that only minor software modifications are necessary for the expanded system.

Detailed consideration of the operation and application of microprocessors and associated ICs concerned with control and interfacing can be found in a variety of texts such as references 14 to 17.

4.8.4 Overview of LSI Development

During the first decade of IC development (1960–70) progressive improvement of the planar fabrication process enabled the chip complexity in terms of the number of circuit elements/chip approximately to double each year (figure 4.41). This exponential relation, known as Moore's law, has been widely used to predict future development. Through the 1970s, chip complexity continued to increase partly by innovations and refinements in the fabrication process, such as the use of ion-implantation, projection lithography and improved image alignment, and partly by the development of technologies such as I^2L

Integrated Circuit Technology

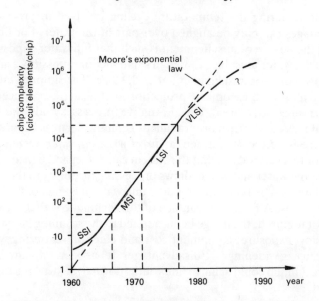

Figure 4.41 Development of the scale of integration

and VMOS which reduced the chip area required/element. By 1980, VLSI chips approximately 6 mm square could be produced supporting more than 10^5 elements such as the 64k dynamic RAM. The capability of an IC fabrication process as regards minimum element size is restricted by processing tolerances and is described in terms of the minimum *feature size* or minimum *line width* attainable while maintaining a reasonable yield and circuit reliability. Minimum feature sizes have been reduced from about 25 μm in 1960 to 3 μm in 1980. It will be noted from figure 4.41 that development during the late 1970s has seen a departure from Moore's law and the rate of development during the 1980s seems likely to reduce progressively.

Developments currently taking place in the lithographic process to improve resolution and to allow line widths to be reduced to 1 μm or less include mask production using electron-beam pattern generators in place of mechanical types, image projection using optical, electron-beam or X-ray methods in place of the standard contact method and a change from wet to dry etching as discussed in section 3.2. Such fabrication equipment is considerably more expensive than traditional equipment and the enormous capital outlay is bound to adversely affect the progressive reduction in function cost or bit cost.

A major difficulty in the continued reduction of element size is the effect known as *run-out* whereby the different thermal expansion coefficients of materials used in IC fabrication, notably silicon and silicon dioxide, cause the wafer to distort during the temperature cycling involved in processing and successive images can only be aligned over part of the wafer. For line widths below 2 μm the degree of misalignment is such that full wafer exposure is not possible and the wafer must be exposed in small sections, possibly a single chip at a time, correcting for misalignment at each stage. The simplest method for single chip exposure uses optical projection of the mask-produced image, stepped over the wafer area. An alternative process avoids the use of a conventional mask and uses a programmed scanning electron-beam system to expose the resist directly, termed a *direct-write-on-wafer* technique. The advantages of this method are that the system can detect and hence correct for distortions in the wafer and also it allows processing flexibility in the sense that different circuits can be fabricated on the same wafer leading possibly to larger integrated systems. A further advantage of this technique is that conventional masks are not required, the image being applied to the scanning beam in digital form, but, slow exposure, equipment cost and electron scattering within the resist layer causing widening, are disadvantages. It has been estimated however that such a technique could achieve line widths down to 0.1 μm in a production environment.

Another aspect of the size reduction (scaling) of electronic devices, notably BJTs and MOSTs, is their operation and characteristics. Prediction of the performance of devices with sub-micron feature sizes based on traditional atomic physics may not be at all reliable; new phenomena due to the extremely small size may cause these devices to perform unexpectedly. One particular uncertainty is the effect of non-uniform dopant distribution within a small region. Dopant densities used in devices are typically in the range 10^{20}–10^{28} atoms/m^3 but as regional volumes would be extremely small it follows that regions would contain only relatively small numbers of dopant atoms. The non-uniform distribution of these atoms within the region seems likely to have a major effect upon device performance. There is also uncertainty about the effect of small region sizes on carrier mobility. Another program concerns fine metallisation tracks in which an effect called *electromigration* can lead to failure. It has been known for some time that if the current density in a metal exceeds a certain threshold there is a progressive drift of metal atoms in the direction of the current. Due to imperfections, the cross-sectional area of metallised tracks varies along the length of the track causing the current density to vary. The current density is highest at a constriction and electromigration at such a point further reduces the metal cross-section leading to escalation and eventual failure. At normal operating current levels, electromigration does not present a problem with present track dimensions but it is likely to become a problem as the scale of integration progresses. References 18 and 19 are particularly useful on scaling problems and limitations.

In the quest for greater packing densities, manufacturers have scaled existing technologies within the limits imposed by the current fabrication process. An important example is the scaled NMOS technology referred to as HMOS (high-density short-channel MOS) which is now widely used in VLSICs. Whereas standard NMOS of the late 1970s provided a minimum power–delay product of the order of 5 pJ, HMOS allows operation below 2 pJ using a cell size which occupies less than half the chip area of the standard NMOS cell. The VMOS derivative of NMOS, although appearing initially to offer major performance and packing density improvements compared with lateral NMOS, is not cost-competitive with scaled NMOS. In the bipolar field, commercial exploitation of I^2L, and its derivatives such as ISL, is expected to develop considerably during the 1980s. Gallium arsenide Schottky-gate FETs (MESFETs) and e-JFETs described in section 4.2.3 are likely to become important in ultra high-speed logic applications, that is, digital circuits operating with clock frequencies above 1 GHz and/or with gate propagation delays below 1 ns. Depletion-type MESFETs using channels lengths down to 0.4 μm have been shown to provide the fastest switching times to date and their use in Schottky-diode FET logic (SDFL) may become the high-performance logic technology of the late 1980s. However, the mature LSI technology based on silicon is still likely to provide strong competition for gallium arsenide circuits with developments such as the use of oxygen-implantation to form silicon dioxide isolation zones providing further performance improvements in bipolar silicon ICs.

It seems certain that during the 1980s, line widths of the order of 1 μm will be used for VLSI production circuits such as memories and advanced processors. Some believe that a combination of economics and processing difficulties will cause the scale of integration to reach a plateau at about the 1 μm level corresponding to between 10^6 and 5×10^6 elements/6 mm square chip for production circuits and that finer widths down to 0.1 μm will be used only for applications demanding the ultimate in performance where cost is of secondary importance, such as military applications. Even at the 1 μm level it is interesting to consider the type of system which could take advantage of such enormous numbers of active devices. One possibility is the production of large-capacity dynamic RAMs with on-chip refresh facilities making them as easy to use as static RAMs. Predictions of such single-chip memories achieving a capacity of 4 Mbit by the late 1980s have been made. While it is confidently predicted that sales of 16-bit μPs will overtake those of 8-bit types in the early 1980s, 32-bit μPs have already become available. Single-chip 8-bit microcomputers have been available for some time using PMOS, NMOS and CMOS technologies. Typical products offer up to 16 kbit (2 kbyte) instruction ROM and up to 512 bit data RAM. Progressive increase of the on-chip memory capacity seems certain in the future while another likely development is a single-chip microcomputer incorporating a 16-bit microprocessor, which would utilise of the order of 10^4 elements, together with high-capacity main memory for data and software storage. A development that is approaching the production stage is the reduction of memory power dissipation by incorporation of on-chip power control circuitry to switch memory cells to standby when they are not being accessed. Another area of interest is the development of a true non-volatile RAM, that is, one that does not require battery back-up to provide non-volatility. A novel mid-range development in this direction is a static RAM incorporating a 'shadow' non-volatile cell for each RAM cell. A single control signal allows data currently stored in the RAM to be transferred to non-volatile storage. The disadvantage of this duplicated cell configuration is the reduced memory capacity possible for a certain chip size. Another important area of development is in EEPROMs which offer considerable performance improvements compared with the older EAROMs and are likely to replace UV EPROMs. If the speed of EEPROMs can be improved and their endurance (allowable read–write cycles) increased, they could provide a general-purpose memory capable of both RAM and ROM application. A current research area that could have far-reaching implications in the future is the use of electrically alterable structures in programmable logic arrays. Such a development would enable the logic functions performed by PLA hardware to be modified electrically.

A problem linked to increased complexity is the generation of 'soft errors' in high-capacity dynamic RAMs due to radiation. The progressive scaling of the memory cell has meant that the capacitance of the bit store and hence the charge representing the bit has been reduced. The level has now been reached where an α-particle emitted during the decay of naturally occurring radioactive material in the vicinity of the memory can cause a bit error, termed a *soft error*, in the memory. One source of minute quantities of radioactive material (for example, uranium and thorium) is the circuit packaging materials and techniques are being introduced to reduce these radioactive traces to acceptably low levels. Another method under consideration is the modification of the cell structure to protect the storage region from α-particles such as the use of a polyimide coating on the chip surface. Increased circuit complexity also introduces problems in testing to check that the IC is functional. Even at present complexities it is not feasible to test completely sequential ICs, where the output depends not only on present inputs but also on the internal state of the circuit, due to the enormous number of possible combinations. With this in mind, designers have already begun to consider the *testability* of a system as being a fundamentally important design factor.

4.9 FILM CIRCUITS

A film circuit is formed by depositing tracks on an insulating ceramic or glass substrate. There are two types termed *thin* and *thick* describing the relative thickness of the deposited film; they are produced by totally different processes. Basic film circuits are usually passive, consisting of conducting tracks, resistive tracks and deposited capacitor structures although in the majority of applications monolithic active devices and/or ICs are added to the basic circuit forming a *hybrid IC*.

4.9.1 Thin-film ICs

The thin-film process involves the deposition of metal and metal-compound tracks typically 0.01–1 μm thick using either vacuum deposition or RF sputtering. In the former process, the substrate is positioned in an evacuated enclosure containing a pellet of the material to be deposited. The pellet is vaporised by electrical heating and the vapour is deposited on the substrate which is masked as required by either a metal foil stencil or photolithographic techniques similar to those used in the fabrication of monolithic circuits (section 3.1.4). The RF sputtering method uses an anode–cathode arrangement in a low-pressure inert gas atmosphere. The cathode is made of the material to be deposited and the masked substrate is mounted on the anode. A high RF voltage is applied between anode and cathode causing the gas to ionise and the resulting cathode bombardment results in atoms of cathode material being emitted and deposited on the substrate.

Conducting interconnection tracks and the lower plates of capacitors are deposited first using aluminium. Resistor tracks are then formed, the resistance values being determined by the sheet resistance of the deposited film and the aspect ratios of the tracks (equation 4.3). Various compounds and metal alloys are used to form resistor tracks such as tin oxide and nichrome, giving sheet resistances in the range 10 Ω/square to 25 kΩ/square. Resistance values from 10 Ω to 1 MΩ can be obtained with a tolerance of 5–20 per cent, the tracks of high-value resistors being folded (figure 4.42a). Although the deposition tolerance is high, resistance values can be adjusted to within one per cent by the use of laser-trimming. The technique is to deposit resistor tracks with a value of resistance lower than required then, by trimming the track by slot cutting or narrowing (figure 4.42a) using a laser while monitoring the resistance value, a close tolerance can be obtained.

Thin-film capacitors are formed by constructing a conductor–dielectric–conductor sandwich. The lower plate is formed at the same time as interconnection tracks and a thin (≈ 0.1 μm) layer of dielectric such as aluminium oxide or hafnium dioxide is then deposited over the lower plate. An aluminium upper plate is deposited on top of the dielectric layer forming a parallel-plate structure. Dielectrics used in this application have relative permittivities up to about 100, enabling capacitance values approaching 10 000 pF/mm^2 to be obtained. The upper plate often incorporates trimming 'fingers' (figure 4.42b) which may be severed from the main plate using a laser to adjust the capacitance value. Since the plates of such capacitors are thin

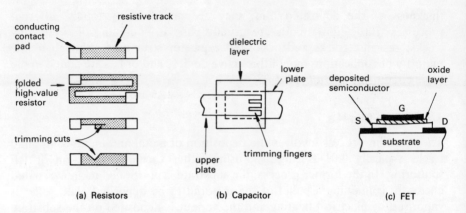

Figure 4.42 Thin-film circuit elements

they introduce series resistance which causes the Q-factor of the device to be low, typically less than 200. Compared with monolithic resistors and capacitors, thin-film elements have the advantages of excellent isolation and closer value tolerance.

Semiconductor junction devices cannot be fabricated using thin-film technology but FET structures such as that shown in figure 4.42c have been produced. Metallised source and drain contacts are first formed on the substrate over which a thin ($0.05-1$ μm) semiconducting layer of cadmium sulphide, cadmium selenide or tellurium is deposited. The surface of the semiconductor is then covered with an insulating layer prior to the deposition of an aluminium gate electrode producing an MOS structure. Channel lengths of 10 μm have been achieved giving good high-frequency performance even though the semiconductor channel region is deposited, not grown, which results in a polycrystalline structure having reduced carrier mobility. Thin-film FETs have not developed significantly due to the economic competition from monolithic active devices which were established first and can be added to a passive thin-film circuit forming a hybrid.

The main applications of thin-film circuits have been as precision resistor networks for measurement equipment, attenuators, decoding circuitry and digital-to-analog converters (DACs) and as the basis of hybrid circuits such as power amplifiers (figure 4.44). Recently, however, thin-film circuits consisting simply of a conductor pattern have found a major application in the interconnection of monolithic LSI chips into electronic systems.

4.9.2 Thick-film ICs

The thick-film fabrication process involves the screen printing of conducting and resistive tracks on to an insulating substrate using conducting inks. The

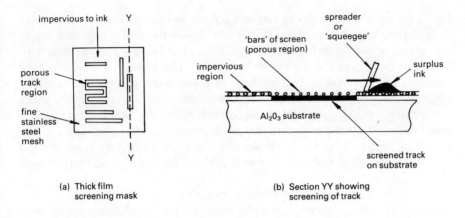

Figure 4.43 Thick-film process

pattern is subsequently fired to dry the ink and to fuse the tracks to the substrate forming an extremely rigid structure. The inks are mixtures of metal and glass particles in an organic 'binder' with a solvent to control the viscosity. The density of metal particles determines the sheet resistance of the resulting film which is typically 10–30 μm thick. The most common substrate material is aluminium oxide containing a small proportion of glass to assist binding of the surface film. Aluminium oxide is popular because it is a good electrical insulator and a good thermal conductor.

The screen is a fine stainless steel mesh with a coating making it impervious to the ink. Using standard photolithographic techniques, areas of the impervious layer are removed corresponding to the track pattern required (figure 4.43a). The screen is positioned over the substrate and a mechanically driven squeegee forces ink through the porous regions forming the tracks on the substrate which is then fired at about 900 °C before the next set of tracks can be printed.

In the first stage of the process, the contact pads and interconnection tracks are formed using conducting inks normally containing powdered mixtures of precious metals (gold, silver and platinum) enabling sheet resistances of less than 0.01 Ω/square to be obtained. After firing, the process is repeated to form the resistor tracks using inks containing metal oxides (for example, palladium silver/palladium oxide) which provide sheet resistances in the range 10 Ω/square to 1 MΩ/square. Close control of the firing temperature is important not only to obtain the sheet resistance required but also to ensure a low temperature coefficient and good stability. The sheet resistance of the film and hence resistance values are inversely proportional to the film thickness (equation 4.2). For example a 1 μm variation in a typical 10 μm thick resistive film would cause a variation in resistance values of about 10 per cent. Due to this sensitivity of resistance value to film thickness, it is usual to 'target' resistance values about 20 per cent below the values required and to use abrasion methods (bombardment with aluminium oxide particles) or laser-trimming to adjust the resistance values. This enables a final tolerance of about one per cent to be obtained. In some cases, particularly in digital applications, multilayer interconnection patterns are necessary using intervening insulating layers of deposited glass. As in other multilayer structures, the conducting patterns are interconnected via windows in the insulating layers. Where a multilayer interconnection structure is required, it is formed before the resistive tracks are added, firing being necessary at each stage.

Although thick-film parallel-plate capacitors can be formed using a screened dielectric layer sandwiched between two conducting layers, it is generally simpler and more economic to bond miniature discrete 'chip' capacitors to contact pads formed in the interconnection pattern. Similarly thick-film FETs can be fabricated having a similar structure to thin-film types (figure 4.42c) although they are relatively expensive and cannot match the performance of monolithic silicon devices. Where active devices are required, therefore, specially packaged 'microminiature' monolithic devices are bonded to the interconnection pattern forming a hybrid circuit.

Thick-film circuits are encapsulated in a variety of package styles, [20] depending on the working environment and the size of the substrate, which have side dimensions typically in the range 1–50 mm and are between 0.5 and 1 mm thick. Since the circuit elements are fused to the substrate, thick-film circuits are extremely robust and find application particularly in hostile environments such as in aircraft and road vehicles where mechanical vibration, wide temperature ranges and humid conditions impose severe stresses. Completed substrates are given a moisture-resistant coating and sealed in a metal package using epoxy resin. For less extreme conditions, thick-film technology provides an economical fabrication method particularly where the precision of thin-film techniques is not required or for relatively small volume production. As with thin-film conductor patterns, thick-film technology is also being used to provide an economic method of connecting LSICs into electronic systems.

4.9.3 Hybrid ICs

Hybrid ICs [21] are thin or thick-film circuits of conductors, resistors and possibly capacitors with the addition of monolithic active devices and/or ICs which are bonded to the conductor pattern. Microminiature transistors, particularly BJTs intended for operation up to UHF, are produced in rectangular packages as small as 3 mm × 1.5 mm × 1 mm with connection 'tabs' specially designed for incorporation in hybrid circuits. To avoid wastage of space by unnecessary encapsulation and bonding wires, some monolithic

Figure 4.44 Hybrid 50 W audio amplifier

ICs designed for use in hybrid circuits are produced unencapsulated although surface passivation is necessary to protect the circuit. *Beam-lead* technology (section 4.1.3), in which the circuit terminals extend laterally from the sides of the rectangular circuit, is used in this application. Another type of structure that is used, known as a *flip-chip* [21] or *leadless inverted device* (LID), has the metallisation on the circuit bonding pads built up to a thickness of about 50 μm. The chip is then inverted and bonded directly to the conductor pattern on the substrate. These techniques are particularly useful for HF circuits where it is important to minimise stray capacitance but the flip-chip method has the disadvantage that since the bonding joints are under the chip they cannot easily be inspected.

It is usually more economic to use miniature discrete capacitors in hybrid circuits instead of film structures. Leadless multilayer ceramic chip capacitors are produced as small rectangular blocks with metallised terminations of palladium–silver. Capacitance values up to about 0.5 μF ± 20 per cent are available in a block size 5 mm × 4 mm × 2 mm having a voltage rating of typically 50 V. Device-bonding using conducting epoxy resin is widely used and has the advantage of being a low-temperature process compared to metal-bonding reducing the stress on components during fabrication. Where necessary, thermocompression wire-bonding is used linking chip contact pads with the conductor pattern. Circuits intended for use in hostile environments are given a moisture-resistant coating prior to final packaging.

Figure 4.44 shows a commercial 50 W audio amplifier produced using thin-film hybrid technology. The main part of the circuit is formed on a single substrate measuring 35 mm × 25 mm while the high-power output transistors are mounted separately on heat sinks. Low-power monolithic BJT chips, ceramic chip capacitors and trimmed thin-film resistors can be readily identified.

Hybrid technology has three main application areas: high-power and high-frequency linear circuits such as amplifiers and oscillators and multi-chip digital systems. The technology offers good heat-sinking for high-power applications and good device isolation/low stray capacitance for high-frequency circuits. In addition, passive component tolerance is far better than for monolithic technology, a feature which is more significant for linear than for digital applications. Hybrid technology also provides a cost-effective method of incorporating monolithic LSICs into electronic systems, certain digital sub-systems being conveniently produced as multi-chip hybrid circuits.

REFERENCES AND FURTHER READING

1. D.J. Hamilton and W.G. Howard, *Basic Integrated Circuit Engineering*, chapter 7 (McGraw-Hill, 1975)
2. A.B. Glaser and G.E. Subak-Sharpe, *Integrated Circuit Engineering*, chapter 4 (Addison-Wesley, 1977)
3. D.J. Hamilton and W.G. Howard, *Basic Integrated Circuit Engineering*, chapter 4 (McGraw-Hill, 1975)
4. J. Millman and C.C. Halkias, *Integrated Electronics*, chapter 15 (McGraw-Hill, 1972)
5. A.B. Glaser and G.E. Subak-Sharpe, *Integrated Circuit Engineering*, chapter 7 (Addison-Wesley, 1977)
6. C.A. Holt, *Electronic Circuits*, chapter 8 (Wiley, 1978)
7. G.S. Hobson, *Charge-Transfer Devices* (Edward Arnold, 1978)
8. M.J. Howes and D.V. Morgan (Eds): *Charge-coupled Devices and Systems* (Wiley, 1979)
9. O. Tomisawa, Y. Horiba, S. Kato, K. Murakami, A. Yasuoka and T. Nakano. Vertical injection logic, *IEEE J. Solid-State Circuits*, **SC-11** (1976) 637–43
10. S.Y. Yu, New V-groove integrated injection logic, *Electron. Lett.*, **13** (1977) 382
11. C.A. Holt, *Electronic Circuits*, chapter 10 (Wiley, 1978)
12. D.J. Hird and D.M. Elliot, Bit-slice microprocessors, *Electronics and Power*, **25** (1979) 179–84
13. D.D. Givone and R.P. Roesser, *Microprocessors/Microcomputers: An Introduction*, chapter 9 (McGraw-Hill, 1980)
14. M. Healey, *Minicomputers and Microprocessors* (Hodder and Stoughton, 1976)
15. D.D. Givone and R.P. Roesser: *Microprocessors/Microcomputers: An Introduction* (McGraw-Hill, 1980)
16. R.J. Tocci and L.P. Laskowski, *Microprocessors and Microcomputers: Hardware and Software* (Prentice-Hall, 1979)
17. A. Veronis, *Microprocessors: Design and Applications* (Reston/Prentice-Hall, 1978)
18. B.L.H. Wilson, Limits to the size of semiconductor devices, *Electronics and Power*, **26** (1980) 704–8
19. C. Mead and L. Conway, *Introduction to VLSI Systems*, chapters 1, 2 and 4 (Addison-Wesley, 1980)

Integrated Circuit Technology

20. J. Hodgkinson, Thick-film circuits for high-reliability applications, *Electronics and Power*, **26** (1980) 709–11
21. D. Roddy, *Introduction to Microelectronics*, 2nd edition, chapter 10 (Pergamon, 1978)
22. C.S. Meyer, D.K. Lynn and D.J. Hamilton, *Analysis and Design of Integrated Circuits* (McGraw-Hill, 1968)
23. J. Allison, *Electronic Integrated Circuits* (McGraw-Hill, 1975)
24. D.H. Navon, *Electronic Materials and Devices*, chapter 11 (Houghton Mifflin, 1975)
25. B.G. Streetman, *Solid State Electronic Devices*, 2nd edition (Prentice-Hall, 1980)
26. G.J. Deboo and C.N. Burrous, *Integrated Circuits and Semiconductor Devices*, 2nd edition, chapters 3 and 5 (McGraw-Hill, 1977)
27. J. Millman, *Microelectronics*, chapters 4, 5, 6 and 9 (McGraw-Hill, 1979)

TUTORIAL QUESTIONS

4.1 A diffused resistor in an integrated circuit is formed by a shallow p-type diffused track in an n-type isolation island. Tests show that the track, which is 100 μm long and 10 μm wide, has a resistance of 2 kΩ. What is the sheet resistance of the diffused track? Mention any assumptions made. If a 20 kΩ resistor is to be fabricated at the same time as the 2 kΩ resistor described above and the permissible range of track widths is 5–25 μm, what is the minimum track length required? Comment on the value obtained and the likely practical layout of this resistor.
(Answers: 200 Ω/sq, 500 μm folded)

4.2 A diffused resistor is formed in an integrated circuit using a constant-source diffusion. The doping profile may be assumed to have the complementary error function form given by equation 3.3. The diffused track is 2.5 μm deep and the dopant concentration of the surrounding isolation island is one-thousandth of the source concentration. Using the information given in figure 3.6, construct the dopant profile of the resistor diffusion. Use a step approximation of the profile to represent the diffused track by five layers of equal thickness each of constant dopant concentration equal to the mean concentration of the layer. Hence produce of simple network model of the diffused resistor and estimate what proportion of the total current flows within the surface layer. Mention any approximations or assumptions made in the solution. List what additional information would be needed to enable the actual value of resistance to be calculated.
(Answers: Model resistance ratios 1.54: 3.33: 12.5: 50: 333. Length and width of diffused region assumed not to vary with depth and minority carrier component assumed negligible. Majority carrier mobility, dopant source concentration, electronic charge, length and width of diffused track.)

4.3 The depletion capacitance of the substrate–epitaxial island junction is used as a capacitor in a monolithic integrated circuit employing basic diode isolation. If the junction has a capacitance of 100 pF/mm^2, the epitaxial layer is 10 μm thick and the surface geometry of the island is a rectangle measuring 400 μm × 200 μm, calculate the capacitance mentioning any approximations made.
(Answer: 9.2 pF. Assuming that sides of isolation diffusion are perpendicular to chip surface.)

4.4 A monolithic capacitor is to be formed by the reverse-biased junction between a p-type diffused region and an n-type epitaxial island. The junction capacitance per unit area is given by

$$C = \sqrt{\left[\frac{\epsilon e N}{2(\psi - V)}\right]}$$

where the contact potential ψ is 0.7 V and N is the dopant concentration on the low-conductivity side of the junction. If $C = 100$ pF/mm^2 at a reverse bias of 1 V and the required value of capacitance is 20 pF at a bias of 5 V, calculate the size of the square mask window required for the 5 μm-deep p-type diffusion. State any simplifying assumptions used.
(Answer: 580 μm side. Assuming that sides of isolation diffusion are perpendicular to chip surface and that lateral and vertical diffusion rates are equal.)

4.5 The following data applies to the fabrication process for a diode-isolated bipolar integrated circuit
  BJT collector island dimensions: 50 μm × 40 μm
  Diode island dimensions: 25 μm × 20 μm
  Sheet resistance of resistor diffusion: 100 Ω/sq
  Minimum diffused resistor track width: 10 μm
  Resistor contact pad dimensions: 15 μm × 15 μm
  Substrate–epitaxial island junction capacitance: 100 pF/mm^2
  Minimum isolation diffusion width: 10 μm
  Peripheral contact pad dimensions: 100 μm × 100 μm.
A bipolar IC is to be produced using this process comprising 40 BJTs, five diodes, one 10 kΩ resistor, four 1 kΩ resistors, one 500 Ω resistor and one 10 pF capacitor. Estimate the absolute minimum *active* chip area. If the circuit requires eight connections to the package, estimate the minimum size of square chip required to accommodate the circuit.
(Answers: 0.209 mm^2, 637 μm side)

4.6 Assuming that a p-type substrate with an n-type epitaxial layer is available, produce a possible chip layout for a single three-input standard TTL NAND gate (figure 4.29b). Sketch a *complete* set of mask patterns necessary to fabricate the layout design, assuming that basic diode isolation is to be used.

Appendix A Terminal Parameter Modelling of Device Characteristics

In general, the current–voltage characteristics of devices are non-linear causing a major complication in the analysis of electronic circuits. A convenient technique is to approximate the real device characteristic by that of a hypothetical linear network. The parameters of the network then approximately represent the terminal properties of the device.

A.1 PIECEWISE-LINEAR MODELS

A piecewise-linear (PWL) model is a hypothetical network representing the performance of a device over a *wide* range by approximating the real characteristic by a linearised characteristic. Each linearised *segment* approximates the variation of current with respect to voltage over a limited range by a constant resistance. The change of gradient as the operating point crosses the *breakpoint* from one segment to the next is represented in the model by the switching action of a hypothetical voltage-controlled ideal switch, resistances being switched in parallel according to the applied voltage, thus modifying the effective resistance of the device. The hypothetical switches are represented by the unblanked diode symbol (figure A.1a) and must not be confused with a real diode as represented by the blanked symbol (figure 2.1b). The hypothetical switch is a perfect short-circuit if the voltage across the switch V_S is such that the current I_S through it is positive or is a perfect open-circuit if V_S is negative.

There are two basic types of non-linear I–V characteristic; those having an increasing gradient (decreasing *slope resistance*) as the applied voltage increases (figure A.1b), and the saturating-type of characteristic (figure A.2a) having an increasing slope resistance with increasing voltage.

The PWL model corresponding to the two-segment linearised characteristic of figure A.1b is shown in figure A.1c. For $0 \leq V \leq V_1$, the ideal switch S_1 is

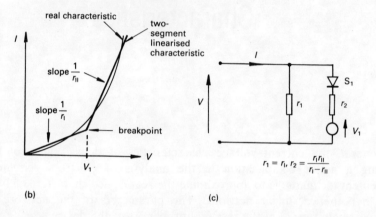

Figure A.1 PWL modelling of an I–V characteristic with increasing gradient

open-circuit and the change of current with voltage over this range, as represented by the slope resistance r_I of the first segment, is represented by resistance r_1 in the model, thus $r_1 = r_I$. For $V > V_1$, S_1 is short-circuit and the current drawn by the model depends on both r_1 and r_2. As far as the *change* of current with voltage is concerned, r_1 is effectively in parallel with r_2 as V_1 is constant and therefore $r_{II} = r_1//r_2$ (where // means 'in parallel with') from which $r_2 = r_1 r_{II}/(r_I - r_{II})$. Note $r_I > r_{II}$.

Figure A.2 shows a two-segment PWL approximation of a saturating I–V characteristic and the corresponding network model. As the slope resistance is *increased* as the applied voltage increases beyond the breakpoint, the technique used in figure A.1 of switching two resistances in parallel at the breakpoint cannot be used, as the combined resistance of two positive resistances in parallel is always less than the resistance of the individuals. Instead, the complementary technique is used whereby two resistances in parallel (r_3 and r_4, figure A.2b) are used to represent the first segment of slope resistance r_{IV} (figure A.2a) and at the breakpoint, switch S_2 becomes an open-circuit so that the current through r_4 can no longer increase and the change of current with voltage above the breakpoint is then described by r_3. Thus $r_3 = r_{III}$ and $r_{IV} = r_3//r_4$ from which $r_4 = r_{III} r_{IV}/(r_{III} - r_{IV})$. Note $r_{III} > r_{IV}$. For voltages below the breakpoint, I_4 is less than the constant source value I_1

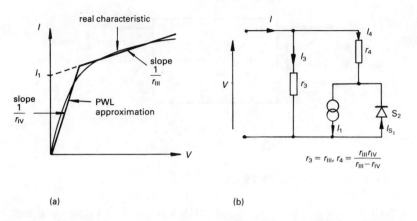

Figure A.2 PWL modelling of an I–V characteristic with decreasing gradient

and so I_{S_2} is positive $(I_1 - I_4)$ and S_2 is therefore short-circuit, connecting r_3 and r_4 in parallel. As V increases, I_4 increases and at the breakpoint $I_4 = I_1$, whereby I_{S_2} is zero. Increase in V above the breakpoint cannot be accompanied by an increase in I_4 as I_1 is constant and I_{S_2} cannot be negative. Thus, as far as *change* in current with voltage is concerned, r_4 has no effect above the breakpoint. Source current I_1 may be viewed as a bias, keeping switch S_2 in the short-circuit state, thus allowing I_4 to pass in the 'reverse' direction, the net current through S_2 then being $(I_1 - I_4)$ in the 'forward' direction. When I_4 reaches the bias value I_1 the switch reverts to its open-circuit state.

The accuracy of a PWL model in representing the real characteristic of a device can be improved by increasing the number of segments although this increases the number of branches in the model and hence complicates analysis using the model.

In circuit analysis using this type of model, the first step is to ascertain the state of each switch, that is, whether short-circuit or open-circuit, which depends on the operating conditions of the device in the real circuit. Once this has been done, the model reduces to a linear network and standard analysis techniques (for example, reduction, loop, mesh) can be used to obtain solution. When the model incorporates more than one switch, it is often not possible to specify the state of some of the switches initially as this requires knowledge of the device operating point which is the object of the analysis. In such cases it is necessary to assume a state for each switch, and after solution of the network based on these assumptions, to check that they are consistent with the solution, if not, the assumption must be altered and the solution repeated until consistency is obtained.

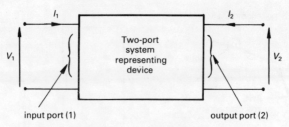

Figure A.3 Two-port representation

A.2 TWO-PORT PARAMETERS

The performance of a three-terminal device over a *narrow* range of operation (that is, applicable to small-signal operation) can be conveniently represented by a set of *parameters* obtained by considering the device as a two-port system (figure A.3) and by establishing mathematical relationships between the input and output variables. The variables V_1, I_1, V_2 and I_2 are interrelated, the actual relations being dependent on the properties of the two-port system. Any two of the four variables can be chosen as the *independent* variables and expressions developed for the other two variables, the *dependent* variables, in terms of them. For example, choosing the input current I_1 and the output voltage V_2 as the independent variables, the changes in input voltage and output current ΔV_1 and ΔI_2 due to changes ΔI_1 and ΔV_2 can be expressed as

$$\Delta V_1 = \left[\frac{\partial V_1}{\partial I_1}\right]_{\Delta V_2=0} \times \Delta I_1 + \left[\frac{\partial V_1}{\partial V_2}\right]_{\Delta I_1=0} \times \Delta V_2 \tag{A.1}$$

$$\Delta I_2 = \left[\frac{\partial I_2}{\partial I_1}\right]_{\Delta V_2=0} \times \Delta I_1 + \left[\frac{\partial I_2}{\partial V_2}\right]_{\Delta I_1=0} \times \Delta V_2 \tag{A.2}$$

The bracketed terms describe the small-signal properties of the two-port system at the particular quiescent operating point and are termed the *terminal parameters* of the system. For this particular choice of independent variables (I_1 and V_2), the units of the four parameters are mixed

$\left[\dfrac{\partial V_1}{\partial I_1}\right]_{\Delta V_2=0}$ is a slope impedance

$\left[\dfrac{\partial V_1}{\partial V_2}\right]_{\Delta I_1=0}$ is a voltage ratio

$\left[\dfrac{\partial I_2}{\partial I_1}\right]_{\Delta V_2=0}$ is a current gain

$\left[\dfrac{\partial I_2}{\partial V_2}\right]_{\Delta I_1=0}$ is a slope admittance

Appendix A

Being 'mixed' parameters they are termed the *hybrid* or '*h*' *parameters* of the system and are given the symbols h_{11}, h_{12}, h_{21} and h_{22} respectively, the suffixes being derived from the suffixes of the variables in each bracketed term.

As ΔV_1, ΔI_1, ΔV_2 and ΔI_2 are the changes in V_1, I_1, V_2 and I_2 respectively, they are the *signal* components v_1, i_1, v_2 and i_2 and equations A.1 and A.2 may therefore be written

$$v_1 = h_{11} i_1 + h_{12} v_2 \qquad (A.3)$$

$$i_2 = h_{21} i_1 + h_{22} v_2 \qquad (A.4)$$

or in matrix form

$$\begin{bmatrix} v_1 \\ i_2 \end{bmatrix} = \begin{bmatrix} h_{11} h_{12} \\ h_{21} h_{22} \end{bmatrix} \begin{bmatrix} i_1 \\ v_2 \end{bmatrix} \qquad (A.5)$$

the corresponding network model being as shown in figure A.4a.

If the input and output voltages are chosen as the independent variables, the small-signal equations are

$$\Delta I_1 = \left[\frac{\partial I_1}{\partial V_1}\right]_{\Delta V_2 = 0} \times \Delta V_1 + \left[\frac{\partial I_1}{\partial V_2}\right]_{\Delta V_1 = 0} \times \Delta V_2 \qquad (A.6)$$

$$\Delta I_2 = \left[\frac{\partial I_2}{\partial V_1}\right]_{\Delta V_2 = 0} \times \Delta V_1 + \left[\frac{\partial I_2}{\partial V_2}\right]_{\Delta V_1 = 0} \times \Delta V_2 \qquad (A.7)$$

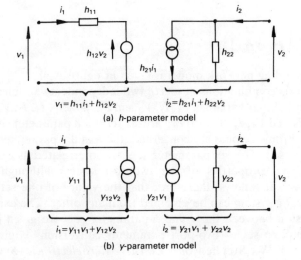

Figure A.4 Two-port parameter models

Table A.1 Admittance — hybrid parameter conversions

$$y_{11} = \frac{1}{h_{11}} \qquad\qquad h_{11} = \frac{1}{y_{11}}$$

$$y_{12} = \frac{-h_{12}}{h_{11}} \qquad\qquad h_{12} = \frac{-y_{12}}{y_{11}}$$

$$y_{21} = \frac{h_{21}}{h_{11}} \qquad\qquad h_{21} = \frac{y_{21}}{y_{11}}$$

$$y_{22} = \frac{|h|}{h_{11}} \qquad\qquad h_{22} = \frac{|y|}{y_{11}}$$

$|\,|$ refers to the matrix determinant, for example, $|h| = h_{11}h_{22} - h_{12}h_{21}$

The four terminal parameters are all slope admittances, given the symbols y_{11}, y_{12}, y_{21} and y_{22} respectively, and the resulting model is termed the y-parameter model. Equations A.6 and A.7 can be written

$$i_1 = y_{11}v_1 + y_{12}v_2 \qquad (A.8)$$

$$i_2 = y_{21}v_1 + y_{22}v_2 \qquad (A.9)$$

or

$$\begin{bmatrix} i_1 \\ i_2 \end{bmatrix} = \begin{bmatrix} y_{11} & y_{12} \\ y_{21} & y_{22} \end{bmatrix} \begin{bmatrix} v_1 \\ v_2 \end{bmatrix} \qquad (A.10)$$

The corresponding network model is given in figure A.4b.

There are six combinations of sets of two independent variables that can be chosen from the four variables V_1, I_1, V_2 and I_2, namely, I_1, I_2; V_1, V_2; I_1, V_2; V_1, I_2; V_2, I_2 and V_1, I_1. Thus there are six terminal parameter sets although, so far as electronic devices are concerned, the h and y-parameter sets are the most widely used. The six parameter sets are interrelated as each describes the small-signal properties of the two-port system although defined in different ways. It follows, therefore, that the values of one set of terminal parameters of a system can be obtained from any other. The expressions for interconversion between the y and h-parameter sets are given in table A.1 while a complete set of two-port parameter conversions is given in A.G. Martin and F.W. Stephenson, *Linear Microelectronic Systems,* p. 8 (Macmillan, 1973).

It should be noted that in use in electronics, the numerical suffixes 11, 12, 21 and 22 are usually replaced by the letters i, r, f and o respectively corresponding to the terms input, reverse, forward and output describing the physical meaning of the parameters.

Appendix B Nomenclature and Terminology

B.1 SYMBOLS, ABBREVIATIONS AND ACRONYMS

Å	angstrom unit ($= 10^{-10}$m)
A	area (m^2)
A	device anode (positive with respect to cathode when device is conducting)
a.c.	alternating current
ACIA	asynchronous communications interface adapter
ADC	analog-to-digital converter
Al$_2$O$_3$	aluminium oxide
ALU	arithmetic/logic unit (section of a CPU)
A_p	power gain
APD	avalanche photodiode
ASCII	American Standard Code for Information Interchange (character set)
ATT	avalanche transit-time device
av	average
A_v	voltage gain
b	width (m); small-signal susceptance (S)
b	(suffix) CB parameter of a BJT
B	base terminal or region of a BJT
B	base transport factor; magnetic flux density (T); bandwidth (Hz)
B'	hypothetical point of current summation within the base of a BJT
BARRITT	barrier transit-time (diode)
BBD	bucket-brigade device
B-C	base–collector port (junction) of a BJT
BCD	binary-coded decimal

B-E	base–emitter port (junction) of a BJT
bit	binary digit
BJT	bipolar junction transistor
BS(I)	British Standards (Institution)
BUS	data communication channel
c	velocity of propagation of electromagnetic radiation in free space ($= 3 \times 10^8$ m/s)
c	(suffix) CC parameter of a BJT
C	collector terminal or region of a BJT
C	capacitance (F)
CATT	controlled-avalanche transit-time triode
$C_{b'c}$	base–collector capacitance (HF Early and hybrid-models) (F)
$C_{b'e}$	base–emitter capacitance (HF Early and hybrid-models) (F)
CB	conduction band, common base connection of a BJT
C-B	collector–base port (junction) of a BJT
CC	common collector connection of a BJT
CCD	charge-coupled device
C_D	diffusion capacitance of a junction (F)
C_{D_c}, C_{D_e}	collector and emitter diffusion capacitances (F)
CDI	collector diffused isolation
CE	common-emitter connection of a BJT
C-E	collector–emitter port of a BJT
C_{gs}, C_{gd}, C_{ds}	capacitances of a FET (F)
C_{in}	input capacitance (F)
$C_{iss}, C_{oss}, C_{rss}$	common-source input, output and reverse transfer capacitances of a FET (F)
CMOS (COSMOS)	complementary (p and n-channel) MOS technology
CPU	central processing unit (of a computing system)
CRT	cathode-ray tube
C_s	shunt capacitance (F)
C_T	depletion (transition) layer capacitance of a junction (F)
C_{T_c}, C_{T_e}	collector and emitter transition (depletion) layer capacitances (F)
CTD	charge-transfer device
d	depletion layer width (m); channel depth of a d-MOST (m); duty cycle ($=t_p/T$)
d–	depletion-type (FET)
D	diffusion coefficient (m^2/s); electric flux density (C/m^2); drain terminal or region of a FET
D_1, D_2	diodes
DAC	digital-to-analog converter
dB	decibel (unit of power ratio) = 10 log P_o/P_i (= 20 log V_o/V_i for symmetrical resistive systems)

428   Semiconductor Device Technology

d.c.	direct current
DIC	dual-in-line ceramic package
DIL	dual-in-line style package
DIP	dual-in-line plastic package
DMOS	double-diffused MOS structure
d_n	width of depletion layer on n-side of junction (m)
D_n	diffusion coefficient for electrons (m^2/s)
DO	diode (package) outline
d_p	width of depletion layer on p-side of junction (m)
D_p	diffusion coefficient for holes (m^2/s)
DTL	diode–transistor logic
e	magnitude of electronic charge (= 1.6×10^{-19} C)
e	(suffix) CE parameter of a BJT
e–	enhancement-type (FET)
E	electric field strength (= negative potential gradient $-\,\mathrm{d}V/\mathrm{d}x$, also = force on a charge of + 1 C) (V/m); illumination (lx)
E	emitter terminal or region of a BJT
EAROM	electrically alterable ROM
E-B	emitter–base port (junction) of a BJT
E-C	emitter–collector port of a BJT
ECL	emitter-coupled logic
EEPROM	electrically erasable PROM
E_H	Hall field (V/m)
EIA	Electronics Industries Association
E_J	electric field at the junction (V/m)
E_n	electric field in depletion region on n-side of junction (V/m)
E_p	electric field in depletion region on p-side of junction (V/m)
epi-	epitaxial (layer)
EPROM	erasable PROM
erfc	complementary error function
eV	electronvolt (unit of energy defined by the change in energy of an electron passing through a p.d. of 1 V = 1.6×10^{-19} J)
E_x, E_y	electric field in the x, y-direction (V/m)
f	frequency (Hz)
$f(0)$	the value of $f(x)$ at $x=0$
F	force (N)
FAMOS	floating-gate avalanche-injection MOS structure
FET	field-effect transistor
f_p	pump frequency in a frequency multiplier (Hz)
FPGA	field-programmable gate array
FPLA	field-programmable logic array
f_T	(= $\beta_0 f_\beta = \alpha_0 f_\alpha \approx f_\alpha$) transition frequency or gain-bandwidth product (figure of merit) for a BJT (= theoretical frequency at which $|h_\mathrm{fe}| = 1$) (Hz)

$f(T_1)$	the value of a variable f at a specific temperature T_1
$f(x)$	a variable which is a function of x
fwd	forward (with particular reference to voltage bias of a junction)
$f_\alpha, f_{h_{fb}}$	α (or h_{fb}) cut-off frequency (= frequency at which $|\alpha| = \alpha_o/\sqrt{2}$) (Hz)
$f_\beta, f_{h_{fe}}$	β (or h_{fe}) cut-off frequency (= frequency at which $|\beta| = \beta_o/\sqrt{2}$) (Hz)
g	small-signal (or slope) conductance (S)
G	gate terminal or region of a FET, SCR or triac
G	conductance (S)
g_m, g_f	mutual conductance (forward transconductance) (S)
GaAs	gallium arsenide
GaP	gallium phosphide
Ge	germanium
GTO	gate turn-off (SCR)
h	hybrid terminal parameter; Planck's constant (= 6.63×10^{-34} J s)
H	irradiance (W/m^2)
$h_{11}, h_{12}, h_{21}, h_{22}$	general hybrid parameters of a system (Ω, —, —, S)
h_f	forward current ratio (gain) h-parameter
HF	high frequency
h_{fb}, α	small-signal CB current gain of a BJT ($h_{fb} = -\alpha$)
h_{fbo}, α_o	low-frequency small-signal CB current gain of a BJT ($h_{fbo} = -\alpha_o$)
h_{FB}, α_{dc}	static CB current gain of a BJT ($h_{FB} = -\alpha_{dc}$)
h_{fe}, β	small-signal CE current gain of a BJT
h_{feo}, β_o	low-frequency small-signal CE current gain of a BJT
h_{FE}, β_{dc}	static CE current gain of a BJT
h_i	input impedance h-parameter (Ω)
h_r	reverse voltage (feedback) ratio h-parameter
h_o	output admittance h-parameter (S)
i	(suffix) input
$i, i_b, i_c, i_e, i_g, i_d, i_s$	signal (varying) current components (A)
$i_B, i_C, i_E, i_G, i_D, i_S$	total (static bias + signal) currents (A)
I	luminous intensity (cd)
$I, I_A, I_B, I_C, I_E, I_G, I_D, I_S$	static (bias) current components (A)
IC	integrated circuit
I_{CBO}	leakage current across the reverse-biased C-B junction with the emitter open circuit (A)
I_{CEO}	C-E leakage current [$= I_{CBO}/(1 - \alpha_{dc})$] (A)
I_D	dark current of a photodetector (A)
IEC	International Electrotechnical Commission
I_{En}	electron component of emitter current (A)

I_{Ep}	hole component of emitter current (A)
I_F	forward current through a diode (A); forward current component due to diffusion across a junction (A); current through E-B junction in the Ebers–Moll BJT model (A)
IFL	integrated fuse logic
IGFET	insulated-gate FET
I_H	holding current of a SCR (A)
I_i	injector current (I^2L gate) (A)
I^2L(IIL)	integrated-injection logic
I^3L	isoplanar integrated-injection logic
IMPATT	impact avalanche transit-time (diode)
I/O	input/output (of a system)
I_p	current due to hole flow (A)
IR	infrared radiation (λ = 780 nm to 300 μm)
I_R	magnitude of reverse current through a diode (A); current through C-B junction in Ebers–Moll BJT model (A)
I_R, I_0	reverse current component due to drift across a junction (= reverse saturation or leakage current) (A)
I_{SC}	short-circuit current of a solar cell (A)
ISL	integrated Schottky logic
I_T	static ON-state current for an SCR (thyristor) (A)
I_x	current flowing in the x-direction (A)
I_Z	magnitude of the reverse current of a voltage-reference (Zener) diode (A)
I_ϕ	photocurrent (A)
j	operator that rotates (advances) a vector by $\pi/2$ radians
J	current density (A/m^2); polar moment of inertia (kg m^2)
J	junction between materials
JEDEC	Joint Electronic Device Engineering Council (USA)
J(diffusion)	component of current density due to carrier diffusion (A/m^2)
J(drift)	component of current density due to carrier drift (A/m^2)
JFET	junction FET
J_n	component of current density carried by electrons (A/m^2)
J_p	component of current density carried by holes (A/m^2)
k	Boltzmann's constant (= 1.38×10^{-23} J/K)
K	e-MOST parameter (= $I_{D(sat)}$ for specific V_{GS}) (A), device cathode (negative with respect to anode when device is conducting)
$K, K_1, K_2, K_3, k_1, k_2$	constants or device parameters
K_H	Hall coefficient (m^3/C)
l	length of sample or region (m)
L	inductance (H); luminance (cd/m^2); length (m)
LDR	light-dependent resistor (photoconductive cell)
LED	light-emitting diode

LF	low frequency
LID	leadless inverted device
L_n	diffusion length of electrons in p-type semiconductor $[= \sqrt{(D_n \tau_n)}]$ (m)
LOCMOS	local oxide isolated CMOS technology
L_p	diffusion length of holes in n-type semiconductor $[= \sqrt{(D_p \tau_p)}]$ (m)
LSIC	large-scale IC
L/W	length to width (aspect) ratio
m	mass (kg)
m^*	effective mass of an electron (kg)
M	collector multiplication factor (collector efficiency)
max	maximum
MESFET	metal-semiconductor (Schottky-barrier) FET
MISFET	metal-insulator-semiconductor FET
MNOS	metal-nitride-oxide-semiconductor structure
MOSR	MOST structure used as a resistor
MOST (MOSFET)	metal-oxide-semiconductor FET
MSIC	medium-scale IC
n	free electron concentration or density (number of electrons or carriers/m^3); integer
n	principal quantum number of an electron orbit
n^+	heavily doped (that is, $> 10^{24}$ dopant atoms/m^3 for Si) n-type semiconductor
n^-	lightly doped (that is, $< 10^{21}$ dopant atoms/m^3) n-type semiconductor
Δn	excess electron concentration (number of electrons/m^3) usually in p-type material (that is, minority carriers) above the equilibrium density of electrons in the semiconductor (due to dopant and electron–hole pairs) due to injection of carriers (for example, at a junction by forward bias) or addition of energy (for example, heat, light)
N	dopant concentration (density) (atoms/m^3); noise power (W)
N_a	acceptor-dopant concentration (density), increasing the free hole concentration in the semiconductor (atoms/m^3)
N_B, N_C, N_E	net dopant density in base, collector and emitter regions (atoms/m^3)
N_c	density of energy levels in CB (number/m^3)
N_d	donor-dopant concentration (density), increasing the free electron concentration in the semiconductor (atoms/m^3)
NF	noise figure (dB)
n_i	intrinsic carrier density (= density of electrons in pure semiconductor) (number of carriers/m^3); n_i^2 = product of *equilibrium* densities of electrons and holes (pn)

NMOS	n-channel MOS technology
n_n	electron density in an n-type semiconductor, that is, majority carriers (electrons/m^3)
n_p	electron density in a p-type semiconductor, that is, minority carriers (electrons/m^3)
n_{po}	total [that is, equilibrium + injected (excess)] electron density at the edge of the depletion region on the p-side of the junction (electrons/m^3)
npn	bipolar transistor structure
n-type	semiconductor containing a majority of donor (*donate* electrons) dopant atoms of density N_d
N_v	density of energy levels in VB (number/m^3)
n_ϕ	rate of incidence of photons (photons/s)
o,O	(suffix) output or open circuit
o/c	open circuit
off, OFF	non-conducting state
on, ON	conducting state
p	free hole concentration or density (number of holes or carriers/m^3)
p^+	heavily doped (that is, $> 10^{24}$ dopant atoms/m^3 for Si) p-type semiconductor
p^-	lightly doped (that is, $< 10^{21}$ dopant atoms/m^3 for Si) p-type semiconductor
$p+jq$	general complex number
Δp	excess hole concentration (number of holes/m^3) usually in n-type material (that is, minority carriers) above the equilibrium density of holes in the semiconductor
P	power (rate of energy flow) handling capability (W)
pcb	printed circuit board
p.d.	potential difference (=voltage) that is, difference in potential between two points (V)
p_i	intrinsic hole density, equal to, and usually denoted by n_i (number of carriers/m^3)
PIA	peripheral interface adaptor
pin	p-type/intrinsic/n-type structure
PLA	programmable logic array
PMOS	p-channel MOS technology
p_n	hole density in an n-type semiconductor, that is, minority carriers (holes/m^3)
p_{no}	total [that is, equilibrium + injected (excess)] hole density at the edge of the depletion region on the n-side of the junction (holes/m^3)
pn	p-type/n-type structure
pnp	bipolar transistor structure

poly	polycrystalline
p_p	hole density in a p-type semiconductor, that is, majority carriers (holes/m^3)
P_{tot}	total power dissipation (W)
p-type	semiconductor containing a majority of acceptor (*accept* electrons, that is, supply holes) dopant atoms of density N_a (atoms/m^3)
PROM	programmable ROM
$P(W)$	probability of an energy level W being occupied
PWL	piecewise-linear
P_ϕ	optical radiation power (W)
q	charge (C)
Q	charged stored or space charge (C); quantity of dopant number of atoms); ($-$factor) quality factor [$= \omega L/R = 1/(\omega CR)$]
Q	quiescent (no signal) operating condition
q_B	stored charge in the base of a BJT (C)
q_{BN}, q_{BI}	normal and inverse components of base charge in a BJT (C)
Q_n	magnitude of space charge in depletion region on n-side of junction (depleted donor atoms) (C)
Q_p	magnitude of space charge in depletion region on p-side of junction (depleted acceptor atoms) (C); charge due to excess holes (C)
q_s	stored charge in the base of a BJT at the onset of saturation (C)
r	radius (m); recombination rate of electron–hole pairs (carriers/s); resistance (Ω) or slope resistance $\Delta V/\Delta I$ or $\partial V/\partial I$ (Ω)
R	resistance (Ω)
RAM	random-access (read–write) memory
R_B	bulk resistance (Ω)
$r_{bb'}, r_c, r_e$	branch resistances in the Early (small-signal, active region, physical) BJT model (Ω)
$r_{bb'}, r_{b'c}, r_{b'e}, r_{ce}$	branch resistances in the hybrid-π (small-signal, active region) BJT model (Ω)
ref	reference
rev	reverse (with particular reference to voltage bias of a junction)
R_L	load resistance (Ω)
ROM	read-only memory
R_s, R_S	source resistance (Ω); series resistance (Ω)
R_{th}	thermal resistance (°C/W)
$R_{th}(hs)$	thermal resistance of a heat sink (°C/W)
$R_{th}(i)$	thermal resistance of an insulating washer (°C/W)
$R_{th(j-amb)}$	thermal resistance between device junction and ambient (°C/W)

$R_{th(j\text{-}case,mb)}$	thermal resistance between device junction and case (mounting base) (°C/W)
$R_{th(case,mb\text{-}amb)}$	thermal resistance between device case (mounting base) and ambient (°C/W)
R_ϕ	responsivity of a photodetector (A/W)
S	signal power (W)
s	(suffix) short circuit; source terminal or region of an FET
S_1, S_2, S_3	hypothetical ideal voltage-controlled switches in PWL device models
sat	saturation mode or condition
s/c	short-circuit
SCR	semiconductor controlled rectifier (thyristor)
SCS	semiconductor controlled switch
SDFL	Schottky-diode FET logic
Si	silicon
Si_3N_4	silicon nitride
SiO_2	silicon dioxide
S/N	signal-to-noise ratio
SOAR	safe operating area
SOS	silicon-on-sapphire technology
SSIC	small-scale IC
sub	substrate on to which semiconductor device is fabricated
S_Z	temperature coefficient of a voltage-reference (Zener) diode
t	time (s); thickness of specimen or region (m)
T	temperature (°C or K); period (s)
t	terminal
T_1, T_2	specific temperatures (°C or K)
T	terminals; transistors
T_{amb}	ambient (environment) temperature (°C or K)
t_B, t_E	time for base and emitter diffusions (s)
t_c	relaxation time, that is, mean time between collisions of free carriers with atomic centres, causing scattering and reducing conductivity (s)
T_C	colour temperature (K); memory column select switch
T_{case}, T_{mb}	case or mounting base temperature (°C or K)
t_d	delay time
t_{fr}, t_{rr}	forward and reverse recovery times for a diode during switching (s)
T_{hs}	heat sink temperature (°C or K)
T_j	junction temperature (°C or K)
TO	transistor (and other) package outline
t_p	pulse width (s)
t_r, t_s, t_f	rise, storage and fall times for a BJT (s)
T_R	memory row select switch

Appendix B

t_s, t_t	storage and transition times for a diode (s)
TTL	transistor–transistor logic (prefixes: H–high speed, S–Schottky, L–low power, LS–low-power Schottky)
TRAPATT	trapped-plasma, avalanche and triggered transit (diode)
u	velocity (m/s)
u_d	drift velocity of carriers under influence of an electric field (m/s)
UHF	ultra high frequency range (300 MHz to 3 GHz)
UJT	unijunction transistor
UV	ultraviolet radiation (λ = 10 to 380 nm)
u_x	velocity in the x-direction (m/s)
v	signal varying alternating voltage (potential difference) (V)
$v_{be}, v_{ce}, v_{cb}, v_{gs}, v_{ds}, v_{dg}$	signal voltages between terminals (V)
$v_{BE}, v_{CE}, v_{CB}, v_{GS}, v_{DS}, v_{DG}$	total (static bias + signal) voltage between terminals (V)
V	static (bias) component of voltage (potential difference) (V); applied voltage across a junction or device (V)
V_1	specific voltage (V)
V_A	potential at point A in material (V)
V_{AK}	static anode–cathode voltage (V)
V_B, V_{BR}	reverse breakdown voltage of a junction (V)
VB	valence band
V_{BB}	static base voltage for a BJT (V)
$V_{BE}, V_{CE}, V_{CB}, V_{GS}, V_{DS}, V_{DG}$	static (bias) voltage between terminals (V)
V_{BO}	forward breakover voltage of a SCR, diac or triac (V)
V_C	contact potential (V)
V_{CC}, V_{EE}	power supply voltages for a BJT circuit (V)
v_D	total voltage across a diode (anode with respect to cathode) (V)
V_{DD}, V_{SS}	power supply voltages for a FET circuit (V)
V_F	applied forward voltage across a diode (V)
V_{GG}	static gate voltage for a FET (V)
$V_{GS(off)}$	value of V_{GS} for a FET to reduce I_D to a specified low value (practical equivalent of V_P) (V)
V_H	Hall voltage (V)
VIL	vertical injection logic
V-I^{2}L	V-groove or vertical I^{2}L
V-JFET	vertical JFET structure
v_L	total load voltage (V)
VLSIC	very large-scale IC
VMOST	V-groove or vertical MOST structure
V_n	voltage across n-type section of depletion layer (V)
v_o	output signal voltage (V)
V_O	static output voltage (V)

V_{OC}	open-circuit voltage of a solar cell (V)
V_p	voltage across p-type section of depletion layer (V)
V_P	pinch-off voltage for a JFET or d-IGFET (V); peak voltage for a tunnel diode (V)
V_R	magnitude of the applied reverse voltage across a diode (V)
V_{RRM}	repetitive peak reverse voltage (V)
$V_T(V_{GS(th)})$	threshold voltage for an e-IGFET (V)
V_V	valley voltage for a tunnel diode (V)
V_Z	magnitude of reverse voltage for a voltage-reference (Zener) diode (V)
V_γ	threshold voltage (for significant E-B conduction) for a BJT (V)
V_σ	saturation voltage of B-E junction ($= V_{BE(sat)}$) (V)
w	width of sample or region (m); metallurgical base width of a BJT (m)
W	energy (J or eV); width (m)
w_B	effective base width of a BJT (m)
W_c	minimum energy of electrons in the conduction band (J or eV)
W_{cs}	minimum energy of electrons in the CB at the semiconductor surface (J or eV)
w_E	distance from E-B junction to external emitter contact in a BJT (m)
W_F	Fermi level (value of energy level having probability of occupation of 0.5) (J or eV)
W_{F1}, W_{F2}	Fermi level of materials 1 and 2 (J or eV)
W_{Fm}	Fermi level of a metal (J or eV)
W_{Fn}	Fermi level of an n-type semiconductor (J or eV)
W_{Fp}	Fermi level of a p-type semiconductor (J or eV)
W_g	energy gap between CB and VB ($= W_c - W_v$) (J or eV)
W_v	maximum energy of electrons in the valance band (J or eV)
W_{vs}	maximum energy of electrons in the VB at the semiconductor surface (J or eV)
W_ϕ	photon energy (J)
x	variable; space direction or spatial position (m)
y	space direction; admittance terminal parameter (S)
Y	admittance (S)
$y_{11}, y_{12}, y_{21}, y_{22}$	general admittance parameters of a system (S)
z	space direction
Z	impedance (Ω); atomic number of an atom (number of protons in the atom)
Z_{Th}	thermal impedance (°C/W)
α, h_{fb}	small-signal CB current gain of a BJT ($\alpha = -h_{fb}$)

α_o, h_{fbo}	low frequency (LF) small-signal CB current gain of a BJT ($\alpha_o = -h_{fbo}$)
α_{dc}, h_{FB}	static CB current gain of a BJT ($\alpha_{dc} = -h_{FB}$)
α_I	static CB current gain of a BJT for inverse operation (Ebers–Moll model)
α_N	static CB current gain of a BJT for normal operation (Ebers–Moll model) ($= \alpha_{dc}$)
β, h_{fe}	small-signal CE current gain of a BJT
β_o, h_{feo}	low frequency (LF) small-signal CE current gain of a BJT
β_{dc}, h_{FE}	static CE current gain of a BJT
β_I	inverse static current gain in CE (Ebers–Moll model)
β_N	normal static current gain in CE (Ebers–Moll model) ($= \beta_{dc}$)
γ	emitter injection efficiency
Δ	small increment (increase) in
Δn	excess electron concentration (number of electrons/m^3)
Δp	excess hole concentration (number of holes/m^3)
ϵ	permittivity of a material (F/m)
ϵ_0	permittivity of free space ($= 8.85 \times 10^{-12}$ F/m)
ϵ_r	relative permittivity of a material ($= \epsilon/\epsilon_0$)
η_q	quantum gain, quantum yield or quantum efficiency (electrons/photon)
θ	angle (° or rad)
θ_c	critical angle (° or rad)
λ	wavelength ($= c/f$) (m)
μ	mobility (m^2/V s); voltage feedback factor in early BJT model ($= h_{rb}$)
μC	microcomputer
μ_H	Hall mobility (m^3/C)
μm	micron ($= 10^{-6}$ m)
μ_n	electron mobility (m^2/Vs)
μ_p	hole mobility (m^2/Vs)
μP	microprocessor
π-type	lightly doped p-type semiconductor ($\equiv p^-$)
ρ	resistivity ($= 1/\sigma$) (Ω m); volume charge density (C/m^3)
ρ_S	sheet resistance (Ω/square)
σ	conductivity ($= 1/\rho$) (S/m)
σ_B, σ_E	conductivities of base and emitter regions of a BJT (S/m)
σ_i	conductivity of an intrinsic (pure, undoped) semiconductor (S/m)
σ_m	angular momentum (kg m^2/s)
τ	lifetime, that is, the mean time carriers exist in free state before recombining (s)
τ_B	lifetime of injected carriers in the base of a BJT (s)
τ_{BI}, τ_{BN}	lifetime of injected carriers in the base of a BJT for inverse

	and normal operation (charge-control model) (s)
τ_n	minority carrier lifetime of electrons (that is, in a p-type semiconductor) (s)
τ_p	minority carrier lifetime of holes (that is, in an n-type semiconductor) (s)
τ_T	transit time of minority carriers across the base of a BJT (charge-control model) (s)
τ_{TI}, τ_{TN}	charge transit time across the base of a BJT for inverse and normal operation (charge-control model) (s)
ϕ	work function, that is, energy addition necessary to cause emission of electrons from the surface of a material (eV); luminous flux (lm)
ϕ_1, ϕ_2	work function of materials 1 and 2 (eV)
ϕ_m	work function of a metal (eV)
ϕ_s	work function of a semiconductor (eV)
χ	range of energy levels in the conduction band (semiconductor affinity) (J or eV)
ψ	contact potential or barrier potential of a junction (V)
ψ_{12}	contact potential of the junction between materials 1 and 2 (V)
ψ_{CB}	collector–base junction contact potential (V)
ψ_{EB}	emitter–base junction contact potential (V)
ω	angular frequency (rad/s)

Mathematical Symbols

$\| \;\|$	matrix determinant; modulus or 'magnitude of'
$\neq$	not equal to
$\sim$	difference between
$\approx$	approximately equal to
$\equiv$	equivalent to
$\leq$	less than or equal to
$\geq$	greater than or equal to
$\ll$	very much less than
$\gg$	very much greater than
$//$	in parallel with
$\rightarrow$	approaches (variable approaching a value); to (in a range of values, or direction of flow)

B.2 DESIGNATION OF VARIABLES

Terminology

The terms *static*, *signal* and *total* are used either to indicate the nature of a variable or to refer to the component being considered. A *static* quantity is a constant (that is, non-time-varying) value. The *signal* is the time-varying component. The *total* value is the actual value of the variable, that is, the sum of constant and time-varying components.

System of Symbols for Electrical Variables

A system of upper and lower case symbols combined with upper and lower case suffixes is used to refer to the static, signal or total components of a variable. Additional suffixes are used to indicate the root-mean-square, average or peak value of a varying quantity.

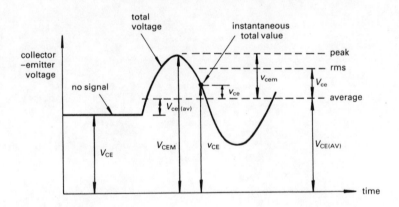

Figure B.1 System of symbols

Considering collector–emitter voltage as an example, figure B.1 shows a variation of voltage over a period of time labelled with the appropriate variables, that is

V_{CE} = static collector–emitter voltage
v_{ce} = instantaneous value of signal component
v_{CE} = instantaneous total value ($=V_{CE(AV)}+v_{ce}$)
V_{ce} = r.m.s. value of signal component
V_{CEM} = peak (maximum) value of total waveform
$V_{CE(AV)}$ = average value of total waveform

V_{cem} = peak (maximum) value of signal component
$V_{ce(av)}$ = average value of signal component.

The order of the suffixes indicates the measurement direction, that is, V_{CE} is the static potential at the collector terminal with reference to the emitter terminal and therefore $V_{CE} = -V_{EC}$. In the case of current flow, I_{CB} would be the static current flowing from collector to base although usually the second suffix is omitted and the convention adopted that positive collector current I_C is the current flowing from collector to base.

Where a variable refers to two terminals of a three terminal device, a third suffix 'O' or 'S' is used to indicate whether the third terminal is open-circuit or short-circuited to the reference terminal, for example

I_{CBO} = static current from collector to base with the emitter (the third terminal of a BJT not indicated in the suffix) open-circuit

V_{CES} = static collector–emitter voltage with the base (the third terminal of a BJT not indicated in the suffix) shorted to the emitter (reference) terminal.

Appendix C Constants, Conversions, Unit Multiples

VALUES OF PHYSICAL CONSTANTS

c (velocity of propagation of electromagnetic radiation in free space) $= 3 \times 10^8$ m/s

e (magnitude of electronic charge) $= 1.6 \times 10^{-19}$ C

h (Planck's constant) $= 6.63 \times 10^{-34}$ J s
$= 4.14 \times 10^{-15}$ eV s

k (Boltzmann's constant) $= 1.38 \times 10^{-23}$ J/K
$= 8.63 \times 10^{-5}$ eV/K

m (electronic rest mass) $= 9.1 \times 10^{-31}$ kg

ϵ_0 (permittivity of free space) $= 8.85 \times 10^{-12}$ F/m

μ_0 (permeability of free space) $= 4\pi \times 10^{-7}$ H/m

e/m (electronic charge/rest mass ratio) $= 1.76 \times 10^{11}$ C/kg

USEFUL VALUES AND CONVERSIONS

kT ≈ 0.026 eV at 300 K

kT/e ≈ 26 mV at 300 K

1 eV $= e$ J $= 1.6 \times 10^{-19}$ J

1 μm (micron) $= 10^{-6}$ m $= 10^{-3}$ mm

1 Å (angstrom) $= 10^{-10}$ m $= 10^{-4}$ μm

Table C.1 Values of physical parameters for silicon, germanium, gallium arsenide, silicon dioxide and silicon nitride at 300 K

	Si	Ge	$GaAs$	SiO_2	Si_3N_4	
W_g (energy gap)	1.1	0.7	1.4			eV
n_i (intrinsic carrier density)	1.5×10^{16}	2.5×10^{19}	10^{13}			carriers/m^3
ϵ_r (relative permittivity)	12	16	12	4	7.5	
μ_n (electron mobility)	0.14	0.39	0.85			m^2/V s
μ_p (hole mobility)	0.05	0.19	0.05			m^2/V s
D_n (electron diffusion coefficient)	3.6×10^{-3}	0.01	2.2×10^{-2}			m^2/s
D_p (hole diffusion coefficient)	1.3×10^{-3}	5×10^{-3}	1.3×10^{-3}			m/s

PREFIXES USED TO INDICATE DECIMAL MULTIPLES

kilo (k)	$\times 10^3$	milli (m)	$\times 10^{-3}$
mega (M)	$\times 10^6$	micro (µ)	$\times 10^{-6}$
giga (G)	$\times 10^9$	nano (n)	$\times 10^{-9}$
tera (T)	$\times 10^{12}$	pico (p)	$\times 10^{-12}$
peta (P)	$\times 10^{15}$	femto (f)	$\times 10^{-15}$
exa (E)	$\times 10^{18}$	atto (a)	$\times 10^{-18}$

PREFIXES USED TO INDICATE BINARY MULTIPLES

kilo (k)	$\times 2^{10}$ ($\times 1024$)
mega (M)	$\times 2^{20}$ ($\times 1.048\ 576 \times 10^6$)
giga (G)	$\times 2^{30}$ ($\times 1.073\ 741\ 824 \times 10^9$)

Appendix D Electromagnetic Spectrum

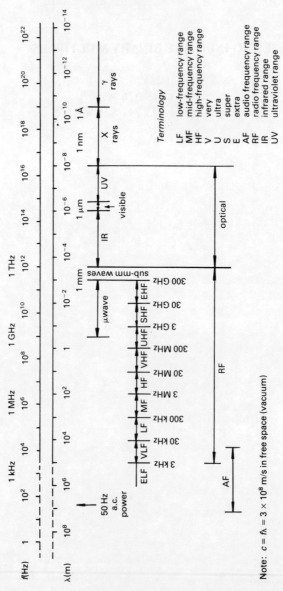

Figure D.1 Electromagnetic spectrum

Appendix E Device Numbering Systems

E.1 DISCRETE SEMICONDUCTOR DEVICES

E.1.1 Original European System

Type number code: OXY nnn where O indicates a semiconductor device, XY is a one or two-letter code indicating the general type of the device

    A diode
  AP photodiode
  AZ voltage–reference diode
    C transistor
  CP phototransistor

nnn is a two or three-digit serial number.
Examples: OA 202 diode, OC 23 transistor.

E.1.2 Present European System (registered by Pro-Electron)

In 1966 an international association called Pro-Electron was established in Brussels to organise the allocation and registration of the type numbers of semiconductor devices. For discrete devices the type number code is of the form XY nnn where X is a letter indicating the *material*

  A germanium
  B silicon
  C compound semiconductors such as GaAs
  D compound semiconductors such as InSb
  R compound semiconductors such as CdS

Y is a letter indicating device *application*

    A switching diode
    B variable capacitance diode
    C low-power AF transistor
    D high-power AF transistor
    E tunnel diode
    F low-power RF transistor
    L high-power RF transistor
    P radiation sensor (for example, photodiode)
    Q radiation emitter (for example, LED)
    R low-power switching device having specific breakdown characteristic (for example, SCR)
    S low-power switching transistor
    T high-power switching device having specific breakdown characteristic (for example, high-power SCR)
    U high-power switching transistor
    X multiplier diode (for example, varactor diode)
    Y rectifier diode
    Z voltage–reference diode

nnn is a three character *serial number* comprising either three digits (devices intended for *consumer* applications) or one letter (Z,Y,X,W etc.) and two digits (devices intended for *industrial/professional* applications).

Range Numbers
Further letters and/or numbers are added to the basic device-type number following a hyphen to identify particular ranges of the same type of device. For example, the maximum peak reverse voltage V_{RRM}max of a rectifier diode or SCR is often given (XY nnn-V_{RRM}max).

An additional letter R for medium and high-power diodes or SCRs (XY nnn-V_{RRM}max R) indicates *reversed* package connections, that is, stud–anode instead of the normal stud–cathode connection.

In the case of voltage–reference diodes, the nominal breakdown voltage and its percentage tolerance is given, where A = ± 1 per cent, B = ± 2 per cent, C = ± 5 per cent, D = ± 10 per cent, E = ± 15 per cent. Where applicable, the letter V replaces the decimal point in the quoted breakdown voltage.

Examples

    BAX 13   industrial grade silicon switching diode
    BC 107   commercial grade low-power silicon AF transistor
    ASY 27   industrial grade low-power germanium switching transistor

Appendix E   447

BYX 42-1200   industrial grade silicon rectifier diode with a
                   maximum repetitive peak reverse voltage of 1200 V
                   and stud–cathode connection
BTW 92-800R   industrial grade high-power silicon SCR with 800 V
                   maximum peak reverse voltage and stud–anode
                   connection
BZY 88-C5V6   industrial grade silicon voltage–reference diode with
                   a 5.6 V ± 5 per cent breakdown voltage
CQY 49   industrial grade gallium arsenide LED
RPY 82   industrial grade cadnium sulphide photoconductive
                   cell.

E.1.3 Housecode

Some manufacturers market devices with type numbers of their own derivation, termed *housecodes*, which usually give some indication of the structure, performance or rating of the devices. Examples
 a TRW power semiconductor SD-51 diode (figure 2.3) is a power *S*chottky *D*iode;
 the Siliconix VN 46/66/88AF transistors (appendix H.7) are n-channel *V*MOS power FETs with maximum drain–source voltages of 40 V, 60 V and 80 V respectively.

E.1.4 British Military Numbering Scheme

Discrete semiconductor devices meeting a Ministry of Defence specification giving approval for use in military applications are given a type number code of the form CV nnnn.

E.1.5 American System

The Joint Electronic Device Engineering Council (JEDEC) approved type number code is of the form MX nnnn where originally

 M number of *pn* junctions in the device (1 — diode, 2 — BJT, 3 — SCR);
 X letter indicating either the semiconductor *material* (G—germanium, S—silicon) or military specification (N—military specification approval);
nnnn three or four digit serial number.

This original coding system is now used less rigorously so that diodes are coded 1Nnnnn and transistors and other devices are coded 2Nnnnn and 3Nnnnn.

E.2 INTEGRATED CIRCUITS

The type number coding of integrated circuits is less unified than that of discrete semiconductor devices.

E.2.1 Housecode

The majority of ICs are coded under systems derived by the individual manufacturers, the code typically comprising two or three letters identifying the company (for example, CA ≡ R.C.A., LM ≡ National, MC ≡ Motorola, NMC ≡ Newmarket, SL ≡ Plessey, SN ≡ Texas, μA ≡ Fairchild) followed by a three or four-digit serial number.

E.2.2 Pro-Electron System

Under the housecode system, ICs produced by different manufacturers to the same specification have different type numbers leading to a proliferation of numbers. The Pro-Electron system attempts to produce a unified system whereby ICs with the same specification from various manufacturers have the same type code number. The Pro-Electron code is of the form XYZ nnnnn

For *solitary* (single type) ICs, X indicates the *mode* of operation

    S   digital IC
    T   linear IC
    U   mixed linear/digital IC.

Y has no special significance, except
    H   hybrid technology

For *family* ICs (for example, groups of logic ICs intended to be used together), XY is an identity code for the group of ICs. Letter Z indicates the operational temperature range

    A   no range specified
    B   0 to +70 °C
    C   −55 to +125 °C
    D   −25 to +70 °C
    E   −25 to +85 °C
    F   −40 to +85 °C

Serial number nnnnn is either, a four digit number assigned by Pro-Electron or, a minimum of four digits of an already widely used housecode number (for example, 7400 derived from the Texas SN 7400 digital IC group, 0741 derived from the Fairchild μA 741 operational amplifier).

In addition, a *version* letter may be added indicating either, *package variations*

⎛ C cylindrical
⎜ D dual-in-line
⎜ F flat-pack
⎝ Q quadruple-in-line

or, *other variations,* such as construction or rating (version letter Z indicates customised wiring).

Examples

> the Mullard/Signetics 'HE' family of LOCMOS small-scale logic ICs includes
> HEF 4012B a dual 4-input NAND gate IC;
> HEF 4737V a quadruple static decade counter IC.
> These ICs have an operational temperature range (F) of -40 to $+85\,°C$. Version letter B indicates the standard power supply voltage range for the family (3–15 V) while version letter V indicates a reduced voltage range (4.5–12.5 V in this case).

> The Mullard/Signetics 'solitary' IC coded TBA 221D is an integrated operational amplifier (that is, a linear IC). No operational temperature range is specified (A) and version letter D indicates encapsulation in an 8-lead plastic flat pack (style SO-8, SOT-96A in this case).

E.2.3 British Military Numbering System

Integrated circuits given approval for use in military applications are given a code of the form CN nnnn analogous to the CV number for approved discrete devices.

Appendix F Component Coding

Colour-coding is widely used to indicate the value and selection tolerance of general-purpose resistors and some plastic–dielectric capacitors, the information being given in a series of coloured bands around the body of the component. Colour-coding is also used to indicate the type number of some devices, particularly low-power diodes.

The colour code conforming to BS 1852:1967 and accepted by the International Electrotechnical Commission (IEC publication 62/1968) and the Electronics Industries Association (EIA) is given in table F.1.

The majority of resistors are produced in the E12 and E24 ranges of *preferred* values (appendix G) and have a four-band code (figure F.1a). Resistors produced in the E96 range have a five-band code (figure F.1a). The code indicates the *nominal* value of resistance of the resistor and the tolerance gives the *selection* tolerance, that is, the range on either side of the nominal value within which the *actual* resistance of an individual resistor is guaranteed to lie. No indication is normally given as to the *stability* of the resistance value in use, although on some older types an additional pink band indicated *high stability* or a coloured band was included to indicate the possible *total excursion* of the resistance value during the *life* of the component. BS 1852 also recommends that in written resistor values, the Ω symbol and decimal point (where applicable) should be omitted and the multiplier (and the position of the decimal point) indicated by a letter

      R   for decimal point
      K   for decimal point and $\times 10^3 \, \Omega$ (that is, kΩ)
      M   for decimal point and $\times 10^6 \, \Omega$ (that is, MΩ)

Examples
    R47 = 0.47 Ω
    4R7 = 4.7 Ω
    47R = 47 Ω
    1K0 = 1 kΩ
    10M = 10 MΩ

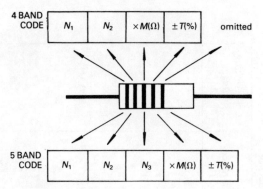

Examples
4 band code: brown, black, red, red ≡ $10 \times 10^2 \Omega \pm 2\%$ = $1 k\Omega \pm 2\%$
5 band code: brown, grey, red, orange, brown ≡ $182 \times 10^3 \Omega \pm 1\%$
= $182 \, k\Omega \pm 1\%$

(a) Resistor coding

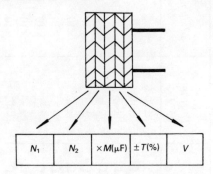

Example
yellow, violet, yellow, white, red ≡ 47×10^4 pF $\pm 10\%$, 250 V d.c.
= 0.47 μF ± 10%, 250 V d.c.

(b) Capacitor coding

Figure F.1 Resistor and capacitor colour-coding

The International Electrotechnical Commission (IEC) have recommended the following tolerance codes (recommendation 62/1968)

F ≡ ± 1 per cent
G ≡ ± 2 per cent
J ≡ ± 5 per cent
K ≡ ± 10 per cent
M ≡ ± 20 per cent

Examples
A resistor coded 390RJ is 390 Ω ± 5 per cent
68KK is 68 k Ω ± 10 per cent
4K7G is 4.7 Ω ± 2 per cent

Table F.1 Component colour code

Colour	Numerical value N	Resistors		Capacitors		
		Multiplier M	Tolerance T	Multiplier M	Tolerance T	d.c. working voltage V
Blank			±20%			
Silver		×10⁻² Ω	±10%			
Gold		×10⁻¹ Ω	± 5%			
Black	0	×1 Ω		×1 pF		
Brown	1	×10 Ω	± 1%	×10 pF		
Red	2	×10² Ω	± 2%	×10² pF	±20%	
Orange	3	×10³ Ω		×10³ pF		
Yellow	4	×10⁴ Ω		×10⁴ pF		100 V
Green	5	×10⁵ Ω		×10⁵ pF	± 5%	250 V
Blue	6	×10⁶ Ω		×10⁶ pF		400 V
Violet	7	×10⁷ Ω				
Grey	8	×10⁸ Ω		×10⁻² pF		
White	9	×10⁹ Ω		×10⁻¹ pF	±10%	

Expressed as $\times 10^M$ where M takes the appropriate sign.

Appendix G Preferred (E-series) Component Values

A series of *preferred* or *standard* nominal values for components was devised (originally for resistors) so that all component values could be covered by the series of nominal values and the associated selection tolerance. The tolerances involved are ± 20 per cent, ± 10 per cent, ± 5 per cent and ± 1 per cent and the corresponding series of preferred values became known as the 20 per cent, 10 per cent, 5 per cent and 1 per cent ranges. The four ranges comprise 6, 12, 24 and 96 nominal values respectively and are now known (BS 2488:1966 and IEC publication 63) as the E6, E12, E24 and E96 series respectively.

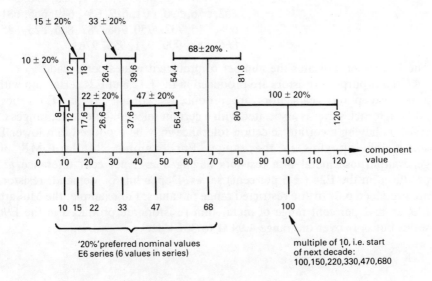

Figure G.1 E6 series of nominal component values

With reference to figure G.1, consider the establishment of the E6 or 20 per cent series. If the first nominal value in the decade is taken as 10, the range of values covered by this value with a selection tolerance of 20 per cent is 10 ± 20 per cent, that is, from 8 to 12. To avoid gaps in the range, the next nominal value (x) in the series must be such that $x - 20$ per cent ≤ 12, from which $x = 15$ is chosen. The next nominal value (y) must be such that $y - 20$ per cent $\leq 15 + 20$ per cent, that is, $0.8y \leq 18$ or $y \leq 22.5$, and so the integer 22 is chosen as the next value. Figure G.1 shows that the ± 20 per cent range of the six values 10, 15, 22, 33, 47 and 68 covers the complete decade from 8 to 80. Similarly the decades 0.8 to 8 and 80 to 800 are covered by the ranges of values 1.0, 1.5, 2.2, 3.3, 4.7, 6.8 and 100, 150, 220, 330, 470, 680 respectively. For a lower selection tolerance, more values are required in the preferred range as shown below.

E6 ($\pm$ 20 per cent) series   10, 15, 22, 33, 47, 68;
E12 ($\pm$ 10 per cent) series   10, 12, 15, 18, 22, 27, 33, 39, 47, 56, 68, 82
E24 ($\pm$ 5 per cent) series   10, 11, 12, 13, 15, 16, 18, 20, 22, 24, 27, 30, 33, 36, 39, 43, 47, 51, 56, 62, 68, 75, 82, 91
E96 ($\pm$ 1 per cent) series   100, 102, 105, 107, 110, 113, 115, 118, 121, 124, 127, 130, 133, 137, 140, 143, 147, 150, 154, 158, 162, 165, 169, 174, 178, 182, 187, 191, 196, 200, 205, 210, 215, 221, 226, 232, 237, 243, 249, 255, 261, 267, 274, 280, 287, 294, 301, 309, 316, 324, 332, 340, 348, 357, 365, 374, 383, 392, 402, 412, 422, 432, 442, 453, 464, 475, 487, 499, 511, 523, 536, 549, 562, 576, 590, 604, 619, 634, 649, 665, 681, 698, 715, 732, 750, 768, 787, 806, 825, 845, 866, 887, 909, 931, 953, 976

The E-number indicates the number of preferred values per decade.

General-purpose resistors are produced in the E12 and E24 series and with a few exceptions, capacitors are produced according to the E6 series. Although each series is associated with a certain selection tolerance, ranges of resistors having a certain selection tolerance may be produced in a lower E series leaving gaps in the range. For example, Mullard MR 30 general-purpose metal–film resistors having a ± 2 per cent tolerance are produced in the E24 (± 5 per cent) series. Depending on demand, resistors are produced only over a restricted range of values. For example, the Mullard MR 25 ± 1 per cent range of metal–film resistors are produced in the E96 series but only over the range 4.99 Ω to 301 kΩ.

Appendix H Manufacturers' Data Sheets Of Selected Devices

The data provided by a manufacturer for a particular type (number) of device includes typically

(1) Description of the device and a brief statement of intended applications;
(2) abridged data giving absolute maximum ratings;
(3) electrical performance for particular operating conditions;
(4) thermal properties;
(5) mechanical details: package style and dimensions;
(6) variation of electrical parameters/properties with operating conditions (for example, current, voltage, frequency, temperature). This information is usually presented graphically.

It must be emphasised that the data provided relates directly to the intended use of the device, thus for a low-power BJT intended for use as an amplifier at audio frequencies, h-parameter information would be supplied in considerable detail but there may be little information as to the switching performance. Alternatively, a large proportion of the data supplied for a high-power device is likely to be concerned with safe operating conditions (SOAR information) while devices intended for switching or high-frequency linear operation have detailed information on the device capacitance and/or frequency response.

The information provided in this section illustrates the properties and performance of typical devices. The data for the BAX13 switching diode, BZY88 voltage–reference diode, BC107 *npn* BJT and BFR29 *n*-channel depletion-type MOST is reproduced by permission of Mullard Limited while that for the 2N 5457 *n*-channel JFET, 3N 163 *p*-channel enhancement-type MOST and VN 46AF *n*-channel enhancement-type VMOS power FET is included with the permission of Siliconix Limited.

H.1 MULLARD LOW-POWER SILICON SWITCHING DIODE TYPE BAX13

Whiskerless diffused silicon diode intended for fast logic applications.

QUICK REFERENCE DATA

V_R max.	50	V
V_{RRM} max.	50	V
V_F max. ($I_F = 20mA$)	1.0	V
I_{FRM} max.	150	mA
t_{rr} max. (when switched from $I_F = 10mA$ to $V_R = 6.0V$, measured at $I_R = 1.0mA$, $R_L = 100\Omega$)	4.0	ns
Q_s max. (when switched from $I_F = 10mA$ to $V_R = 5.0V$, $R_L = 500\Omega$)	45	pC

RATINGS

Limiting values of operation according to the absolute maximum system.

Electrical

V_R max.	Continuous reverse voltage	50	V
V_{RRM} max.	Repetitive peak reverse voltage	50	V
$I_{F(AV)}$ max.	Average forward current, ($t_{av} = 20ms$ 75 (for sinusoidal operation see page 6)		mA
I_F max.	Forward current (d.c.)	75	mA
I_{FRM} max.	Repetitive peak forward current	150	mA
I_{FSM} max.	Non-repetitive peak forward current $t = 1\mu s$	2000	mA
	$t = 1s$	500	mA

Temperature

T_{stg}	Storage temperature	−65 to +200	°C
T_j max.	Junction temperature	200	°C

THERMAL CHARACTERISTIC

$R_{th(j-amb)}$ Thermal resistance in free air 0.60 degC/mW

ELECTRICAL CHARACTERISTICS ($T_j = 25°C$ unless otherwise stated)

		Max.	
V_F	Forward voltage		
	$I_F = 2.0mA$	0.7	V
	$I_F = 10mA$, $T_j = 100°C$	0.8	V
	$I_F = 20mA$	1.0*	V
	$I_F = 75mA$	1.53*†	V
I_R	Reverse current		
	$V_R = 10V$	25	nA
	$V_R = 10V$, $T_j = 150°C$	10	μA
	$V_R = 25V$	50	nA
	$V_R = 50V$	200*	nA
	$V_R = 50V$, $T_j = 150°C$	25	μA
C_d	Diode capacitance $V_R = 0$, $f = 1.0MHz$	3.0	pF

*These are the characteristics which are recommended for acceptance testing purposes.

†Measured under pulsed conditions to prevent excessive dissipation.

OUTLINE AND DIMENSIONS

The diodes may be either type branded or colour coded

Dimensions in mm

DO-35 (type branded)

SOD-17 (colour coded)

Cathode indicated by coloured band

min. mounting width 7.6

brown (cathode) orange black

BAX13

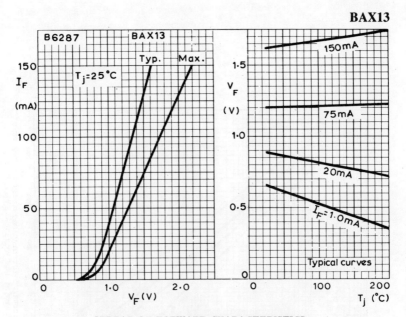

SPREAD OF FORWARD CHARACTERISTICS
FORWARD VOLTAGE DROP PLOTTED AGAINST JUNCTION TEMPERATURE
WITH FORWARD CURRENT AS A PARAMETER

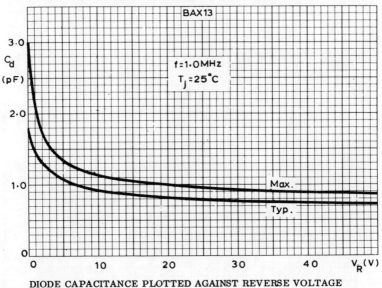

DIODE CAPACITANCE PLOTTED AGAINST REVERSE VOLTAGE

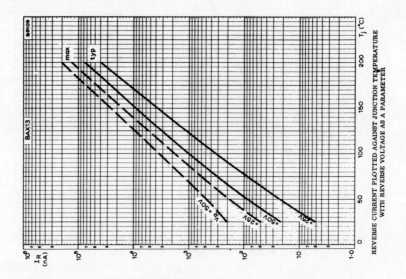
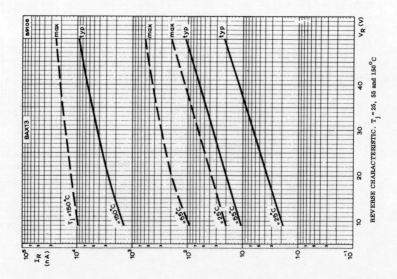

Appendix H

BAX13

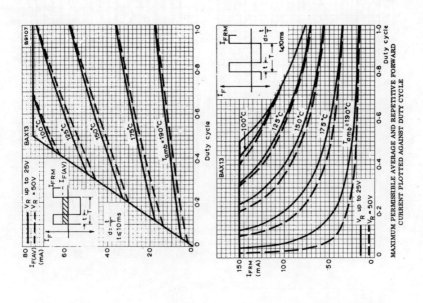

MAXIMUM PERMISSIBLE AVERAGE AND REPETITIVE FORWARD CURRENT PLOTTED AGAINST DUTY CYCLE

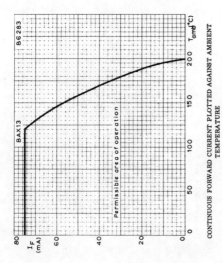

CONTINUOUS FORWARD CURRENT PLOTTED AGAINST AMBIENT TEMPERATURE

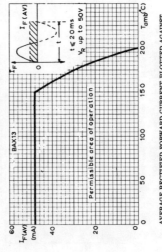

AVERAGE RECTIFIED FORWARD CURRENT PLOTTED AGAINST AMBIENT TEMPERATURE (SINUSOIDAL OPERATION)

H.2 MULLARD VOLTAGE-REFERENCE DIODE TYPE BZY88

Also available to BS 9305—N041

Silicon alloy junction voltage regulator diodes in DO-7 construction having a "5%" voltage tolerance. Intended for use as low current stabilisers or voltage reference diodes.

In addition to the standard range offered, devices can also be supplied to specific requirements. (See page 17 for details of the "Select" Service).

QUICK REFERENCE DATA

BZY88-C2V7 to C33

V_Z (I_Z = 5.0mA) = 2.7 to 33V with limits of voltage in a "5 PERCENT" range (Ref. B.S. 3494 appendix C)

I_{ZM} max.	250	mA
P_Z max. (T_{amb} = 50°C)	400	mW
T_j max.	200	°C
$R_{th(j-amb)}$ (in free air)	0.37	degC/mW

Unless otherwise shown data is applicable to all types in the series

OUTLINE AND DIMENSIONS

Conforming to J.E.D.E.C. DO-7
B.S. 3934 SO-6

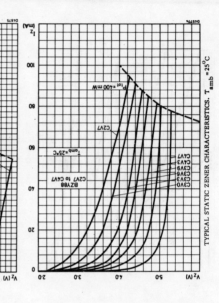

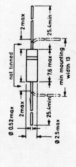

For operation as a zener diode the positive voltage is connected to the lead adjacent to the broad coloured band.

Appendix H

BZY88

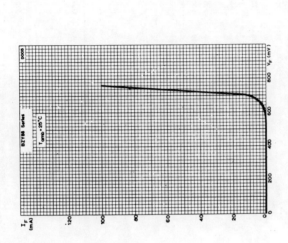

TYPICAL FORWARD CHARACTERISTIC. $T_{amb} = 25°C$

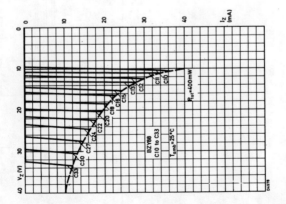

TYPICAL STATIC ZENER CHARACTERISTICS. $T_{amb} = 25°C$

H.3 MULLARD LOW-POWER SILICON NPN BJT TYPE BC107/8/9

Also available to BS9365—F112

N-P-N silicon planar epitaxial transistors in TO-18 encapsulation.
The BC107 is primarily intended for use in audio driver stages and television signal processing circuits.
The BC108 is a general purpose l.f. transistor.
The BC109 is primarily intended for low noise audio input stages.

QUICK REFERENCE DATA

		BC107	BC108	BC109	
V_{CES} max.		50	30	30	V
V_{CEO} max.		45	20	20	V
I_{CM} max.		200	200	200	mA
P_{tot} max. ($T_{amb} \leq 25°C$)		300	300	300	mW
T_j max.		175	175	175	°C
h_{fe} ($I_C = 2mA$, $V_{CE} = 5V$, $f = 1kHz$)	min.	125	125	240	
	max.	500	900	900	
f_T ($I_C = 10mA$, $V_{CE} = 5V$, $f = 35MHz$) typ.		300	300	300	MHz
N ($I_C = 200\mu A$, $V_{CE} = 5V$, $R_S = 2k\Omega$)					
$f = 30Hz$ to $15kHz$	typ.	-	-	1.4	dB
	max.	-	-	4.0	dB
$f = 1kHz$, $B = 200Hz$	typ.	2.0	2.0	1.2	dB
	max.	10	10	4.0	dB

OUTLINE AND DIMENSIONS
Conforms to B.S.3934 SO-132A
J.E.D.E.C. TO-18

Collector connected to case
Accessories available: 56246, 56263

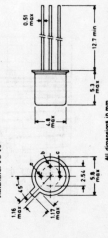

All dimensions in mm

RATINGS

Limiting values of operation according to the absolute maximum system.

Electrical

	BC107	BC108	BC109	
V_{CBO} max.	50	30	30	V
V_{CES} max.	50	30	30	V
V_{CEO} max.	45	20	20	V
V_{EBO} max.	6.0	5.0	5.0	V
I_C max.	100	100	100	mA
I_{CM} max.	200	200	200	mA
$-I_{EM}$ max.	200	200	200	mA
I_{BM} max.	200	200	200	mA
P_{tot} max. ($T_{amb} \leq 25°C$)	300	300	300	mW

Temperature

T_{stg} range	−65 to +175	°C
T_j max.	175	°C

THERMAL CHARACTERISTICS

$R_{th(j-amb)}$	0.5	°C/mW
$R_{th(j-case)}$	0.2	°C/mW

ELECTRICAL CHARACTERISTICS ($T_j = 25°C$ unless otherwise stated)

		Min.	Typ.	Max.	
I_{CBO}	Collector cut-off current $V_{CB} = 20V$, $I_E = 0$, $T_j = 150°C$	-	-	15	μA
V_{BE}	Base-emitter voltage				
	$I_C = 2.0mA$, $V_{CE} = 5.0V$	550	620	700	mV
	$I_C = 10mA$, $V_{CE} = 5.0V$	-	-	770	mV
$V_{CE(sat)}$	Collector-emitter saturation voltage				
	$I_C = 10mA$, $I_B = 0.5mA$	-	90	250	mV
	$I_C = 100mA$, $I_B = 5.0mA$	-	200	600	mV
$V_{BE(sat)}$	Base-emitter saturation voltage				
	$I_C = 10mA$, $I_B = 0.5mA$	-	700	-	mV
	$I_C = 100mA$, $I_B = 5.0mA$	-	900	-	mV

Appendix H

BC107

ELECTRICAL CHARACTERISTICS (contd.)

The following supplementary gain groups are available on request:—

			BC107A BC108A	BC107B BC108B BC109B	BC108C BC109C		
h_{FE}	Static forward current transfer ratio $I_C = 10\mu A, V_{CE} = 5.0V$	typ.	90	150	270		
	$I_C = 2.0mA, V_{CE} = 5.0V$	min. typ. max.	110 180 220	200 290 450	420 520 800		
h parameters	$I_C = 2.0mA, V_{CE} = 5.0V, f = 1.0kHz$						
h_{ie}	Input impedance	min. typ. max.	1.6 2.7 4.5	3.2 4.5 8.5	6.0 8.7 15	kΩ kΩ kΩ
h_{re}	Voltage feedback ratio	typ.	1.5	2.0	3.0	$\times 10^{-4}$
h_{fe}	Small signal current gain	min. typ. max.	125 220 260	240 330 500	450 600 900	
h_{oe}	Output admittance	typ. max.	18 30	30 60	60 110	$\mu A/V$ $\mu A/V$

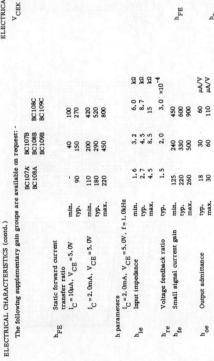

ELECTRICAL CHARACTERISTICS (contd.)		Min.	Typ.	Max.	
V_{CEK}	Collector knee voltage $I_C = 10mA$, I_B = the value for which $I_C = 11mA$ at $V_{CE} = 1.0V$	-	300	600	mV
h_{FE}	Static forward current transfer ratio $I_C = 2.0mA, V_{CE} = 5.0V$ BC107 BC108 BC109	110 110 200	- - -	450 800 800	
h_{fe}	Small signal forward current transfer ratio $I_C = 2.0mA, V_{CE} = 5.0V, f = 1.0kHz$ BC107 BC108 BC109	125 125 240	- - -	500 900 900	
f_T	Transition frequency $I_C = 10mA, V_{CE} = 5.0V, f = 35MHz$	-	300	-	MHz
C_{Tc}	Collector capacitance $I_E = 0, V_{CB} = 10V, f = 1.0MHz$	-	2.5	4.5	pF
C_{Te}	Emitter capacitance $I_C = I_c = 0, V_{EB} = 0.5V, f = 1.0MHz$	-	9.0	-	pF
N	Noise figure $I_C = 0.2mA, V_{CE} = 5.0V, R_S = 2.0k\Omega$ BC109 BC107/108 $f = 30Hz$ to $15kHz$ BC109 $f = 1.0kHz, B = 200Hz$	- - -	1.4 2.0 1.2	4.0 10 4.0	dB dB dB

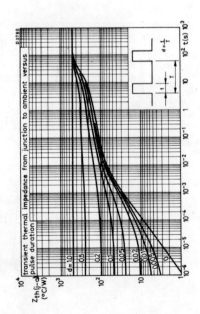

BC107

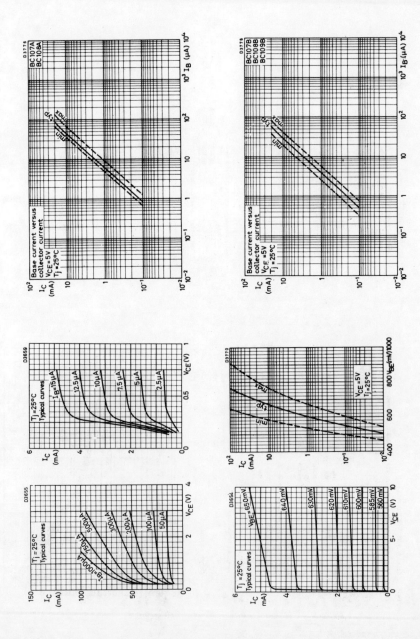

Appendix H

BC107

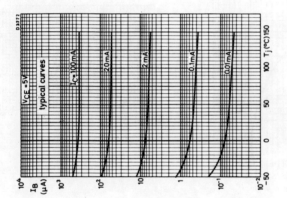

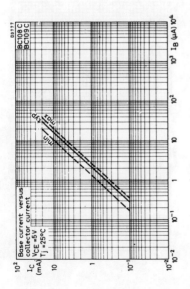

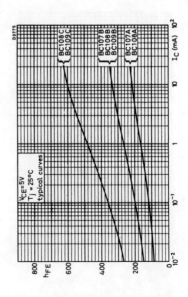

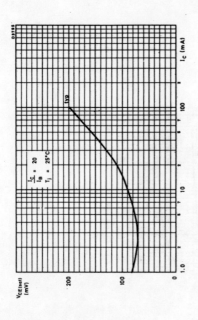
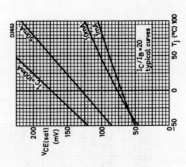
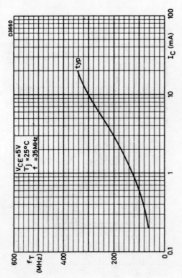
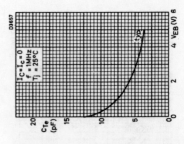
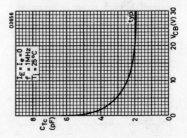

Appendix H

BC107

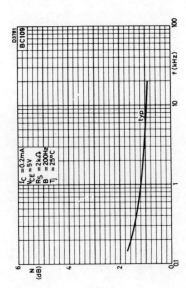

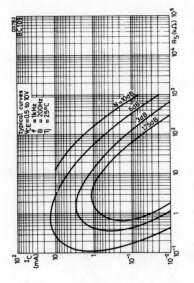

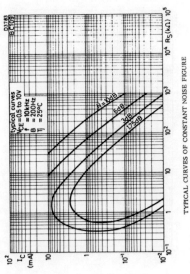

TYPICAL CURVES OF CONSTANT NOISE FIGURE

468 Semiconductor Device Technology

BC107

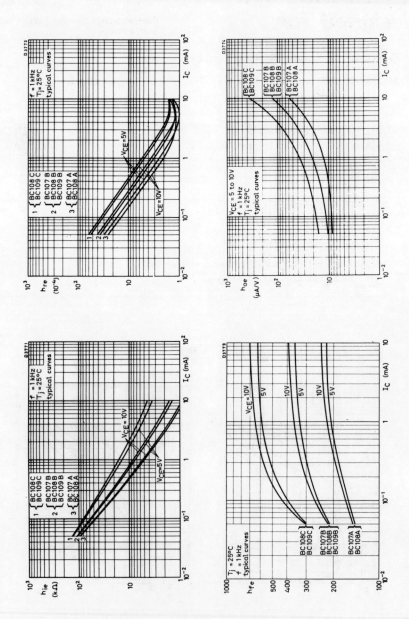

H.4 SILCONIX LOW-POWER SILICON N-CHANNEL JFET TYPE 2N5457/8/9

n-channel JFETs
designed for...

Siliconix

Performance Curves NRL
See Section 5

- **General Purpose Amplifiers**
- **Switches**

BENEFITS
- Low Cost
- Automated Insertion Package

*ABSOLUTE MAXIMUM RATINGS (25°C)

Drain-Source Voltage	25 V
Drain-Gate Voltage	25 V
Source-Gate Voltage	25 V
Total Device Dissipation at 25°C	310 mW
Derate above 25°C	2.82 mW/°C
Operating Junction Temperature	135°C
Storage Temperature Range	−65 to +150°C

TO-92
See Section 7

Bottom View

*ELECTRICAL CHARACTERISTICS (25°C unless otherwise noted)

		Characteristic	2N5457			2N5458			2N5459			Unit	Test Conditions	
			Min	Typ	Max	Min	Typ	Max	Min	Typ	Max			
1	S	I_{GSS} Gate Reverse Current		−.01	−1.0		−.01	−1.0		−.01	−1.0	nA	V_{GS} = −15 V, V_{DS} = 0	
2	T				−200			−200			−200			T_A = +100°C
3	A	BV_{GSS} Gate-Source Breakdown Voltage	−25	−60		−25	−60		−25	−60		V	I_G = −10 μA, V_{DS} = 0	
4	T I	$V_{GS(off)}$ Gate-Source Cutoff Voltage	−0.5		−6.0	−1.0		−7.0	−2.0		−8.0		V_{DS} = 15 V, I_D = 10 nA	
5	C	I_{DSS} Saturation Drain Current	1.0	*	5.0	2.0		9.0	4.0		16	mA	V_{DS} = 15 V, V_{GS} = 0 (Note 1)	
6		g_{fs} Common-Source Forward Transconductance	1,000		5,000	1,500		5,500	2,000		6,000	μmho		f = 1 kHz
7	D Y	g_{os} Common-Source Output Conductance		10	50		15	50		20	50		V_{DS} = 15 V, V_{GS} = 0	
8	N A	C_{iss} Common-Source Input Capacitance		4.5	7.0		4.5	7.0		4.5	7.0	pF		f = 1 MHz
9	M I	C_{rss} Common-Source Reverse Transfer Capacitance		1.0	3.0		1.0	3.0		1.0	3.0			
10	C	NF Noise Figure		.04	3.0		.04	3.0		.04	3.0	dB	V_{DS} = 15 V, V_{GS} = 0, R_G = 1 MΩ, NBW = 1 Hz	f = 1 kHz

*JEDEC registered data

NOTE:
1. Pulse test pulsewidth = 2 ms.

NRL

© 1979 Siliconix incorporated

2N5457

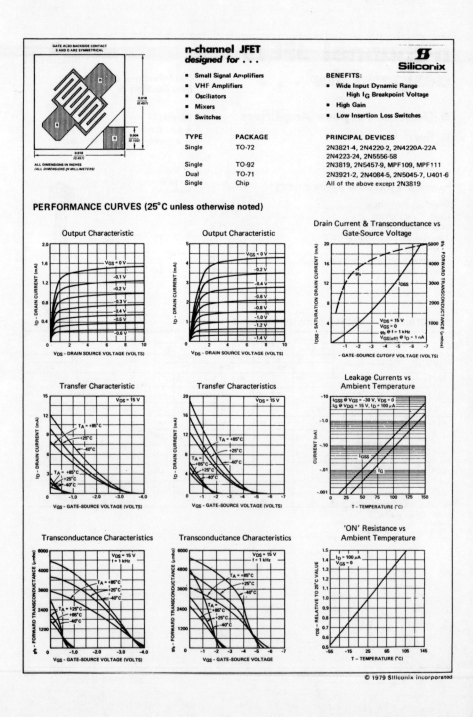

Appendix H

2N5457

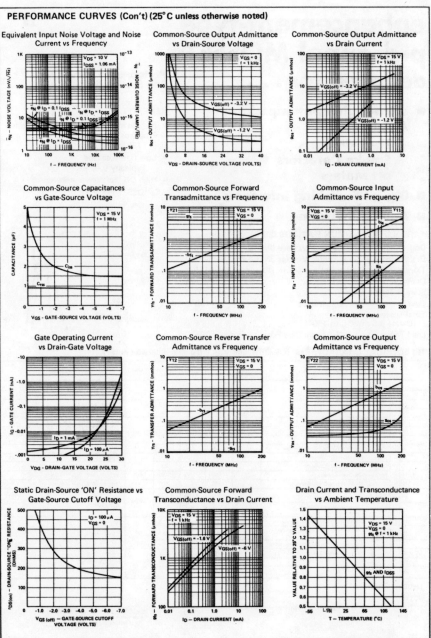

H.5 SILICONIX LOW-POWER P-CHANNEL ENHANCEMENT-TYPE MOST TYPE 3N163/4

enhancement-type p-channel MOSFETs designed for...

Siliconix

Performance Curves MRA
See Section 5

- **Ultra-High Input Impedance Amplifiers**
  - Electrometers
  - Smoke Detectors
  - pH Meters
- **Digital Switching Interfaces**
- **Analog Switching**

BENEFITS
- Rugged MOS Gate Minimizes Handling Problems
  ±150 V Transient Capability
- Low Gate-Leakage
  Typically 0.02 pA
- High Off-Isolation as a Switch
  $I_{DSS} <$ 200 pA

*ABSOLUTE MAXIMUM RATINGS (25°C)

Drain-Source or Gate-Source Voltage 3N163	–40 V
Drain-Source or Gate-Source Voltage 3N164	–30 V
Transient Gate-Source Voltage (Note 1)	±150 V
Drain Current	–50 mA
Storage Temperature	–65 to +200°C
Operating Junction Temperature	–55 to +150°C
Total Device Dissipation (Derate 3.0 mW/°C to 150°C)	375 mW
Lead Temperature 1/16" From Case For 10 Seconds	265°C

TO-72
See Section 7

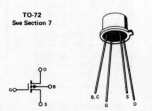

*ELECTRICAL CHARACTERISTICS (25°C and V_{BS} = 0 unless otherwise noted)

		Characteristic	3N163 Min	3N163 Max	3N164 Min	3N164 Max	Unit	Test Conditions	
1		I_{GSS} Gate-Body Leakage Current		–10			pA	V_{GS} = –40 V, V_{DS} = 0	
2				–25					T_A = 125°C
3						–10		V_{GS} = –30 V, V_{DS} = 0	
4						–25			T_A = 125°C
5	S	BV_{DSS} Drain-Source Breakdown Voltage	–40		–30			I_D = –10 μA, V_{GS} = 0	
6	T	BV_{SDS} Source-Drain Breakdown Voltage	–40		–30			I_S = –10 μA, V_{GD} = V_{BD} = 0	
7	A	V_{GS} Gate Source Voltage	–3	–6.5	–2.5	–6.5	V	V_{DS} = –15 V, I_D = –0.5 mA	
8	T	$V_{GS(th)}$ Gate-Source Threshold Voltage	–2	–5	–2	–5		V_{DS} = V_{GS}, I_D = –10 μA	
9	I	I_{DSS} Drain Cutoff Current		–200		–400	pA	V_{DS} = –15 V, V_{GS} = 0	
10	C	I_{SDS} Source Cutoff Current		–400		–800		V_{SD} = –20 V, V_{GD} = 0, V_{DB} = 0	
11		$I_{D(on)}$ ON Drain Current	–5	–30	–3	–30	mA	V_{DS} = –15 V, V_{GS} = –10 V	
12		$r_{DS(on)}$ Drain-Source ON Resistance		250		300	Ω	V_{GS} = –20 V, I_D = –100 μA	
13	D	g_{fs} Common-Source Forward Transconductance	2,000	4,000	1,000	4,000	μmho	V_{DS} = –15 V, I_D = –10 mA	f = 1 kHz
14	Y N	g_{os} Common-Source Output Conductance		250		250			
15	A M	C_{iss} Common-Source Input Capacitance		2.5		2.5			
16	I C	C_{rss} Common-Source Reverse Transfer Capacitance		0.7		0.7	pF	V_{DS} = –15 V, V_{GS} = –10 mA	f = 1 MHz
17		C_{oss} Common-Source Output Capacitance		3		3			
18	S	$t_{d(on)}$ Turn-ON Delay Time		12		12		V_{DD} = –15 V	
19	W	t_r Rise Time		24		24	ns	$I_{D(on)}$ = –10 mA	
20		t_{off} Turn-OFF Time		50		50		R_G = R_L = 1.5 kΩ	

*JEDEC registered data

NOTE:
1. Transient gate-source voltage JEDEC registered as ±125 V.

MRA

© 1979 Siliconix incorporated

Appendix H

3N163

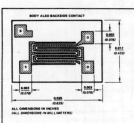

enhancement-type p-channel MOSFET designed for...

- Analog and Digital Switching
- General Purpose Amplifiers
- Smoke Detectors

Siliconix

BENEFITS:
- High Gate Transient Voltage Breakdown Eliminates Need for Gate Protective Diode
- Ultra-High Input Impedance
- Low Leakage
- Normally OFF

TYPE	PACKAGE	PRINCIPAL DEVICES
Single	TO-18	MFE823
Single	TO-72	3N163-64
Single	Chip	3N163-64CHP, MFE823CHP

PERFORMANCE CURVES (25°C unless otherwise noted)

Output Characteristics

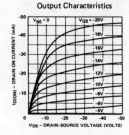

Low-Level Output Characteristics

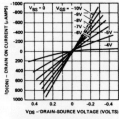

Common-Source, Short-Circuit, Output Admittance vs Drain Voltage

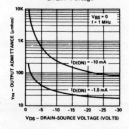

Transfer Characteristic

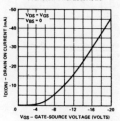

Common-Source, Short-Circuit, Forward Transadmittance vs Drain Current

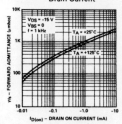

Common-Source, Short-Circuit, Output Admittance vs Drain Current

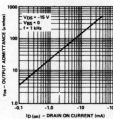

© 1979 Siliconix incorporated

3N163

PERFORMANCE CURVES (Cont'd) (25°C unless otherwise noted)

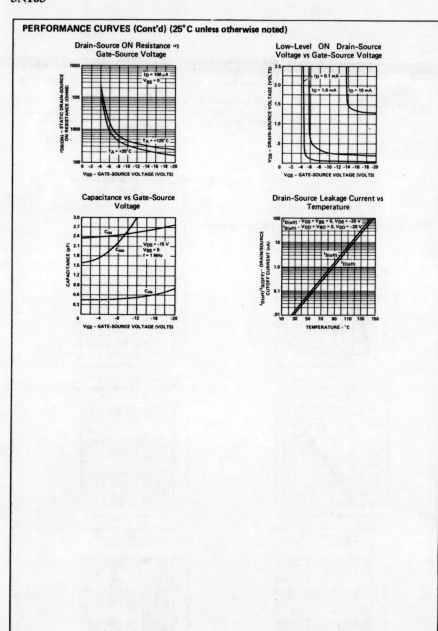

H.6 MULLARD LOW-POWER N-CHANNEL DEPLETION TYPE MOST TYPE BFR29

Depletion type, insulated gate, field effect transistor in a TO-72 metal envelope with the substrate connected to the case. It is intended for linear applications in the audio as well as the i.f. and v.h.f. frequency region, and in cases where high input impedance, low gate leakage currents and low noise figures are of importance.

QUICK REFERENCE DATA

V_{DB} max.	30	V
$\pm V_{GB}$ max.	10	V
I_{DSS} ($V_{DS} = 15V$, $V_{GS} = 0$)	10 to 40	mA
$|y_{fs}|$ min. ($I_D = 5mA$, $V_{DS} = 15V$, $f = 1kHz$)	6.0	mA/V
$-C_{rs}$ max. ($I_D = 5mA$, $V_{DS} = 15V$, $f = 1MHz$)	0.7	pF
N max. ($I_D = 5mA$, $V_{DS} = 15V$, $f = 200MHz$), $G_S = 1mA/V$, $B_S = $ optimum)	5.0	dB
$V_n/\sqrt{B}$ typ. ($I_D = 5mA$, $V_{DS} = 15V$, $f = 1kHz$)	100	nV/$\sqrt{Hz}$

OUTLINE AND DIMENSIONS

Conforms to BS 3934 SO-12A/SB4-3
J.E.D.E.C. TO-72

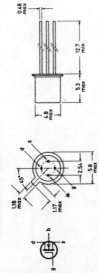

* Substrate connected to envelope

All dimensions in mm

NOTE

To exclude the possibility of damage to the gate oxide layer by an electrostatic charge building up on the high resistance gate electrode, the device is fitted with a conductive rubber ring around the leads. This ring should not be removed until after the device has been mounted in the circuit.

RATINGS

Limiting values of operation according to the absolute maximum system.

Electrical

V_{DB} max.	Drain-substrate voltage	30	V
V_{SB} max.	Source-substrate voltage	30	V
$\pm V_{GB}$ max.	Gate-substrate voltage (continuous)	10	V
$\pm V_{G-N}$ max.	Repetitive peak gate voltage (gate to all other terminals) $V_{SB} = V_{DB} = 0$, $f > 100kHz$	15	V
I_D max.	Drain current (d.c.)	20	mA
I_{DM} max.	Peak drain current $t_r = 20ns$, $\delta \leq 0.1$	50	mA
P_{tot} max.	Total power dissipation $T_{amb} \leq 25°C$	200	mW

Temperature

T_{stg}	Storage temperature	-65 to +125	°C
T_j max.	Junction temperature	125	°C

THERMAL CHARACTERISTIC

$R_{th(j-amb)}$	Thermal resistance, junction to ambient, in free air	0.5	degC/mW

ELECTRICAL CHARACTERISTICS ($T_j = 25°C$ unless otherwise stated)

		Min.	Typ.	Max.	
$-I_{GSS}$	Gate current, $V_{BS} = 0$ $-V_{GS} = 10V$, $V_{DS} = 0$	-	-	10	pA
$-I_{GSS}$	$V_{GS} = 10V$, $V_{DS} = 0$	-	-	10	pA
$-I_{GSS}$	$-V_{GS} = 10V$, $V_{DS} = 0$, $T_j = 125°C$	-	-	200	pA
I_{GSS}	$V_{GS} = 10V$, $V_{DS} = 0$, $T_j = 125°C$	-	-	200	pA
	Substrate current, $V_{GB} = 0$				
$-I_{BDO}$	$-V_{BD} = 30V$, $I_S = 0$	-	-	10	µA
$-I_{BSO}$	$-V_{BS} = 30V$, $I_D = 0$	-	-	10	µA
I_{DSS}	Drain current $V_{DS} = 15V$, $V_{GS} = 0$	10	-	40	mA
$-V_{GS}$	Gate-source voltage $I_D = 100mA$, $V_{DS} = 15V$	0.5	-	3.5	V

BFR29

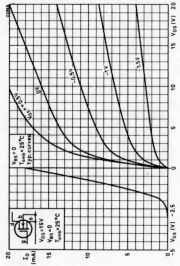
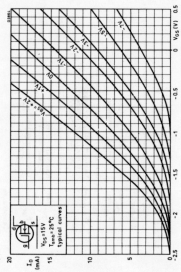

ELECTRICAL CHARACTERISTICS (contd.)

		Min.	Typ.	Max.	
$-V_{(P)GS}$	Gate-source cut-off voltage $I_D = 100nA$, $V_{DS} = 15V$	-	-	4.0	V
N	Noise figure at f = 200MHz $I_D = 5mA$, $V_{DS} = 15V$, $T_{amb} = 25°C$ $G_S = 1mA/V$, $B_S = $ optimum	-	-	5.0	dB
$V_n/\sqrt{B}$	Equivalent noise voltage, $T_{amb} = 25°C$ $I_D = 5mA$, $V_{DS} = 15V$, f = 120Hz	-	300	-	$nV/\sqrt{Hz}$
	f = 1kHz	-	100	-	$nV/\sqrt{Hz}$
	f = 10kHz	-	35	-	$nV/\sqrt{Hz}$
y-parameters	$I_D = 5mA$, $V_{DS} = 15V$, $T_{amb} = 25°C$				
$\|y_{fs}\|$	Transfer admittance at f = 1kHz	6.0	-	-	mA/V
$\|y_{os}\|$	Output admittance at f = 1kHz	-	-	0.4	mA/V
C_{is}	Input capacitance at f = 1MHz	-	-	5.0	pF
$-C_{rs}$	Feedback capacitance at f = 1MHz	-	-	0.7	pF
C_{os}	Output capacitance at f = 1MHz	-	-	3.0	pF

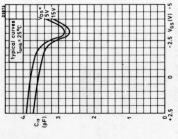
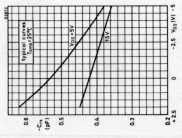

Appendix H

BFR29

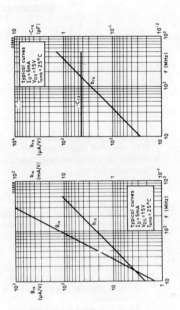
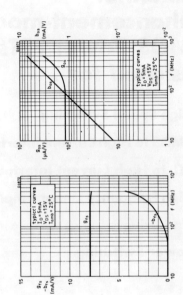
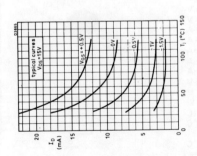
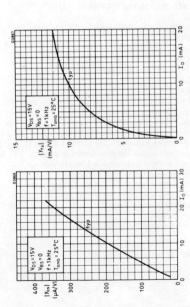

H.7 SILCONIX MEDIUM-POWER N-CHANNEL ENHANCEMENT-TYPE VMOST TYPE VN46/66/88AF

n-channel enhancement-mode VMOS Power FETs designed for...

- High Speed Switching
- CMOS to High Current Interface
- TTL to High Current Interface
- High Frequency Linear Amplifiers
- Line Drivers
- Power Switching

Siliconix

Performance Curves VNAZ
See Section 3

BENEFITS

- Directly Interfaces to CMOS, TTL, DTL and MOS Logic Families
- Permits More Efficient and Compact Switching Designs
- Reduces Component Count and Design Time/Effort
  Drives Inductive Loads Directly
  Fan Out From CMOS Logic > 100
  Easily Paralleled with Inherent Current Sharing Capability
  High Gain
- Improves Reliability
  Free From Secondary Breakdown Failures and Voltage Derating
  Current Decreases as Temperature Increases
  Input Protected From Static Discharge

F- Package
TO-202
See Section 5

ABSOLUTE MAXIMUM RATINGS

```
Maximum Drain-Source Voltage
  VN46AF ................................ 40 V
  VN66AF ................................ 60 V
  VN88AF ................................ 80 V
Maximum Drain-Gate Voltage
  VN46AF ................................ 40 V
  VN66AF ................................ 60 V
  VN88AF ................................ 80 V
Maximum Continuous Drain Current ............... 2.0 A
Maximum Pulsed Drain Current ................... 3.0 A
Maximum Continuous Forward Gate Current ...... 2.0 mA
Maximum Pulsed Forward Gate Current (Note 1)... 100 mA
Maximum Continuous Reverse Gate Current ....... 100 mA
Maximum Forward Gate-Source (Zener) Voltage ..... 15 V
Maximum Reverse Gate-Source Voltage ........... -0.3 V
Maximum Dissipation at 25°C Case Temperature ..... 15 W
Linear Derating Factor ..................... 120 mW/°C
Temperature (Operating and Storage) ....... -40 to +150°C
Lead Temperature
  (1/16" from case for 10 seconds) .............. 300°C
```

NOTE:
1. Pulse test — 80 μsec pulse, 1% duty cycle.

© 1979 Siliconix incorporated

Appendix H

VN46AF

ELECTRICAL CHARACTERISTICS (25°C unless otherwise noted)

		Characteristic	VN46AF Min	VN46AF Typ	VN46AF Max	VN66AF Min	VN66AF Typ	VN66AF Max	VN88AF Min	VN88AF Typ	VN88AF Max	Unit	Test Conditions
1		Drain-Source	40			60			80				$V_{GS} = 0$, $I_D = 500\mu A$
2	BV_{DSS}	Breakdown	40			60			80			V	$V_{GS} = 0$, $I_D = 2.5$ mA
3	$V_{GS(th)}$	Gate Threshold Voltage	0.8	1.7		0.8	1.7		0.8	1.7			$V_{DS} = V_{GS}$, $I_D = 1$ mA
4				0.01	10		0.01	10		0.01	10		$V_{GS} = 10$ V, $V_{DS} = 0$
5	I_{GSS}	Gate Body Leakage			100			100			100		$V_{GS} = 10$ V, $V_{DS} = 0$, $T_A = 125°C$ (Note 2)
6	S				10			10			10	μA	V_{DS} = Max. Rating, $V_{GS} = 0$
7	T A T I C	I_{DSS}	Zero Gate Voltage Drain Current			100			100			100		$V_{DS} = 0.8$ Max. Rating, $V_{GS} = 0$, $T_A = 125°C$ (Note 2)
8				100			100			100		nA	$V_{DS} = 25$ V, $V_{GS} = 0$
9	$I_{D(on)}$	ON-State Drain Current (Note 1)	1.0	2		1.0	2		1.0	2		A	$V_{DS} = 25$ V, $V_{GS} = 10$ V
10				0.3			0.3			0.4			$V_{GS} = 5$ V, $I_D = 0.1$ A
11		Drain-Source Saturation		1.0	1.5		1.0	1.5		1.4	1.7	V	$V_{GS} = 5V$, $I_D = 0.3$ A
12	$V_{DS(on)}$	Voltage (Note 1)		1.0			1.0			1.3			$V_{GS} = 10$ V, $I_D = 0.5$ A
13				2.0	3.0		2.0	3.0		3.0	4.0		$V_{GS} = 10$ V, $I_D = 1.0$ A
14	g_m	Forward Transconductance (Note 1)	150	250		150	250		150	250		m℧	$V_{DS} = 24$ V, $I_D = 0.5$ A
15	C_{iss}	Input Capacitance (Note 2)		50			50			50			
16	D Y C_{rss}	Reverse Transfer Capacitance (Note 2)		10			10			10		pF	$V_{GS} = 0$, $V_{DS} = 25$ V, $f = 1.0$ MHz
17	N A M C_{oss}	Common-Source Output Capacitance (Note 2)		50			50			50			
18	I C $t_{d(on)}$	Turn-ON Delay Time (Note 2)		2	5		2	5		2	5		
19	t_r	Rise Time (Note 2)		2	5		2	5		2	5	ns	See Switching Time Test Circuit VNAZ, Section 3
20	$t_{r(off)}$	Turn-OFF Delay Time (Note 2)		2	5		2	5		2	5		
21	t_f	Fall Time (Note 2)		2	5		2	5		2	5		

NOTES: 1. Pulse test — 80 μs pulse, 1% duty cycle.
2. Sample test.

VNAZ

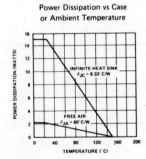

Power Dissipation vs Case or Ambient Temperature

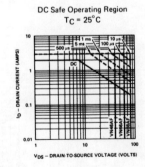

DC Safe Operating Region $T_C = 25°C$

VN46AF

**n-channel
enhancement-mode
VMOS Power FETs**

Siliconix

TYPE	PACKAGE	PRINCIPAL DEVICES
Single	TO-3	2N6656, 2N6657, 2N6658, VN30AA, VN35AA, VN67AA, VN89AA, VN90AA
	TO-39	2N6659, 2N6660, 2N6661, VN30AB, VN35AB, VN67AB, VN89AB, VN90AB
	TO-202	VN40AF, VN46AF, VN66AF, VN67AF, VN88AF, VN89AF

TYPICAL PERFORMANCE CURVES (25°C unless otherwise noted)

Output Characteristics   Saturation Characteristics   Transfer Characteristic

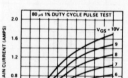
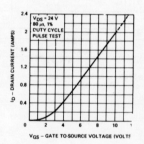

Normalized Drain-to-Source ON Resistance vs Temperature   Capacitance vs Drain-to-Source Voltage   Drain-to-Source ON Resistance vs Gate-to-Source Voltage

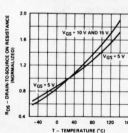
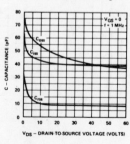
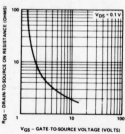

© 1979 Siliconix incorporated

Appendix H

VN46AF

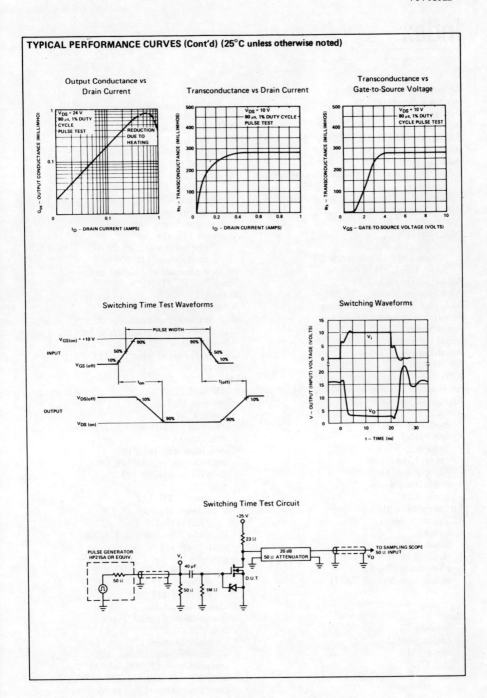

Index

Absorption of energy 273
Acceptor dopant 19, 20
Access time, memory 389, 392
Activator 287
Active mode (BJT) 112–32, 148, 149
Active pull-up 371, 377
Address, memory 388
Admittance parameter model (see y-parameter model)
Affinity 55
Allowed energy band 11
Alloy junction 309
Alpha (BJT)
   small-signal 143, 209
      variation with frequency 209
   static 120, 129, 158
Alpha rays (particles) 3
Alphanumeric display 279, 280
ALU (see Arithmetic/logic unit)
Amorphous material 11
Amplification 119
Angle-lap measurement 327
Angstrom unit 274, 310
Anode 3, 97
APD (see Avalanche photodiode)
Architecture, microprocessor 403, 404
Arithmetic/logic unit 404
Atom 2, 14
   models of 3–9, 14
Atomic model
   Bohr 6
   energy band 11–4
   particle 4–8, 14
   Rutherford 4
   wave 9, 14
Atomic number 10
ATT devices 268–71
Avalanche
   diode 104–6
   multiplication 83, 121, 157
   photodiode 288–92
   transit-time devices 268–71

Band
   bending 50
   conduction 13
   energy 11
   gap 13, 29
   model 11–4, 18, 22, 25
   valence 13

Barrier
   potential (see also Contact potential) 53
   transit-time diode 271
BARRITT diode 271
Base (BJT)
   charge distribution 124
   current 118, 137
   drive 134
      reverse 136
   resistance 175, 200
   stored charge 126
   transport factor 118, 128
   width 115, 118, 200
      control in manufacture 315
      modulation 124, 200
BBD (see Bucket-brigade device)
Beam-lead IC 333, 415
Beta (BJT)
   small-signal 153, 208
      variation with current 155
      variation with frequency 208
   static 145–7, 153–5
      variation with current 155
Beta rays 3
Bias
   forward 57, 71
   reverse 57, 72
   voltage 55
Bipolar junction transistor (see Bipolar transistor)
Bipolar transistor 107–212
   active mode 112–32, 148, 149
   amplification using 119
   base
      current 118, 137
      transport factor 118, 128
      width 115, 118, 200
   base-width modulation 124, 200
   bottoming 151
   carrier
      collection 115
      diffusion 113
      injection 113
   charge, stored 126
   charge-control model 194–9
   collector
      multiplication 121
      transition capacitance 175
   common-base 120, 138–44, 184
   common-collector 138, 184
   common-emitter 138, 144–55, 184

Index

current gain 120, 129, 140, 143, 145–7, 153, 158, 208, 209
cut-off
  mode 113, 135, 151
  frequency 177, 208, 209
Early
  effect 124
  model 199–201
Ebers–Moll model 191–4
emitter
  crowding 172
  injection efficiency 117, 126–8
  slope resistance 176, 200
energy model 129–32
fall time 199
frequency response 175–7, 207–12
gain–bandwidth product 177, 210
high-frequency properties 175–7
high-power types 170–4
h-parameter model 182–9
h-parameters, typical values 187
hybrid-π model 201–12
input
  characteristics 138, 139, 147
  resistance 139, 148
integrated 335–9
interdigitated geometry 172
inverse active operation 151, 191
leakage current 118, 120, 140, 145, 161
model, thermal 168
models
  electrical, 179–212
    summary of 212
noise in 177–9
npn type 109, 112–212
operation 112–9
output
  characteristics 138, 140–3, 148–53
  conductance 140, 142, 148, 150
piecewise-linear model 180-2
planar epitaxial structure 109, 111, 112, 116, 137, 309–21
pnp type 109, 112, 137
power
  dissipation 164–74
  type 164–74
punch-through 157
ratings 157–60
rise time 199
saturation
  line 151
  mode 113, 132–5
  voltages 148, 151
stability factors 163
storage time 199
stored charge 126
structure 111, 112
switching 132–7
symbols 108

temperature effects 160–74
transit time 177
thermal aspects 160–74
threshold voltage 148
transfer characteristics 138, 143, 153–5
transition frequency 116, 177, 210
y-parameter model 189–91
Bit 388
  kilo 388, 443
  mega 388, 443
Bit-slice microprocessor 406
BJT (see Bipolar transistor)
Blocking state (SCR) 256
Bohr atomic model 6
Boltzmann
  constant 20
  energy distribution 20
Bond 10
  covalent 10
  ionic 10
  metallic 11
Bottoming (BJT) (see also Saturation) 151
Breakdown
  diode 104–6
  junction 83–5
  second (BJT) 170
Breakover voltage (SCR) 256
Bucket-brigade device 376
Buried layer 316, 335
Burn-in 320
Burrus LED 281
Bus 404
Byte 388

Cadmium sulphide cell (see Photoconductive cell)
Capacitance
  BJT 175–7, 192, 200–12
  FET 241, 242, 247–50
  junction 80–3
Capacitors, integrated 352
Capacity, memory 388, 393, 397, 400, 401
Capture of carriers 30
Carriers
  diffusion of 36–8, 41, 113
  drift of 33–6, 39–41
  excess 31, 113, 118
  free 28
Cathode 3, 97
  ray 3
Cathodoluminescence 287
CATT 271
CCD (see Charge-coupled device)
CDI (see Collector-diffused isolation)
Cell
  memory 388, 391, 393
  NMOS RAM
    dynamic 391, 393

Index

static 391
NMOS ROM 395, 396
Central processing unit 403
Channel (FET) 212
  stop 373
  types 213
Channelling 316
Characteristics
  BJT
    common-base 138–44
    common-emitter 144–55
    input 138, 139, 147
    output 138, 140–3, 148–53
    transfer 138, 143, 153–5
  diode 93–6
  d-MOST 233
  e-MOST 226
  FET 217, 226, 233
  JFET 217
Charge 2
Charge-control model
  BJT 194–9
  diode 252–4
  pn junction 86, 252
Charge-coupled device 374–6, 402
  buried-channel type 375
  (in) imaging 376
  (as) serial memory 376, 402
  surface-channel type 375
Charge-transfer device 376
Chip 310, 317, 330, 403
CMOS (see Complementary MOS technology)
Coactivator 287
Code
  colour 452
  conversion 397
Coding of components 450–2
Coherent emission 286
COLDFET 242
Collector multiplication 121
  factor 122
Collector transition capacitance 175
Collector-diffused isolation 357
Colour
  component code 450–2
  temperature 276
Common-base (BJT) 120, 138–44, 184
Common-collector (BJT) 138, 184
Common-emitter (BJT) 138, 144–55, 184
Compensation 277, 315
  self- 277
Complementary
  error function 322, 324
  MOS technology 357, 372–4
Compound semiconductors 28, 29
Computer 403
Conduction band 13
Conductivity 13, 15, 34–6
  conductor 34

intrinsic 15, 35
semiconductor 34–6
Conductor 13
Constants, physical, values 441
Contact
  deposition 316
  potential 53, 55, 65, 66, 129
Contacts (see Junction)
Continuity equations 39–42
Controlled-avalanche transit-time triode 271
Controlled-conductance diode 105
Core levels 13
Covalent bond 10
CPU (see Central processing unit)
Critical angle 279
Crossovers 354
Crystal 11
  pulling 309, 311
CTD (see Charge-transfer device)
Current 2
Current gain (BJT) 120, 129, 140, 143, 145–7, 153, 158, 208, 209
  common-base, small-signal 143, 209
    static 120, 129, 140, 158
  common-emitter, small-signal 153, 208
    static 145–7, 153
  variation with frequency 208, 209
Current hogging 246
Current mirror 362
Cut-off (BJT) 113, 135, 151
  frequency, common-base 177, 209
    common-emitter 177, 208
Czochralski process 311

Darlington
  pair 156
  photo- 293
Data sheets
  general 455
  (for) selected devices 456–81
Defect, crystal 30
Degenerate semiconductor 26
Depletion layer 50, 55, 63
  capacitance 80, 175, 252
  width 67–9, 77
Depletion mode (FET) 217, 222, 232
Depletion type FET (see also Metal–oxide–semiconductor FET) 213
Derating factor 166
Design of ICs 357–65
  circuit design rules 360–3
  layout 363–5
  system selection 357
  technology choice 358
Destructive read operation 394
Device numbering systems 445–9
Diac 260
Diagnostic keys 317

Index

Dielectric isolation 332
Diffusion
  capacitance 80, 82, 175, 251
  (of) carriers 36–8, 41, 113
  coefficient 37
  constant-source 315, 324
  current 37, 38
  (of) dopant 315, 323–7
  lateral 316
  length 42–4
  limited-source 315, 324
Diode 93–107
  a.c. switch (see Diac)
  applications 94
  BARRITT 271
  Burrus 281
  controlled-conductance 105
  equation 59, 76, 93
  Gunn 265
  hot carrier 255
  IMPATT 270
  integrated 347
  isolation 331
  laser 281–7
  light-emitting 276–81
  models 106, 252
  photo- 288–92
  pn junction
    fast-switching 251–4
    forward recovery 252
    high-power 103
    low-power 93–103
    reverse recovery 251, 252
  Schottky-barrier 254
  step-recovery 268
  temperature effects 101–3
  thermal model 103
  TRAPATT 270
  tunnel 263–5
  varactor 266
  voltage reference 104–6
Dipole layer (see also Depletion layer) 63
Direct-write-on-wafer technique 408
Discretionary wiring 367
Display
  alphanumeric 279, 280
  electroluminescent 288
  light-emitting diode 276–81
d-MESFET (see Metal–semiconductor FET)
DMOS (see Double-diffused MOS structure)
d-MOST (see Metal–oxide–semiconductor FET, depletion type)
Dominant transition 277
Donor
  dopant 16, 20
  level 18
Dopant 15, 20
  acceptor 19, 20
  donor 16, 20

drive-in 315, 324
  profile 324
    graded-base 336
Doping 15–20, 314–6
  heavy 19, 344
  light 19, 344
Double-diffused MOS structure 242, 343
Drain resistance (FET) 218, 250
Drift
  (of) carriers 33–36, 39–41
  current 34
  velocity 33
Dual-in-line package 317–20
Duty cycle 103, 170
Dynamic RAM 390, 393–5

Early
  effect 124
  model (BJT) 199–201
EAROM (see Electrically alterable ROM)
Ebers–Moll model (BJT) 191–4
ECL (see Emitter-coupled logic)
EEPROM (see Electrically erasable PROM)
Einstein 1, 8, 38
  relation 38
e-JFET (see Junction FET, enhancement type)
Electric field 2, 16, 33, 63
Electrically alterable ROM 341, 390, 400
Electrically erasable PROM 390, 401
Electroluminescence 275, 276, 288
  injection 276
Electromagnetic spectrum 444
Electromigration 408
Electron 3
  conduction 13
  valence 10
Electron–hole pair 15
Electronvolt 8
e-MESFET (see Metal–semiconductor FET)
Emission
  coherent 286
  monochromatic 283
  spectral width 282, 283
  spontaneous 282
  stimulated 283
Emitter, crowding 172
  injection efficiency 117, 126–8
  slope resistance 176, 200
Emitters, optical 277
Emitter-coupled logic 380
e-MOST (see Metal–oxide–semiconductor FET, enhancement type)
Encoder 397
Endurance, memory 390, 401
Energy 1, 14
  absorption 273
  band 11
    allowed 11

Index

atomic model 11–4
forbidden 11
Boltzmann distribution 20
conduction band 13
core level 13
Fermi 21, 23
gap 11, 13, 273
values 29
kinetic 1, 4, 6
level 6
potential 1, 4
valence band 13
Enhancement mode (FET) 217, 222, 225, 232
Enhancement type FET (see also
Metal–oxide–semiconductor FET) 213
Epitaxial
layer 312
planar structure 96, 98, 109, 111, 112, 116, 137, 309–21
Epitaxy 312
EPROM (see Erasable PROM)
Erasable PROM 343, 390, 398–400
Esaki diode (see Tunnel diode)
E-series component values 450, 453
Etching 314
Excess carrier 31, 113, 118
Extrinsic semiconductor 16, 24–6

f_α (BJT) 209
f_β (BJT) 208
Fabrication (see Manufacture)
Fall time (BJT) 199
FAMOS structure 343, 398, 400
Fan-out 359
Feature size 407
Fermi
energy 21, 23
level 21, 23
Fermi–Dirac function 21
FET (see Field-effect transistor)
Fick's law 36
Field
electric 2, 16, 33, 63
emission 85
oxide (FET) 239
Field-effect transistor 109, 212–50
capacitance 241, 242, 247–50
COLDFET 242
double-diffused MOS 242, 343
gallium arsenide 241
high-frequency performance 240–4
high-power types 244–7
insulated-gate type (see also
Metal–oxide–semiconductor FET) 213
integrated 339–47, 409, 412
ion-implantation, use of 242
junction type (see Junction FET)
lateral type 215, 223

metal–insulator–semiconductor type 240
metal–oxide–semiconductor type 213, 223–40, 339–44, 412
metal–semiconductor type 240, 242, 346, 409
models 247–50
MOS 213, 223–40, 339–44, 412
noise in 246
parameter values 250
photofet 294
power considerations 240
ratings 240
safe operating area 240
Schottky-barrier type 240, 242
symbols 214
thin film type 412
V-groove MOS type 242, 243, 343, 344
V-JFET 244
Field-programmable
gate array 402
logic array 402
Figure of merit (BJT) 177, 210
Film ICs 330, 410–6
Firmware 403
Flaw, crystal 30
Flicker noise 178
Flip-chip 415
Flip-flop 390
Floating-gate avalanche-injection MOS
structure 343, 398, 400
Fluorescence 287
Forbidden energy band 11
Forward recovery 252
FPGA (see Field-programmable gate array)
FPLA (see Field-programmable logic array)
Free carrier 28
Frequency 274
optical range 271

g_{fs} (FET) 218, 222, 227, 230, 235, 250
g_m (FET) (see also g_{fs}) 218
g_{os} (FET) 218, 226, 231, 250
Gain-bandwidth product 177, 210
Gallium arsenide 28, 29, 442
parameter values 29, 442
Gamma rays 3
Gate oxide (FET) 239
Gate turn-off SCR 259
Generation 15, 30, 39
Germanium 28, 29, 309, 442
parameter values 29, 442
Gettering 51
Graded-base dopant profile 336
Ground state 7
Grown junction 309
GTO SCR (see Gate turn-off SCR)
Guard ring 51, 99, 290, 373
Gunn

Index

diode 265
effect 265

h_{fb} (BJT) 143, 209
    variation with frequency 209
h_{FB} (BJT) 121, 129, 140, 147, 157
h_{fe} (BJT) 153, 208
    variation with current 155
    variation with frequency 208
h_{FE} (BJT) 146, 153
    variation with current 153–5
    variation with temperature 162, 163
h_{ib} (BJT) 139
h_{ie} (BJT) 148
h_{ob} (BJT) 140, 142
h_{oe} (BJT) 148, 150
Hall
    coefficient 47
    effect 45
    mobility 47
    voltage 45
Hardware 403
Hardwired memory 397
Haynes–Shockley measurements 44
Heat sink 167
Heterojunction 281, 283
High-density MOS technology 409
High-level injection (BJT) 153
High-speed TTL 379
HMOS (see High-density MOS technology)
Hockey puck package 103, 256
Holding current (SCR) 256
Hole 15, 19
Hologram 286
Hot spot 172
Hot-carrier diode 255
h-parameter model
    BJT 182–9
    general 422, 423
h-parameters, typical BJT values 187
H-TTL (see High-speed TTL)
Hybrid active devices 347
Hybrid ICs 330, 411, 414–6
Hybrid parameter model (see h-parameter model)
Hybrid-π
    model, BJT 201–12
    parameters, typical BJT values 202

I_{CBO} (BJT), 118, 120, 140, 161, 162
    variation with temperature 161, 162
I_{CEO} (BJT) 145, 163
    variation with temperature 163
$I_{D(sat)}$ (FET) 217, 218, 222, 226, 230, 233, 235
I_{DSS} (FET) 216, 217, 224, 226, 233, 235
IC (see Integrated circuit)
IFL (see Integrated fuse logic)

IGFET (see Insulated-gate FET; Metal–oxide–semiconductor FET)
I^2L (see Integrated-injection logic)
I^3L (see Isoplanar integrated-injection logic)
Illumination 276
Imaging, using CCDs 376
Impact avalanche transit-time diode 270
IMPATT diode 270
Implantation of ions 315, 320
Inductors, integrated 349
Infrared radiation 271, 274, 275, 444
Injection
    (of) carriers (BJT) 113
    high-level 153
    low-level 153
Injection electroluminescence 276
Input characteristic, BJT 138, 139, 147
Input resistance
    BJT, common-base 139
        common-emitter 148
    FET 218, 227, 232, 236
Insulated-gate FET (see also Metal-oxide-semiconductor FET) 213
Insulator 13
Integrated
    BJTs 335–9
    capacitors 352–4
    circuit 329–416
        advantages of 329
        beam-lead 333, 415
        BJTs 335–9
        buried layer 335
        capacitors 352–4
        CCD 374–6
        circuit elements 335–55
        classification 330
        CMOS 357, 372–4
        collector-diffused isolation 357
        cost–complexity relation 368
        design 357–65
        diodes 347
        discretionary wiring of 367
        DMOS transistors 343
        ECL 380
        FAMOS transistors 343, 398, 400
        feature size 407
        film types 330, 410–6
        gallium arsenide 346
        HMOS 409
        H-TTL 379
        hybrid 330, 411, 414–6
        hybrid devices in 347
        I^2L 368, 381–5, 409
        I^3L 385
        inductors 349
        inverter, CMOS 372
        inverter, I^2L 382, 383
        ISL 384, 385, 409
        isolation 330–4, 356–7

Isoplanar process 355
JFETs 344–7, 409
large-scale 388–410
line width 407
loads, logic gates 370, 371
logic arrays 402
LSI development 406–10
LS-TTL 379
L-TTL 379
manufacture 309–27
memories 388–402
MESFETs 346, 409
metallisation 354
microprocessor 403–6
MNOS transistors 341, 400
monolithic type 330–410
Moore's law 406, 407
MOSTs 339–44, 412
MTL (see I^{2}L)
NAND gate, CMOS 372
NAND gate, LS-TTL 378, 379
NAND gate, NMOS 370, 371
NAND gate, TTL 377, 378
NMOS 371, 409
NOR gate, I^{2}L 382, 383
NOR gate, PMOS 370
packing density (see also Scale of integration) 369
performance comparison 385–7, 392
performance parameters 369
PMOS 370, 409
power–delay product 369
power dissipation 369
propagation delay 369
resistors 349–52
scale of integration 365–8
Schottky clamping in 378, 379
Schottky I^{2}L 384, 385
Schottky-barrier FETs 346, 409
silicon-on-sapphire 334, 373
SOS 334, 373
SOS-CMOS 373
S-TTL 379
superintegration 347, 368, 383
surface passivation 354
testability 410
thick film 410, 412–4
thin film 410–2
TTL 377–9
VIL 384, 385
V-I^{2}L 384, 385
VMOS transistors 343, 344, 371, 409
yield 366
diodes 347
fuse logic 402
inductors 349
-injection logic 368, 381–5, 409
JFETs 344–7, 409
MOSTs 339–44, 412

resistors 349–52
Schottky logic 384, 385, 409
Interdigitated geometry (BJT) 172
Intrinsic
  carrier density 15, 24
    values 442
  conduction 15, 35
  conductivity 35
  semiconductor 14, 21–4
Inverse active operation (BJT) 151, 191
Inversion layer 50, 224, 239, 244, 340
Inverter, digital
  CMOS 372
  I^{2}L 382, 383
Ion 3
  implantation 315, 320
  migration 236, 237
Ionic bond 10
Irradiance 276
ISL (see Integrated Schottky logic)
Isolation in ICs 330–4, 356, 357
Isoplanar integrated-injection logic 385
Isoplanar process 355

JFET (see Junction FET)
Johnson noise 178
Junction 51–88
  abrupt 61
  breakdown 83–5
  FET 213, 215–23, 344–7, 409
    analysis 218–23
    characteristics 217
    depletion mode operation 217
    drain resistance 218
    enhancement mode operation 217
    enhancement type 346, 409
    forward transconductance 218, 222
    gate current 218
    $I_{D(sat)}$ 217, 218, 222
    I_{DSS} 216, 217
    input resistance 218
    integrated 344–7, 409
    mutual conductance 218
    ohmic region 216, 221
    operation 215–8
    output conductance 218, 250
    pinch-off region 216, 217, 221
    pinch-off voltage 216, 217
    saturation region (see Pinch-off region)
    structure 215, 236–40
    symbols 214
    $V_{GS(off)}$ 217, 218
    V_P 216, 217
    vertical type 244
  field 69–71, 77, 78
  field–effect transistor (see Junction FET)
  forward bias 57, 71
  forward current 57, 71

Index

hetero- 281, 283
isolation (see Diode isolation)
metal–semiconductor 53–61
ohmic 53, 59–61
pn 61–83
pn^+ 69, 71, 85, 99
p^+n 69, 71, 85, 99
rectifying 53–9
reverse bias 57, 72
reverse current 57, 72
step 61
transistor (see Bipolar transistor)

Kilobit 388, 443
Kinetic energy 1, 4, 6

Large-scale integration (see Scale of integration)
Laser, semiconductor 281–7
   heterojunction 283
   strip-geometry 283
   threshold 282, 286
Lasing action 283
Latch 390
Lateral
   diffusion 316
   FET 215, 223
   pnp transistor 338
LDR (see Light-dependent resistor)
Leadless inverted device 415
Leakage current
   BJT 118, 120, 140, 145, 161–3
   diode 95, 96, 101, 104
   junction 59, 76, 78, 95, 96, 101, 104
   variation with temperature 101, 161–3
LED (see Light-emitting diode)
LID (see Leadless inverted device)
Lifetime 31, 32
   killer 252
Light (see Visible radiation)
Light pipe 280
Light-activated switch 295
Light-dependent resistor 296
Light-emitting diode 276–81
   Burrus 281
   edge-emitting 286
   face-emitting 281
Line width 407
Lithography (photomasking) 314, 407
   projection 407
Load
   depletion type 371
   enhancement type 370
Local oxide isolated CMOS 357, 373
LOCMOS (see Local oxide isolated CMOS)
Low-level injection (BJT) 153
Low-power

Schottky TTL 379
TTL 379
LSI (see Scale of integration)
LS-TTL (see Low-power Schottky TTL)
L-TTL (see Low-power TTL)
Luminance 276
Luminescence
   cathodol- 287
   electro- 275, 276, 288
   photo- 278, 287
Luminescent coatings 287
Luminous
   flux 275
   intensity 275

Majority carriers 16, 19, 48
Manufacture of devices 309–27
   contact deposition 316
   crystal pulling 311
   current developments 320–3
   diffusion 315, 323–7
   direct-write-on-wafer 408
   doping 314–6
   electromigration 408
   epitaxy 312
   etching 314
   feature size 407
   ion-implantation 315, 320
   line width 407
   LSI development 406–10
   metallisation 316, 354
   oxidation 313
   packaging 317–20
   photomasking 314, 407
   planar process 309–21
   projection lithography 407
   run-out 407
   scaling 408
   silicon, choice of 309, 314, 323
   testing 317, 320
   zone refining 311
Mask-programmed memories 395
Medium-scale integration (see Scale of integration)
Megabit 388, 443
Memory 388–402
   access time 389, 392
   address 388
      decoder 388
   battery back-up 395
   capacity 388, 393, 397, 400, 401
   CCD 376, 402
   cell 388, 391, 393
   code conversion 397
   dynamic 390, 393–5
   EAROM 341, 390, 400
   EEPROM 390, 401
   electrically alterable ROM 341, 390, 400

Index

electrically erasable PROM 390, 401
encoder 397
endurance 390, 401
EPROM 343, 390, 398–400
erasable PROM 343, 390, 398–400
hardwired 397
mask programmed 395
organisation 388
performance comparison 392
power dissipation 389, 392
programmable ROM 390, 397
PROM 390, 397
RAM 389–395
random-access 389–95
read operation 389, 391
read-only 390, 395–401
refresh 393, 394
ROM 390, 395–401
shift register 376, 401
soft error 410
static 390–3
volatility 389, 390, 397
write operation 389, 391,
Merged-transistor logic (see
  Integrated-injection logic)
Mesa structure 51, 104
MESFET (see Metal–semiconductor FET)
Metal–insulator–semiconductor FET 240
Metallic bond 11
Metallisation 316
Metal–nitride–oxide–semiconductor
  structure 238, 341, 400
Metal–oxide–semiconductor FET (see also
  Field-effect transistor) 213, 223–36,
  339–44, 412
  depletion type 213, 231–6, 341
    analysis 233–6
    characteristics 233
    depletion mode operation 232
    enhancement mode operation 232
    forward transconductance 235
    gate current 236
    handling precautions 232
    $I_{D(sat)}$ 233, 235
    I_{DSS} 233, 235
    input resistance 232, 236
    offset gate geometry 231, 233
    operation 231–3
    pinch-off 232
    pinch-off voltage 232, 233
    structure 231, 236–40
    $V_{GS(off)}$ 232
    V_P 232, 233
  enhancement type 213, 223–31, 339–41
    analysis 227–31
    characteristics 226
    double-diffused structure 242, 343
    enhancement mode operation 225
    forward transconductance 227, 230
    gate current 227
    handling precautions 227, 341
    $I_{D(sat)}$ 226, 230
    I_{DSS} 224, 226
    input resistance 227
    ohmic region 225, 230
    operation 224–7
    output conductance 226, 231, 250
    pinch-off (saturation) region 226, 230
    silicon gate 227, 340
    structure 223, 236–40
    threshold voltage 224–6
    $V_{GS(th)}$ 224
    V_T 224–6
    V-groove structure 242, 243, 343, 344
  field oxide 239
  gate oxide 239
  integrated 339–44, 412
  inversion layer 224, 239, 244, 340
  ion migration 236, 237
  MNOS structure 238, 341, 400
  parameter spread 239
  pin holes 239
  sealed junction 238
  silicon dioxide 223, 227, 236, 239
  silicon nitride 238
  surface passivation 238, 319, 354
  temperature variations 239
  thin film 412
Metal–semiconductor
  FET 240, 242, 346, 409
  junctions 53–61
Micro (see Microprocessor; Prefixes, decimal)
Micro-alloying 354
Microcomputer 403
Microminiature transistors 414
Micron 274, 310
Microprocessor 403–6
  architecture 403
  bit-slice 406
Miller
  effect 212, 249
  capacitance 212
Minority carriers 16, 19, 48
MISFET (see Metal–insulator–semiconductor
  FET)
MNOS structure (see
  Metal–nitride–oxide–semiconductor
  structure)
Mobility 34
  values 29
Model
  atomic 3–9, 14
  BJT 179–212
    summary 212
  charge-control
    BJT 194–9
    diode 252–4
    pn junction 86, 252

Index

diode 106, 252
Early, BJT 199–201
Ebers–Moll, BJT 191–4
energy, BJT 129–32
FET 247–50
h-parameter
   BJT 182–9
   general 422, 423
hybrid-π, BJT 201–12
photodiode 289, 291
physical 191
piecewise-linear 86, 180–2, 419–21
rectifying junction 86–8
SCR 259
small-signal
   BJT 179–91, 199–212
   diode 106, 107
   FET 247–50
   pn junction 86
solar cell 298, 299
terminal parameter 180, 419–25
thermal
   diode 103
   transistor 168
two-port 422–5
y-parameter
   BJT 189–91
   FET 247
   general 423, 424
Monochromatic emission 283
Monolithic IC 330–410
Moore's law 406, 407
MOSFET (see Metal–oxide–semiconductor FET)
MOST (see Metal–oxide–semiconductor FET)
Mounting base 165
MSI (see Scale of integration)
MTL (see Integrated-injection logic)
Mutual conductance (see also Transconductance, forward) 201, 218

NAND gate
   CMOS 372
   LS-TTL 378, 379
   NMOS 370, 371
   TTL 377, 378
n-channel
   FET (see also Field-effect transistor) 213
   MOS technology 371, 409
Neutral regions 63
Neutrality 47
Neutron 4
NMOS (see n-channel MOS technology)
Noise
   (in) BJTs 177–9
   (in) FETs 246
   figure 178
   flicker 178
   Johnson 178
   shot 178
   surface 178
   thermal 178
   white 178
NOR gate
   I^2L 382, 383
   PMOS 370
npn transistor (see Bipolar transistor)
n-type semiconductor 16–9
Nucleus 4
Numbering of devices 445–9

Offset gate geometry (FET) 231, 233
Ohmic
   junction 53, 59–61
   region (FET) 216, 221, 225, 230
Optical
   devices 271–300
   emitters 277
   isolator 293
   radiation 271, 274, 444
Optocoupler 293
Optoelectronic pair 294
Outdiffusion 344
Output
   characteristic, BJT 138, 140–3, 148–53
   conductance
      BJT, common-base 140, 142
      BJT, common-emitter 148, 150
      FET 218, 226, 231, 250
Overdriving (BJT) 133
Oxidation of silicon 313
Oxide isolation (see Dielectric isolation)

Package
   dual-in-line 317–20
   hockey puck 103, 256
   outlines (see Package styles)
   styles
      diodes 96, 97
      transistors 109, 110
Packaging of devices 317–20
Packing density 369
Parametric amplifier 267
Particle
   atomic model 4–8, 14
   detector 292
Passivation 51, 319, 354
Pauli exclusion principle 11
p-channel
   FET (see also Field-effect transistor) 213
   MOS technology 370
Performance comparison
   logic technologies 385–7
   memories 392
Periodic table 10, 17

Permittivity 16, 69, 341, 353
   values 442
Persistence 287
Phosphor 287
Phosphorescence 287
Photoconductive cell 296
Photocoupler 293
Photodarlington 293
Photodetector 288
Photodiode 288–92
   avalanche 288, 290
   model 289, 291
   pin 288
   quantum gain 290
   responsivity 290
   spectral response 289, 291
Photofet 294
Photoluminescence 278, 287
Photomasking (see also Lithography) 314
Photon 6, 273
Photoresist 314
Photothyristor 295
Phototransistor 292
Photovoltaic
   cell 297–300
   potential 299
Physical models
   BJT 180, 191–212
   diode 106, 107
Pi (π)-type semiconductor 244, 344
Piecewise-linear modelling 419–21
   (of) BJT 180–2
   (of) junction diode 86, 107
Pin diode (see Impact avalanche transit-time diode; Photodiode; Step-recovery diode)
Pin holes (FET) 239
Pinch-off (FET) 216–8, 221, 226, 230–3
   voltage 216, 217, 232, 233
PLA (see Programmable logic array)
Planar
   epitaxial structure 96, 98, 109, 111, 112, 116, 137, 309–21
   process 309–21
Planck 6
PMOS (see p-channel MOS technology)
pn junction 61–83
   alloy 309
   applied bias 71–8
   breakdown 83–5
   capacitance 80–3
   depletion layer 63
      width 67–9, 77
   diode 93–107
      fast switching 251–4
      high-power 103
      low-power 93–103
      voltage reference 104–6
   field 69–71, 77, 78
   grown 309
   planar 96, 98, 309
   pn^+ structure 69, 71, 85, 99
   p^+n structure 69, 71, 85, 99
   temperature effects 78–80
pnp transistor (see Bipolar transistor)
Point-contact transistor 108
Poisson's equation 64
Polycrystalline structure 11
Population
   (of) energy band 282
   inversion 282
Potential 2
   barrier (see also Contact potential) 53
   contact 53, 55, 65, 66, 129
   energy 1, 4
   well 8
Power dissipation
   BJT 164–74
   diode 102, 103
   FET 240
   logic gates 369
   memories 389, 392
Power–delay product 369
Preferred values of components (see E-series component values)
Prefixes
   binary 443
   decimal 443
Probability function 20, 21
Processor 403
Pro-Electron system 445, 448
Program 403
Programmable
   logic array 402
   ROM 390, 397
Projection lithography 407
PROM (see Programmable ROM)
Propagation delay 369
Proton 4
p-type semiconductor 19
Punch-through 99, 157
PWL models (see Piecewise-linear modelling)

Quality control in manufacture 320
Quantum 6, 273
   gain 290
   number 7
Quaternary semiconductors 28, 29
Quinternary semiconductors 28

r_{ds}(FET) 218, 250
Radiation
   infrared 271, 274, 275, 444
   optical 271, 275, 444
   ultraviolet 271, 274, 275, 444
   visible 271, 274, 275, 444
RAM (see Random-access memory)

Index

Random-access memory 389–95
Rate effect (SCR) 258
Ratings
　BJT 157–60
　FET 240
Rating system, absolute maximum 160
Read (memory) 389, 391
Read diode 268
Read-only memory 390, 395–401
Read/write memory (see Random-access memory)
Recombination 15, 30, 39, 118
　centre 31
Rectifier (see also Diode) 57, 72
　equation (see also Diode equation) 59
Rectifying junction 53–9
　equation 59, 76
　model 86–8
Refresh, dynamic memory 393, 394
Relativity 8
Reliability testing 320
Resistivity 34
Resistors, integrated 349–52
Reverse
　base drive 136
　breakdown 83–5
　recovery 251–4
　saturation current 59, 76
Rise time (BJT) 199
ROM (see Read-only memory)
Run-out 407
Rutherford atomic model 4

Safe operating area
　BJT 165, 167, 172
　diode 103
　FET 240
Saturation (BJT) 113, 132–5
　line (BJT) 151
　(pinch-off) region (FET) 216–18, 221, 226, 230–3
　voltages (BJT) 148, 151
Scale of integration 365–8
Scaling of devices 408
Schottky clamping 378, 379
Schottky
　effect 254
　I^2L 384, 385
　TTLK 379
Schottky barrier,
　diode 53, 57, 254
　FET 240, 242
Schottky-diode FET logic 409
SCR (see Semiconductor controlled rectifier)
SCS (see Semiconductor controlled switch)
SDFL (see Schottky-diode FET logic)
Sealed junction technology 238, 354
Second breakdown (BJT) 170

Seed crystal 312
Self-aligned gate 340
Self-compensation 277
Self-isolation 340
Semiconductor 14, 20
　compound 28, 29
　conductivity 35
　controlled rectifier 255–60
　　characteristics 257
　　forward blocking state 256
　　forward breakover voltage 256
　　gate turn-off type 259
　　holding current 256
　　model 259
　　operation 256–60
　　rate effect 258
　　structure 257
　　symbols 257
　controlled switch 259
　degenerate 26
　extrinsic 16, 24–6
　intrinsic 14, 21–4
　n-type 16–9
　n^+-type 19
　n^--type 19
　pi (π)-type 244, 344
　p-type 19
　p^+-type 19
　p^--type 19
　quaternary 28, 29
　quinternary 28
　ternary 28, 29
Sheet resistance 350, 351, 411
Shell 10
Shot noise 178
Signal-to-noise ratio 178
Silicon 28, 29, 309, 442
　advantages 309, 314, 323
　dioxide 223, 227, 236, 239, 311, 313, 314
　　permittivity 442
　gate 227, 340
　nitride 238
　　permittivity 442
　parameter values 442
　purification 311
Silicon-on-sapphire ICs 334
　CMOS 373
Slope resistance 86, 419
Small-scale integration (see Scale of integration)
Small-signal model
　BJT 179–91, 199–212
　diode 106, 107
　FET 247–50
　pn junction 86
Snap diode (see Step-recovery diode)
SOAR (see Safe operating area)
Soft error 410
Software 403

Solar cell 297–300
Solid angle 275
SOS (see Silicon-on-sapphire ICs)
SOS-CMOS (see Silicon-on-sapphire ICs, CMOS)
Space charge 47, 64
  layer (see also Depletion layer) 50
Spectral width 282, 283
Spectrum, electromagnetic 444
Speed–power product (see Power–delay product)
Spontaneous emission 282
Sputtering 411
SSI (see Scale of iuntegration)
Stability factors (BJT) 163
Stabistor 105
Stacked-gate MOS structure 343, 401
Static characteristics
  BJT 138–55
  diode 93–6
  FET 217, 226, 233
Static RAM 390–3
Step-recovery diode 268
  in frequency multiplication 268
Steradian 275
Stimulated emission 283
Storage time
  BJT 199
  diode 253
S-TTL (see Schottky TTL)
Substrate 99, 109, 116, 312
Superintegration 347, 368, 383
Surface
  conditions 48
  noise 178
  passivation 238, 319, 354
  states 48
Switching devices 250–62

Temperature coefficient
  breakdown diode 105
  junction forward voltage 102
  of resistance 36
Terminal parameter models
  BJT 180–91, 212
  FET 247–50
  general 419–25
Ternary semiconductors 28, 29
Testability of LSICs 410
Testing of devices 317, 320
Thermal
  model, diode 103
  model, transistor 168
  noise 178
  resistance 102, 166–70
  runaway 164
Thick-film IC 410, 412–4
Thin-film IC 410–2

Threshold
  lasing 282, 286
  voltage
    BJT 148
    FET 224–6
Thyristor (see Semiconductor controlled rectifier)
Totem-pole configuration 377
Transadmittance 190
Transconductance 201
  forward (FET) 218, 222, 227, 230, 235, 250
Transfer admittance 190
Transfer characteristic (BJT) 138, 143, 153–5
Transferred-electron device (see Gunn diode)
Transistor (see Bipolar transistor; Field-effect transistor)
  bipolar (see Bipolar transistor)
  field-effect (see Field-effect transistor)
  integrated 335–47, 409, 412
  microminiature 414
  models
    electrical 179–212, 247–50
    thermal 168
  photo- 292
  unijunction 262
Transistor–transistor logic 377–9
Transit time (BJT) 177
Transition
  frequency (BJT) 116, 177, 210
  region (layer) 63
  temperature 27
  time (diode) 254
Trap 30
TRAPATT diode 270
Trapped plasma avalanche and triggered transient diode 270
Triac 261
Triode a.c. switch (see Triac)
TTL (see Transistor–transistor logic)
Tunnel diode 263–5
Tunnelling 85, 264
Two-port models 422–5
Type branding (see also Device numbering systems) 99, 109

Ultraviolet radiation 271, 274, 275, 444
Unijunction transistor 262
Unipolar transistor (see also Field-effect transistor) 212

V_{BE} (BJT), variation with temperature 161, 162
$V_{BE(sat)}$ (BJT) 148
$V_{CE(sat)}$ (BJT) 151
$V_{GS(off)}$ (FET) 217, 232
$V_{GS(th)}$ (FET) 224
V_P (FET) 216, 217, 232, 233

Index

V_T (FET) 224–6
Valence
   band 13
   electron 10
Varactor diode 266
   in frequency multiplication 267
   in parametric amplifier 267
   in up-converter 267
   in voltage-controlled oscillator 266
Variables, terminology 439
Vertical
   injection logic 384, 385
   junction FET 244
Very large-scale integration (see Scale of integration)
V-groove
   integrated-injection logic 384, 385
   MOS structure 242, 243, 343, 344
VIL (see Vertical injection logic)
V-I^{2}L (see V-groove integrated-injection logic)
Visible radiation 271, 174, 275, 444
V-JFET (see Vertical junction FET)
VLSI (see Scale of integration)

VMOS (see V-groove MOS structure)
Volatility, memory 389, 390, 397
Voltage 2
Voltage reference diode 104–6
Voltage-controlled oscillator 266

Wave model 9, 14
Wavelength 274
White noise 178
Word 388
Work function 52
Write (memory) 389, 391

Yield 366
y-parameter model
   BJT 189–91
   FET 247
   general 423, 424
Zener
   diode 104–6
   effect 85
Zone refining 311